Contents

Chapter 1: Describing Data with Graphs..1

Chapter 2: Describing Data with Numerical Measures..17

Chapter 3: Describing Bivariate Data..36

Chapter 4: Probability..49

Chapter 5: Discrete Probability Distributions..65

Chapter 6: The Normal Probability Distribution..83

Chapter 7: Sampling Distributions...98

Chapter 8: Large-Sample Estimation..112

Chapter 9: Large-Sample Test of Hypotheses..128

Chapter 10: Inference from Small Samples..144

Chapter 11: The Analysis of Variance..172

Chapter 12: Linear Regression and Correlation..193

Chapter 13: Multiple Regression Analysis..218

Chapter 14: The Analysis of Categorical Data..230

Chapter 15: Nonparametric Statistics..246

© 2020 Cengage Learning, Inc. May not be scanned, copied or duplicated, or posted to a publicly accessible website, in whole or in part.

Student Solutions Manual

Introduction to Probability and Statistics

FIFTEENTH EDITION

Barbara M. Beaver
University of California, Riverside, Emerita

CENGAGE

Australia • Brazil • Mexico • Singapore • United Kingdom • United States

© 2020 Cengage Learning

Unless otherwise noted, all content is © Cengage

ALL RIGHTS RESERVED. No part of this work covered by the copyright herein may be reproduced or distributed in any form or by any means, except as permitted by U.S. copyright law, without the prior written permission of the copyright holder.

For product information and technology assistance, contact us at
**Cengage Customer & Sales Support,
1-800-354-9706 or support.cengage.com.**

For permission to use material from this text or product, submit all requests online at **www.cengage.com/permissions**.

ISBN: 978-1-337-55828-0

Cengage
20 Channel Center Street
Boston, MA 02210
USA

Cengage is a leading provider of customized learning solutions with employees residing in nearly 40 different countries and sales in more than 125 countries around the world. Find your local representative at:
www.cengage.com.

Cengage products are represented in Canada by Nelson Education, Ltd.

To learn more about Cengage platforms and services, register or access your online learning solution, or purchase materials for your course, visit **www.cengage.com.**

Printed in the United States of America
1 2 3 4 5   23 22 21 20 19

1: Describing Data with Graphs

Section 1.1

1.1.1 The experimental unit, the individual or object on which a variable is measured, is the student.

1.1.3 The experimental unit is the patient.

1.1.5 The experimental unit is the car.

1.1.7 "Number of students" is a *quantitative* variable because a numerical quantity (1, 2, etc.) is measured.

1.1.9 "State of residence" is a *qualitative* variable since a quality (CA, MT, AL, etc.) is measured.

1.1.11 "Weight" is a *continuous* variable, taking on any values associated with an interval on the real line.

1.1.13 "Number of consumers" is a *discrete* variable because it can take on only integer values.

1.1.15 "Time" is a *continuous* variable, taking on any values associated with an interval on the real line..

1.1.17 "Number of brothers and sisters" is integer-valued and hence *discrete*.

1.1.19 The statewide database contains a record of all drivers in the state of Michigan. The data collected represents the *population* of interest to the researcher.

1.1.21 The researcher is interested in the weight gain of all animals that might be put on this diet, not just the twenty animals that have been observed. The responses of these twenty animals is a *sample*.

1.1.23   **a**   The experimental unit, the item or object on which variables are measured, is the vehicle.

     b   Type (qualitative); make (qualitative); carpool or not? (qualitative); one-way commute distance (quantitative continuous); age of vehicle (quantitative continuous)

     c   Since five variables have been measured, this is *multivariate data.*

1.1.25   **a**   The population of interest consists of voter opinions (for or against the candidate) <u>at the time of the election</u> for all persons voting in the election.

     b   Note that when a sample is taken (at some time prior or the election), we are not actually sampling from the population of interest. As time passes, voter opinions change. Hence, the population of voter opinions changes with time, and the sample may not be representative of the population of interest.

1.1.27   **a**   The variable "reading score" is a quantitative variable, which is probably integer-valued and hence discrete.

     b   The individual on which the variable is measured is the student.

     c   The population is hypothetical – it does not exist in fact – but consists of the reading scores for all students who could possibly be taught by this method.

Section 1.2

1.2.1 The pie chart is constructed by partitioning the circle into five parts, according to the total contributed by each part. Since the total number of students is 100, the total number receiving a final grade of A represents $31/100 = 0.31$ or 31% of the total. Thus, this category will be represented by a sector angle of $0.31(360) = 111.6°$. The other sector angles are shown next, along with the pie chart.

Final Grade	Frequency	Fraction of Total	Sector Angle
A	31	.31	111.6
B	36	.36	129.6
C	21	.21	75.6
D	9	.09	32.4
F	3	.03	10.8

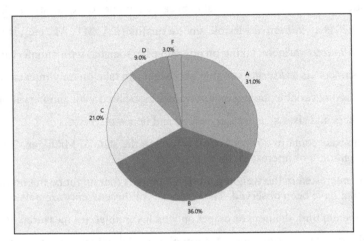

The bar chart represents each category as a bar with height equal to the frequency of occurrence of that category and is shown in the figure that follows.

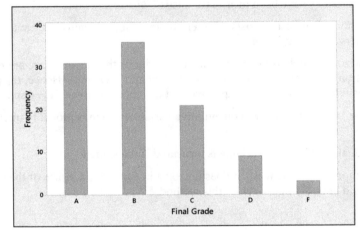

1.2.3 Construct a statistical table to summarize the data. The pie and bar charts are shown in the figures that follow.

Status	Frequency	Fraction of Total	Sector Angle
Humanities, Arts & Sciences	43	.43	154.8
Natural/Agricultural Sciences	32	.32	115.2
Business	17	.17	61.2
Other	8	.08	28.8

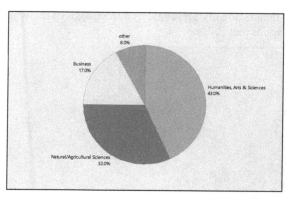 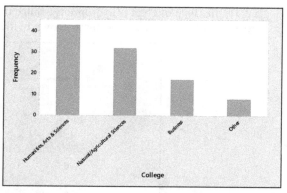

1.2.5 a-b Construct a statistical table to summarize the data. The pie and bar charts are shown in the figures that follow.

State	Frequency	Fraction of Total	Sector Angle
CA	9	.36	129.6
AZ	8	.32	115.2
TX	8	.32	115.2

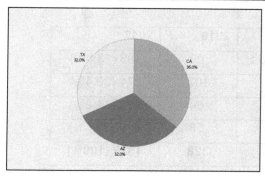 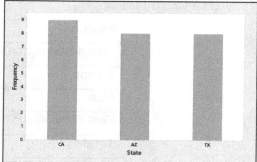

c From the table or the chart, Texas produced $8/25 = 0.32$ of the jeans.

d The highest bar represents California, which produced the most pairs of jeans.

e Since the bars and the sectors are almost equal in size, the three states produced roughly the same number of pairs of jeans.

1.2.7, 9 The bar charts represent each category as a bar with height equal to the frequency of occurrence of that category.

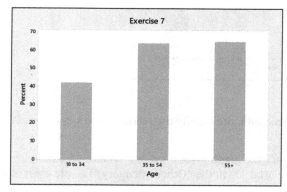 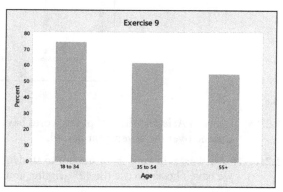

1.2.11 a The percentages given in the exercise only add to 94%. We should add another category called "Other", which will account for the other 6% of the responses.

b Either type of chart is appropriate. Since the data is already presented as percentages of the whole group, we choose to use a pie chart, shown in the figure that follows.

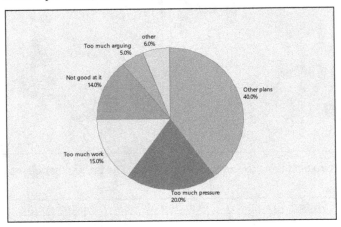

c-d Answers will vary.

1.2.13 The percentages falling in each of the four categories in 2017 are shown below (in parentheses), and the bar chart for 2010 follow.

Region	2010	2017
United States/Canada	99	183 (13.8%)
Europe	107	271 (20.4%)
Asia	64	453 (34.2%)
Rest of the World	58	419 (31.6%)
Total	328	1326 (100%)

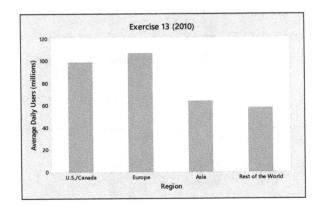

1.2.15 Users in Asia and the rest of the world have increased more rapidly than those in the U.S., Canada or Europe over the seven-year period.

1.2.17-18 Answers will vary from student to student. Since the graph gives a range of values for Zimbabwe's share, we have chosen to use the 13% figure, and have used 3% in the "Other" category. The pie chart is shown next.

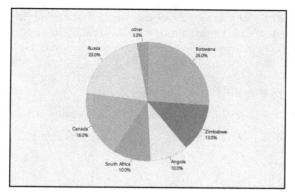

1.2.19 The Pareto chart is shown next, with the bars arranged from largest to smallest percent share.

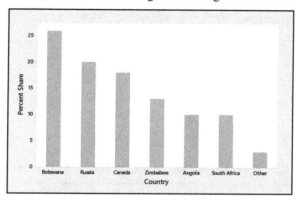

1.2.21 The data should be displayed with either a bar chart or a pie chart. The pie chart is shown next.

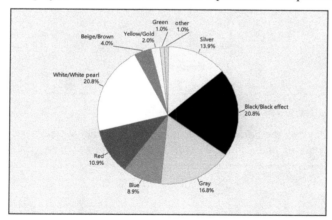

Section 1.3

1.3.1 The dotplot is shown next; the data is skewed right, with one outlier, $x = 2.0$.

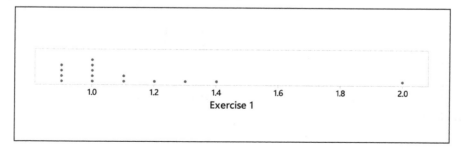

1.3.3, 5 The most obvious choice of a stem is to use the ones digit. The portion of the observation to the right of the ones digit constitutes the leaf. Observations are classified by row according to stem and also within each stem according to relative magnitude. The stem and leaf display is shown next.

```
1   6 8
2   1 2 5 5 5 7 8 8 9 9
3   1 1 4 5 5 6 6 6 7 7 7 7 8 9 9 9        leaf digit = 0.1
4   0 0 0 1 2 2 3 4 5 6 7 8 9 9 9          1  2 represents 1.2
5   1 1 6 6 7
6   1 2
```

3. The stem and leaf display has a mound shaped distribution, with no outliers.

5. The eight and ninth largest observations are both 4.9 (4 9).

1.3.7 The stems are split, with the leaf digits 0 to 4 belonging to the first part of the stem and the leaf digits 5 to 9 belonging to the second.

```
3 | 2 3 4
3 | 5 5 5 6 6 7 9 9 9 9        leaf digit = 0.1   1  2 represents 1.2
4 | 0 0 2 2 3 3 3 4 4
4 | 5 8
```

1.3.9 The scale is drawn on the horizontal axis and the measurements are represented by dots.

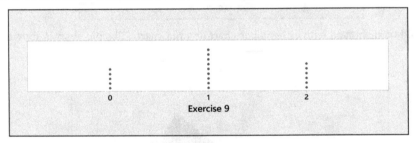

1.3.11 The distribution is relatively mound-shaped, with no outliers.

1.3.13 The line chart plots "day" on the horizontal axis and "time" on the vertical axis. The line chart shown next reveals that learning is taking place, since the time decreases each successive day.

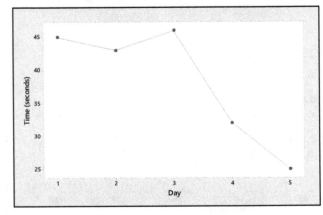

1.3.15 The dotplot is shown next.

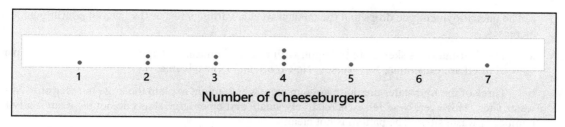

    a   The distribution is somewhat mound-shaped (as much as a small set can be); there are no outliers.

    b $2/10 = 0.2$

1.3.17 **a**   We choose a stem and leaf plot, using the ones and tenths place as the stem, and a zero digit as the leaf. The *Minitab* printout is shown next.

```
Stem-and-leaf of RBC Count   N = 15
  1    49   0
  2    50   0
  3    51   0
 (5)   52   00000
  7    53   000
  4    54   000
  1    55   0

Leaf Unit = 0.01
```

  b   The data set is relatively mound-shaped, centered at 5.2.

  c   The value $x = 5.7$ does not fall within the range of the other cell counts, and would be considered somewhat unusual.

1.3.19 **a** Stem and leaf displays may vary from student to student. The most obvious choice is to use the tens digit as the stem and the ones digit as the leaf.

```
 7 | 8 9
 8 | 0 1 7
 9 | 0 1 2 4 4 5 6 6 6 8 8
10 | 1 7 9
11 | 2
```

  b The display is fairly mound-shaped, with a large peak in the middle.

1.3.21 **a-b** The bar charts for the median weekly earnings and unemployment rates for eight different levels of education are shown next.

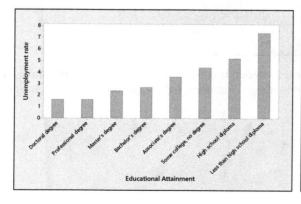 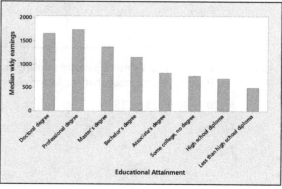

c The unemployment rate drops and the median weekly earnings rise as the level of educational attainment increases.

1.3.23 a The distribution is skewed to the right, with a several unusually large measurements. The five states marked as HI are California, New Jersey, New York and Pennsylvania.

b Three of the four states are quite large in area, which might explain the large number of hazardous waste sites. However, New Jersey is relatively small, and other large states do not have unusually large number of waste sites. The pattern is not clear.

1.3.25 a Answers will vary.

b The stem and leaf plot is constructed using the tens place as the stem and the ones place as the leaf. Notice that the distribution is roughly mound-shaped.

```
Stem-and-leaf of Ages   N = 38

   2    4    69
   7    5    36778
  19    6    003344567778
  19    7    0112347889
   9    8    01358
   4    9    0033

   Leaf Unit = 1
```

c-d Three of the five youngest presidents – Kennedy, Lincoln and Garfield – were assassinated while in office. This would explain the fact that their ages at death were in the lower tail of the distribution.

Section 1.4

1.4.1 The relative frequency histogram displays the relative frequency as the height of the bar over the appropriate class interval and is shown next. The distribution is relatively mound-shaped.

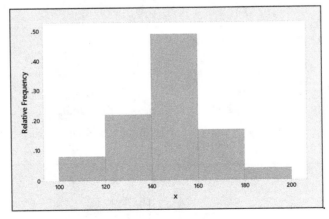

1.4.3, 5, 7 The proportion of measurements falling in each interval is equal to the sum of the heights of the bars over that interval. Remember that the lower class boundary is included, but not the upper class boundary.

    3. .20 + .40 + .15 = .75

    5. .05

    7. .15

1.4.9 Answers will vary. The range of the data is $110 - 10 = 90$ and we need to use seven classes. Calculate $90/7 = 12.86$ which we choose to round up to 15. Convenient class boundaries are created, starting at 10: 10 to < 25, 25 to < 40, ..., 100 to < 115.

1.4.11 Answers will vary. The range of the data is $1.73 - .31 = 1.42$ and we need to use ten classes. Calculate $1.42/10 = .142$ which we choose to round up to .15. Convenient class boundaries are created, starting at .30: .30 to < .45, .45 to < .60, ..., 1.65 to < 1.80.

1.4.13, 15 The table containing the classes, their corresponding frequencies and their relative frequencies and the relative frequency histogram are shown next.

Class i	Class Boundaries	Tally	f_i	Relative frequency, f_i/n
1	1.6 to < 2.1	11	2	.04
2	2.1 to < 2.6	11111	5	.10
3	2.6 to < 3.1	11111	5	.10
4	3.1 to < 3.6	11111	5	.10
5	3.6 to < 4.1	11111 11111 1111	14	.28
6	4.1 to < 4.6	11111 11	7	.14
7	4.6 to < 5.1	11111	5	.10
8	5.1 to < 5.6	11	2	.04
9	5.6 to < 6.1	111	3	.06
10	6.1 to < 6.6	11	2	.04

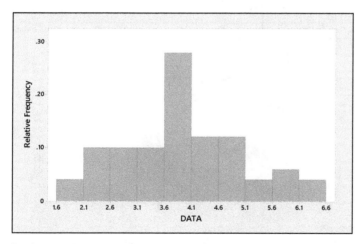

13. The distribution is roughly mound-shaped.

15. The fraction larger than 3.6 lies in classes 5-10, or $(14 + 7 + \cdots + 3 + 2)/50 = 33/50 = 0.66$.

1.4.17, 19 Since the variable of interest can only take the values 0, 1, or 2, the classes can be chosen as the integer values 0, 1, and 2. The table shows the classes, their corresponding frequencies and their relative frequencies. The relative frequency histogram follows the table.

Value	Frequency	Relative Frequency
0	5	.25
1	9	.45
2	6	.30

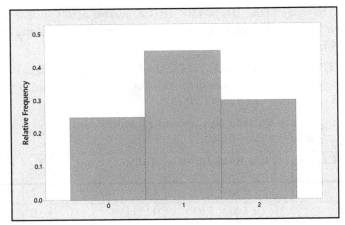

17. Using the table above, the proportion of measurements greater than 1 is the same as the proportion of "2"s, or 0.30.

19. The probability of selecting a "2" in a random selection from these twenty measurements is $6/20 = .30$.

1.4.21, 23 Answers will vary. The range of the data is $94 - 55 = 39$ and we choose to use 5 classes. Calculate $39/5 = 7.8$ which we choose to round up to 10. Convenient class boundaries are created, starting at 50 and the table and relative frequency histogram are created.

Class Boundaries	Frequency	Relative Frequency
50 to < 60	2	.10
60 to < 70	6	.30
70 to < 80	3	.15
80 to < 90	6	.30
90 to < 100	3	.15

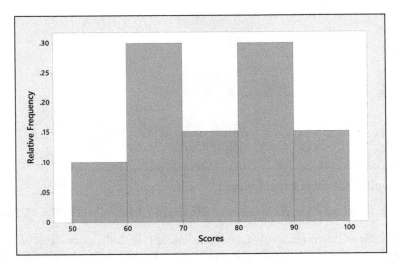

21. The distribution has two peaks at about 65 and 85. Depending on the way in which the student constructs the histogram, these peaks may or may not be clearly seen.

23. The shapes are roughly the same, but this may not be the case if the student constructs the histogram using different class boundaries.

1.4.25 a The range of the data $32.3 - 0.2 = 32.1$. We choose to use eleven class intervals of length 3 ($32.1/11 = 2.9$, which when rounded to the next largest integer is 3). The subintervals 0.1 to < 3.1, 3.1 to < 6.1, 6.1 to < 9.1, and so on, are convenient and the tally and relative frequency histogram are shown next.

Class i	Class Boundaries	Tally	f_i	Relative frequency, f_i/n
1	0.1 to < 3.1	11111 11111 11111	15	15/50
2	3.1 to < 6.1	11111 1111	9	9/50
3	6.1 to < 9.1	11111 11111	10	10/50
4	9.1 to < 12.1	111	3	3/50
5	12.1 to < 15.1	1111	4	4/50
6	15.1 to < 18.1	111	3	3/50
7	18.1 to < 21.1	11	2	2/50
8	21.1 to < 24.1	11	2	2/50
9	24.1 to < 37.1	1	1	1/50
10	27.1 to < 30.1		0	0/50
11	30.1 to < 33.1	1	1	1/50

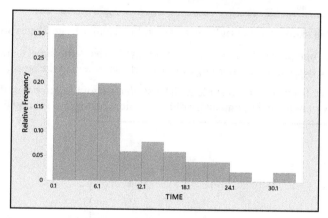

b The data is skewed to the right, with a few unusually large measurements.

c Looking at the data, we see that 36 patients had a disease recurrence within 10 months. Therefore, the fraction of recurrence times less than or equal to 10 is $36/50 = 0.72$.

1.4.27 **a** The data ranges from .2 to 5.2, or 5.0 units. Since the number of class intervals should be between five and twelve, we choose to use eleven class intervals, with each class interval having length 0.50 ($5.0/11 = .45$, which, rounded to the nearest convenient fraction, is .50). We must now select interval boundaries such that no measurement can fall on a boundary point. The subintervals .1 to < .6, .6 to < 1.1, and so on, are convenient and a tally is constructed.

Class i	Class Boundaries	Tally	f_i	Relative frequency, f_i/n
1	0.1 to < 0.6	11111 11111	10	.167
2	0.6 to < 1.1	11111 11111 11111	15	.250
3	1.1 to < 1.6	11111 11111 11111	15	.250
4	1.6 to < 2.1	11111 11111	10	.167
5	2.1 to < 2.6	1111	4	.067
6	2.6 to < 3.1	1	1	.017
7	3.1 to < 3.6	11	2	.033
8	3.6 to < 4.1	1	1	.017
9	4.1 to < 4.6	1	1	.017
10	4.6 to < 5.1		0	.000
11	5.1 to < 5.6	1	1	.017

The relative frequency histogram is shown next.

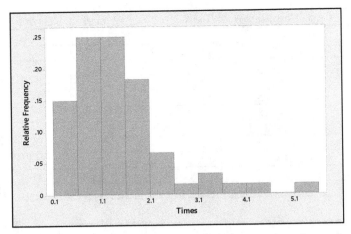

b The distribution is skewed to the right, with several unusually large observations.

c For some reason, one person had to wait 5.2 minutes. Perhaps the supermarket was understaffed that day, or there may have been an unusually large number of customers in the store.

1.4.29 a-b Answers will vary from student to student. The students should notice that the distribution is skewed to the right with a few pennies being unusually old. A typical histogram is shown next.

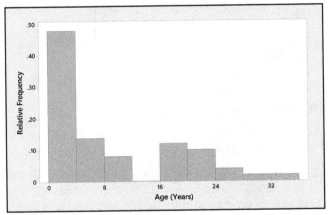

1.4.31 a Answers will vary from student to student. The relative frequency histogram below was constructed using classes of length 1.0 starting at $x = 4$. The value $x = 35.1$ is not shown in the table but appears on the graph shown next.

Class i	Class Boundaries	Tally	f_i	Relative frequency, f_i/n
1	4.0 to < 5.0	1	1	1/54
2	5.0 to < 6.0	0	0	0/54
3	6.0 to < 7.0	11111 1	6	6/54
4	7.0 to < 8.0	11111 11111 11111	15	15/54
5	8.0 to < 9.0	11111 111	8	8/54
6	9.0 to < 10.0	11111 11111 111	13	13/54
7	10.0 to < 11.0	11111 11	7	7/54
8	11.0 to < 12.0	111	3	3/54

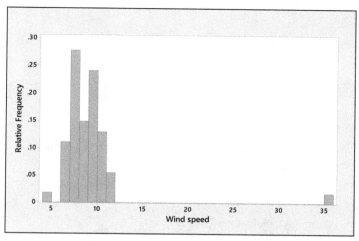

b Since Mt. Washington is a very mountainous area, it is not unusual that the average wind speed would be very high.

c The value $x = 9.9$ does not lie far from the center of the distribution (excluding $x = 35.1$). It would not be considered unusually high.

1.4.33 a The relative frequency histogram below was constructed using classes of length 1.0 starting at $x = 0.0$.

Class i	Class Boundaries	Tally	f_i	Relative frequency, f_i/n
1	0.0 to < 1.0	11	2	2/39
2	1.0 to < 2.0	11	2	2/39
3	2.0 to < 3.0	1	1	1/39
4	3.0 to < 4.0	111	3	3/39
5	4.0 to < 5.0	111	4	4/39
6	5.0 to < 6.0	11111	5	5/39
7	6.0 to < 7.0	111	3	3/39
8	7.0 to < 8.0	11111	5	5/39
9	8.0 to < 9.0	11111 111	8	8/39
10	9.0 to < 10.0	11111 1	6	6/39

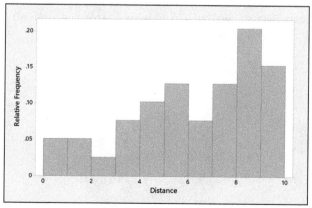

a The distribution is skewed to the left, with slightly higher frequency in the first two classes (within two miles of UCR).

b As the distance from UCR increases, each successive area increases in size, thus allowing for more Starbucks stores in that region.

Reviewing What You've Learned

1.R.1 **a** "Ethnic origin" is a *qualitative variable* since a quality (ethnic origin) is measured.

b "Score" is a *quantitative variable* since a numerical quantity (0-100) is measured.

c "Type of establishment" is a *qualitative variable* since a category (Carl's Jr., McDonald's or Burger King) is measured.

d "Mercury concentration" is a *quantitative variable* since a numerical quantity is measured.

1.R.3 **a** The length of time between arrivals at an outpatient clinic is a continuous random variable, since it can be any of the infinite number of positive real values.

b The time required to finish an examination is a continuous random variable as was the random variable described in part **a**.

c Weight is continuous, taking any positive real value.

d Body temperature is continuous, taking any real value.

e Number of people is discrete, taking the values 0, 1, 2, …

1.R.5 **a**   **b**

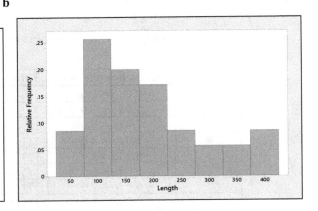

```
Stem-and-leaf of Length   N = 35

    7    0   6779999
  (11)   1   00122334444
   17    1   5799
   13    2   0004
    9    2   669
    6    3   0
    5    3   5679
    1    4   2

Leaf Unit = 10
```

c These data are skewed right.

1.R.7 **a** The popular vote within each state should vary depending on the size of the state. Since there are several very large states (in population) in the United States, the distribution should be skewed to the right.

b-c Histograms will vary from student to student but should resemble the histogram generated by **Minitab** in the next figure. The distribution is indeed skewed to the right, with three "outliers" – California, Florida and Texas.

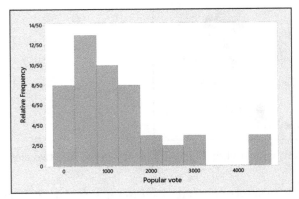

1.R.9 **a-b** Popular vote is skewed to the right while the percentage of popular vote is roughly mound-shaped. While the distribution of popular vote has outliers (California, Florida and Texas), there are no outliers in the distribution of percentage of popular vote. When the stem and leaf plots are turned 90°, the shapes are very similar to the histograms.

c Once the size of the state is removed by calculating the percentage of the popular vote, the unusually large values in the set of "popular votes" will disappear, and each state will be measured on an equal basis. The data then distribute themselves in a mound-shape around the average percentage of the popular vote.

1.R.11 a-b Answers will vary from student to student. A typical histogram is shown next—the distribution is skewed to the right, with an extreme outlier (Texas).

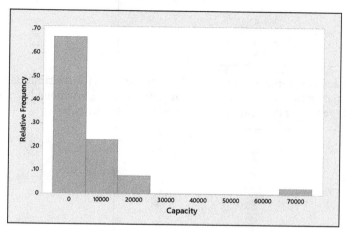

c Answers will vary.

1.R.13 a-b The *Minitab* stem and leaf plot is shown next. The distribution is slightly skewed to the right.

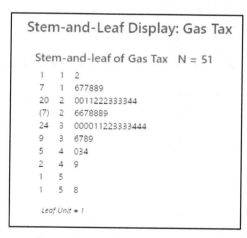

c Pennsylvania (58.20) has an unusually high gas tax.

1.R.15 a-b The distribution is approximately mound-shaped, with one unusual measurement, in the class with midpoint at 100.8°. Perhaps the person whose temperature was 100.8 has some sort of illness coming on?

c The value 98.6° is slightly to the right of center.

On Your Own

1.R.17 a The line chart is shown next. The year in which a horse raced does not appear to have an effect on his winning time.

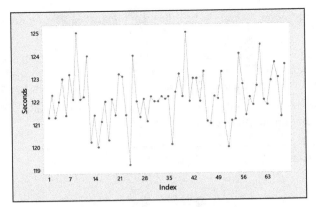

b Since the year of the race is not important in describing the data set, the distribution can be described using a relative frequency histogram. The distribution that follows is roughly mound-shaped with an unusually fast ($x = 119.2$) race times the year that *Secretariat* won the derby.

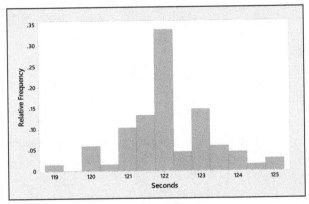

1.R.19 Answers will vary. A typical relative frequency histogram is shown next. There is an unusual bimodal feature.

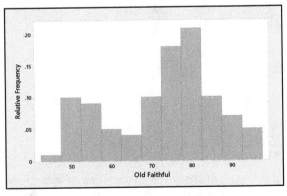

2: Describing Data with Numerical Measures

Section 2.1

2.1.1 The mean is the sum of the measurements divided by the number of measurements, or

$$\bar{x} = \frac{\sum x_i}{n} = \frac{0+5+1+1+3}{5} = \frac{10}{5} = 2$$

To calculate the median, the observations are first ranked from smallest to largest: 0, 1, 1, 3, 5. Then since $n = 5$, the position of the median is $0.5(n+1) = 3$, and the median is the 3rd ranked measurement, or $m = 1$. The mode is the measurement occurring most frequently, or mode = 1. The three measures are located on the dotplot that follows.

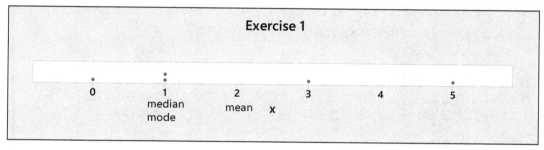

2.1.3 The mean is $\bar{x} = \frac{\sum x_i}{n} = \frac{57}{10} = 5.7$. The ranked observations are: 2, 3, 4, 5, 5, 5, 6, 8, 9, 10. Since $n = 10$, the median is halfway between the 5th and 6th ordered observations, or $m = (5+5)/2 = 5$. The measurement $x = 5$ occurs three times. Since this is the highest frequency of occurrence for the data set, the *mode* is 5. The dotplot is shown next.

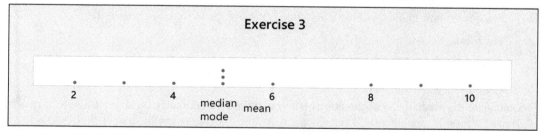

2.1.5 Since the mean is less than the median, the data set is skewed left.

2.1.7 Since the mean is greater than the median, the data set is skewed right.

2.1.9 The mean is

$$\bar{x} = \frac{\sum x_i}{n} = \frac{94.3}{24} = 3.929$$

The data set is ordered from smallest to largest below. The position of the median is $.5(n+1) = 12.5$ and the median is the average of the 12th and 13th measurements, or $(3.9 + 3.9)/2 = 3.9$.

3.2	3.5	3.7	3.9	4.2	4.4
3.3	3.5	3.9	4.0	4.3	4.4
3.4	3.6	3.9	4.0	4.3	4.5
3.5	3.6	3.9	4.2	4.3	4.8

2.1.11 The mean is $\bar{x} = \dfrac{\sum x_i}{n} = \dfrac{862}{15} = 57.467$. The ranked observations are: 53, 54, 54, 56, 56, 56, 58, 58, 58, 58, 58, 60, 60, 61, 62. Since $n = 15$, the median is the $.5(15+1) = 8^{\text{th}}$ ordered observation, or $m = 58$. The measurement $x = 58$ occurs five times. Since this is the highest frequency of occurrence for the data set, the *mode* is 58.

2.1.13 Similar to previous exercises. The mean is

$$\bar{x} = \dfrac{\sum x_i}{n} = \dfrac{0.99 + 1.92 + \cdots + 0.66}{14} = \dfrac{12.55}{14} = 0.896$$

To calculate the median, rank the observations from smallest to largest. The position of the median is $0.5(n+1) = 7.5$, and the median is the average of the 7th and 8th ranked measurement or $m = (0.67 + 0.69)/2 = 0.68$. The mode is .60, which occurs twice. Since the mean is slightly larger than the median, the distribution is slightly skewed to the right.

2.1.15 Similar to previous exercises. The mean is

$$\bar{x} = \dfrac{\sum x_i}{n} = \dfrac{2150}{10} = 215$$

The ranked observations are: 175, 185, 190, 190, 200, 225, 230, 240, 250, 265. The position of the median is $0.5(n+1) = 5.5$ and the median is the average of the 5th and 6th observation or

$$\dfrac{200 + 225}{2} = 212.5$$

and the mode is 190 which occurs twice. The mean is slightly larger than the median, and the data may be slightly skewed right.

2.1.17

a Although there may be a few households who own more than one DVR, the majority should own either 0 or 1. The distribution should be slightly skewed to the right.

b Since most households will have only one DVR, we guess that the mode is 1.

c The mean is

$$\bar{x} = \dfrac{\sum x_i}{n} = \dfrac{1 + 0 + \cdots + 1}{25} = \dfrac{27}{25} = 1.08$$

To calculate the median, the observations are first ranked from smallest to largest: There are six 0s, thirteen 1s, four 2s, and two 3s. Then since $n = 25$, the position of the median is $0.5(n+1) = 13$, which is the 13th ranked measurement, or $m = 1$. The mode is the measurement occurring most frequently, or mode = 1.

d The relative frequency histogram is shown next, with the three measures superimposed. Notice that the mean falls slightly to the right of the median and mode, indicating that the measurements are slightly skewed to the right.

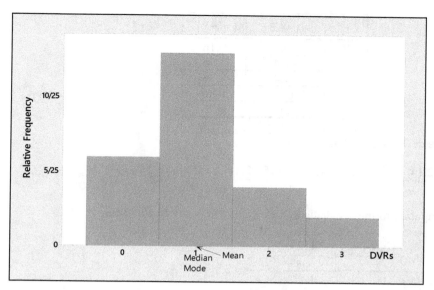

2.1.19 It is obvious that any one family cannot have 2.5 children, since the number of children per family is a quantitative discrete variable. The researcher is referring to the average number of children per family calculated for all families in the United States during the 1930s. The average does not necessarily have to be integer-valued.

2.1.21 a The mean is

$$\bar{x} = \frac{\sum x_i}{n} = \frac{10.4 + 10.7 + \cdots + 8.8}{21} = \frac{203.5}{21} = 9.690$$

To calculate the median, rank the observations from smallest to largest.

6.4	6.5	6.8	6.8	8.2	8.2
8.6	8.8	8.9	9.0	9.2	
9.7	10.0	10.0	10.4	10.4	
10.7	12.0	13.0	14.3	15.6	

Then since $n = 21$, the position of the median is $0.5(n+1) = 11$, the 11th ranked measurement or $m = 9.2$. There are four modes—6.8, 8.2, 10.0 and 10.4—all of which occur twice.

b-c Since the mean is larger than the median, the data set is skewed right. The dotplot is shown next.

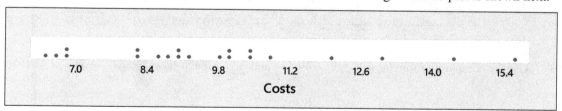

Section 2.2

2.2.1 Calculate $\bar{x} = \frac{\sum x_i}{n} = \frac{12}{5} = 2.4$. Then create a table of differences, $(x_i - \bar{x})$ and their squares, $(x_i - \bar{x})^2$.

x_i	$x_i - \bar{x}$	$(x_i - \bar{x})^2$
2	−0.4	0.16
1	−1.4	1.96
1	−1.4	1.96
3	0.6	0.36
5	2.6	6.76
Total	0	11.20

Then, using the definition formula,

$$s^2 = \frac{\Sigma(x_i - \bar{x})^2}{n-1} = \frac{(2-2.4)^2 + \cdots + (5-2.4)^2}{4} = \frac{11.20}{4} = 2.8$$

To use the computing formula, calculate $\Sigma x_i^2 = 2^2 + 1^2 + \cdots + 5^2 = 40$. Then

$$s^2 = \frac{\Sigma x_i^2 - \frac{(\Sigma x_i)^2}{n}}{n-1} = \frac{40 - \frac{(12)^2}{5}}{4} = \frac{11.2}{4} = 2.8.$$

The sample standard deviation is the positive square root of the variance or

$$s = \sqrt{s^2} = \sqrt{2.8} = 1.673$$

2.2.3 Calculate $\bar{x} = \frac{\Sigma x_i}{n} = \frac{31}{8} = 3.875$. Then create a table of differences, $(x_i - \bar{x})$ and their squares, $(x_i - \bar{x})^2$.

x_i	$x_i - \bar{x}$	$(x_i - \bar{x})^2$	x_i	$x_i - \bar{x}$	$(x_i - \bar{x})^2$
3	−.875	.765625	4	.125	.015625
1	−2.875	8.265625	4	.125	.015625
5	1.125	1.265625	3	−.875	.765625
6	2.125	4.515625	5	1.125	1.265625

Then, using the definition formula,

$$s^2 = \frac{\Sigma(x_i - \bar{x})^2}{n-1} = \frac{(3-3.875)^2 + \cdots + (5-3.875)^2}{7} = \frac{16.875}{7} = 2.4107$$

To use the computing formula, calculate $\Sigma x_i^2 = 3^2 + 1^2 + \cdots + 5^2 = 137$. Then

$$s^2 = \frac{\Sigma x_i^2 - \frac{(\Sigma x_i)^2}{n}}{n-1} = \frac{137 - \frac{(31)^2}{8}}{7} = \frac{16.875}{7} = 2.4107 \text{ and } s = \sqrt{s^2} = \sqrt{2.4107} = 1.553.$$

2.2.5 Use the computing formula, with $\Sigma x_i^2 = 49{,}634$ and $\Sigma x_i = 862$. Then $\bar{x} = \frac{\Sigma x_i}{n} = \frac{862}{15} = 57.467$,

$$s^2 = \frac{\Sigma x_i^2 - \frac{(\Sigma x_i)^2}{n}}{n-1} = \frac{49634 - \frac{(862)^2}{15}}{14} = \frac{97.73333}{14} = 6.98095$$

and $s = \sqrt{s^2} = \sqrt{6.98095} = 2.642$.

The range is calculated as $R = 62 - 53 = 9$ and $R/s = 9/2.642 = 3.41$, so that the range is approximately 3.5 standard deviations.

2.2.7 The range is $R = 1.92 - .53 = 1.39$. Use the computing formula, with $\Sigma x_i^2 = 13.3253$ and $\Sigma x_i = 12.55$. Then,

$$s^2 = \frac{\sum x_i^2 - \frac{(\sum x_i)^2}{n}}{n-1} = \frac{13.3253 - \frac{(12.55)^2}{14}}{13} = \frac{2.07512}{13} = .15962 \text{ and } s = \sqrt{s^2} = \sqrt{.15962} = .3995.$$

2.2.9 The range is $R = 265 - 175 = 90$. Use the computing formula, with $\sum x_i^2 = 470,900$ and $\sum x_i = 2150$. Then,

$$s^2 = \frac{\sum x_i^2 - \frac{(\sum x_i)^2}{n}}{n-1} = \frac{470,900 - \frac{(2150)^2}{10}}{9} = \frac{8650}{9} = 961.11111 \text{ and } s = \sqrt{s^2} = \sqrt{961.11111} = 31.002.$$

2.2.11   **a**   The range is $R = 459.21 - 233.97 = 225.24$.   **b** $\bar{x} = \frac{\sum x_i}{n} = \frac{3777.30}{12} = 314.775$

     c   Calculate $\sum x_i^2 = 243.92^2 + 233.97^2 + \cdots + 286.41^2 = 1,267,488.3340$. Then

$$s^2 = \frac{\sum x_i^2 - \frac{(\sum x_i)^2}{n}}{n-1} = \frac{1267488.334 - \frac{(3777.30)^2}{12}}{11} = 7135.338773$$

and $s = \sqrt{s^2} = \sqrt{7135.338773} = 84.471$.

2.2.13   **a**   Max = 27, Min = 20.2 and the range is $R = 27 - 20.2 = 6.8$.

     b   Answers will vary. A typical histogram is shown next. The distribution is slightly skewed to the left.

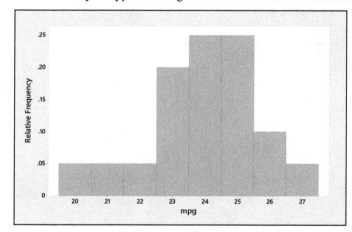

     c   Calculate $n = 20$, $\sum x_i = 479.2$, $\sum x_i^2 = 11532.82$. Then $\bar{x} = \frac{\sum x_i}{n} = \frac{479.2}{20} = 23.96$

$$s = \sqrt{\frac{\sum x_i^2 - \frac{(\sum x_i)^2}{n}}{n-1}} = \sqrt{\frac{11532.82 - \frac{(479.2)^2}{20}}{19}} = \sqrt{2.694} = 1.641$$

2.2.15 They are probably referring to the average number of times that men and women go camping per year.

Section 2.3

2.3.1 The range of the data is $R = 6 - 1 = 5$ and the range approximation with $n = 10$ is $s \approx \frac{R}{3} = 1.67$. The standard deviation of the sample is

$$s = \sqrt{s^2} = \sqrt{\frac{\sum x_i^2 - \frac{(\sum x_i)^2}{n}}{n-1}} = \sqrt{\frac{130 - \frac{(32)^2}{10}}{9}} = \sqrt{3.0667} = 1.751$$

which is very close to the estimate.

2.3.3 The range of the data is $R = 7.1 - 4.0 = 3.1$ and the range approximation with $n = 15$ is $s \approx \frac{R}{3} = 1.033$. The standard deviation of the sample is

$$s = \sqrt{s^2} = \sqrt{\frac{\sum x_i^2 - \frac{(\sum x_i)^2}{n}}{n-1}} = \sqrt{\frac{531.17 - \frac{(88.1)^2}{15}}{14}} = \sqrt{.98067} = .990$$

which is very close to the estimate.

2.3.5 From the dotplot below, you can see that the data set is relatively mound-shaped. Hence you can use both Tchebysheff's Theorem and the Empirical Rule to describe the data.

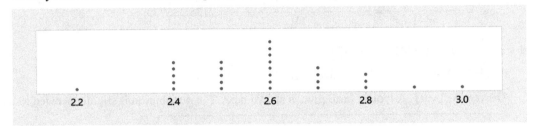

2.3.7, 9, 11 Since the distribution is relatively mound-shaped, use the Empirical Rule.

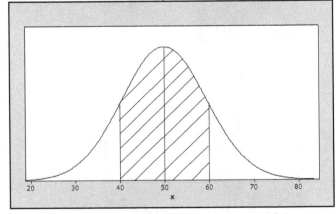

7. Refer to the figure above. Using the Empirical Rule, the interval $\mu \pm 2\sigma = 50 \pm 2(10)$ or between 30 and 70 contains approximately .95 (or 95%) of the measurements.

9. From the figure above, the interval from 40 to 60 represents $\mu \pm \sigma = 50 \pm 10$. Since the distribution is relatively mound-shaped, the proportion of measurements between 40 and 60 is .68 (or 68%) according to the Empirical Rule. The proportion of the measurements between 50 and 60 is then half of .68 or .34 and the proportion of the measurements which are greater than 50 is 0.50. Therefore, the proportion that is greater than 60 must be $0.5 - 0.34 = 0.16$.

11. From the figure above the proportion of the measurements between 40 and 50 is 0.34 and the proportion of the measurements which are greater than 50 is 0.50. Therefore, the proportion that is 40 or more must be $0.5 + 0.34 = 0.84$.

2.3.13 Since nothing is known about the shape of the data distribution, you must use Tchebysheff's Theorem to describe the data. The interval from 65 to 85 represents $\mu \pm 2\sigma$ which will contain at least 3/4 of the measurements.

2.3.15 The range of the data is $R = 1.1 - 0.5 = 0.6$ and the approximate value of s is

$$s \approx \frac{R}{3} = 0.2$$

Calculate $\sum x_i = 7.6$ and $\sum x_i^2 = 6.02$. The sample variance is

$$s^2 = \frac{\sum x_i^2 - \frac{(\sum x_i)^2}{n}}{n-1} = \frac{6.02 - \frac{(7.6)^2}{10}}{9} = \frac{0.244}{9} = .027111$$

and the standard deviation of the sample is $s = \sqrt{.027111} = 0.165$ which is very close to the range approximation.

2.3.17 The range is $R = 125 - 25 = 100$ and the approximate value of s is

$$s \approx \frac{R}{3} = 33.33$$

Use the computing formula, with $\sum x_i^2 = 131,125$ and $\sum x_i = 1305$. Then,

$$s^2 = \frac{\sum x_i^2 - \frac{(\sum x_i)^2}{n}}{n-1} = \frac{131,125 - \frac{(1305)^2}{15}}{14} = \frac{17,590}{14} = 1256.42857 \text{ and } s = \sqrt{s^2} = \sqrt{1256.42857} = 35.446.$$

which is very close to the range approximation.

2.3.19 a The stem and leaf plot generated by *Minitab* shows that the data is roughly mound-shaped. Note however the gap in the center of the distribution and the two measurements in the upper tail.

```
Stem-and-leaf of Weight   N = 27
 1    7   5
 2    8   3
 6    8   7999
 8    9   23
13    9   66789
13   10
(3)  10   688
11   11   2244
 7   11   788
 4   12   4
 3   12   8
 2   13
 2   13   8
 1   14   1
Leaf Unit = 0.01
```

b Calculate $\sum x_i = 28.41$ and $\sum x_i^2 = 30.6071$, the sample mean is $\bar{x} = \frac{\sum x_i}{n} = \frac{28.41}{27} = 1.052$ and the standard deviation of the sample is

$$s = \sqrt{s^2} = \sqrt{\frac{\sum x_i^2 - \frac{(\sum x_i)^2}{n}}{n-1}} = \sqrt{\frac{30.6071 - \frac{(28.41)^2}{27}}{26}} = 0.166$$

c The following table gives the actual percentage of measurements falling in the intervals $\bar{x} \pm ks$ for $k = 1, 2, 3$.

k	$\bar{x} \pm ks$	Interval	Number in Interval	Percentage
1	1.052 ± 0.166	0.866 to 1.218	21	78%
2	1.052 ± 0.332	0.720 to 1.384	26	96%
3	1.052 ± 0.498	0.554 to 1.550	27	100%

d The percentages in part **c** do not agree too closely with those given by the Empirical Rule, especially in the one standard deviation range. This is caused by the lack of mounding (indicated by the gap) in the center of the distribution.

e The lack of any one-pound packages is probably a marketing technique intentionally used by the supermarket. People who buy slightly less than one-pound would be drawn by the slightly lower price, while those who need exactly one-pound of meat for their recipe might tend to opt for the larger package, increasing the store's profit.

2.3.21 a The stem and leaf plots are shown next. The second set has a slightly higher location and spread.

```
Stem-and-leaf of Titanium content    Method = 1    N = 10
  1    10   0
  3    11   00
  4    12   0
 (4)   13   0000
  2    14   0
  1    15   0

Leaf Unit = 0.0001
```

```
Stem-and-leaf of Titanium content    Method = 2    N = 10
  1    11   0
  3    12   00
  5    13   00
  5    14   0
  4    15   00
  2    16   0
  1    17   0

Leaf Unit = 0.0001
```

b *Method 1*: Calculate $\sum x_i = 0.125$ and $\sum x_i^2 = 0.001583$. Then $\bar{x} = \dfrac{\sum x_i}{n} = 0.0125$ and

$$s = \sqrt{s^2} = \sqrt{\dfrac{\sum x_i^2 - \dfrac{(\sum x_i)^2}{n}}{n-1}} = \sqrt{\dfrac{0.001583 - \dfrac{(0.125)^2}{10}}{9}} = 0.00151$$

Method 2: Calculate $\sum x_i = 0.138$ and $\sum x_i^2 = 0.001938$. Then $\bar{x} = \dfrac{\sum x_i}{n} = 0.0138$ and

$$s = \sqrt{s^2} = \sqrt{\frac{\sum x_i^2 - \frac{(\sum x_i)^2}{n}}{n-1}} = \sqrt{\frac{0.001938 - \frac{(0.138)^2}{10}}{9}} = 0.00193$$

The results confirm the conclusions of part **a**.

2.3.23 **a** Similar to previous exercises. The intervals, counts and percentages are shown in the table.

k	$\bar{x} \pm ks$	Interval	Number in Interval	Percentage
1	4.586 ± 2.892	1.694 to 7.478	43	61%
2	4.586 ± 5.784	−1.198 to 10.370	70	100%
3	4.586 ± 8.676	−4.090 to 13.262	70	100%

b The percentages in part **a** do not agree with those given by the Empirical Rule. This is because the shape of the distribution is not mound-shaped, but flat.

2.3.25 **a** The value of x is $\mu - \sigma = 32 - 36 = -4$.

b The interval $\mu \pm \sigma$ is 32 ± 36 or −4 to 68. This interval should contain approximately 68% of the survival times, so that 16% will be longer than 68 days and 16% will be less than −4 days.

c The latter is clearly impossible. Therefore, the approximate values given by the Empirical Rule are not accurate, indicating that the distribution cannot be mound-shaped.

2.3.27 **a-b** Calculate $R = 93 - 51 = 42$ so that $s \approx R/4 = 42/4 = 10.5$.

c Calculate $n = 30$, $\sum x_i = 2145$ and $\sum x_i^2 = 158{,}345$. Then

$$s^2 = \frac{\sum x_i^2 - \frac{(\sum x_i)^2}{n}}{n-1} = \frac{158{,}345 - \frac{(2145)^2}{30}}{29} = 171.6379 \text{ and } s = \sqrt{171.6379} = 13.101$$

which is fairly close to the approximate value of s from part **b**.

d The two intervals are calculated below. The proportions agree with Tchebysheff's Theorem but are not too close to the percentages given by the Empirical Rule. (This is because the distribution is not quite mound-shaped.)

k	$\bar{x} \pm ks$	Interval	Fraction in Interval	Tchebysheff	Empirical Rule
2	71.5 ± 26.20	45.3 to 97.7	30/30 = 1.00	at least 0.75	≈ 0.95
3	71.5 ± 39.30	32.2 to 110.80	30/30 = 1.00	at least 0.89	≈ 0.997

2.3.29 **a** Answers will vary. A typical stem and leaf plot is generated by **Minitab**.

```
Stem-and-leaf of Completed Passes   N = 16
 1    1   8
 3    2   01
 8    2   22223
 8    2   5
 7    2   66777
 2    2   99
     Leaf Unit = 1
```

b Calculate $n = 16$, $\sum x_i = 386$ and $\sum x_i^2 = 9476$. Then $\bar{x} = \frac{\sum x_i}{n} = \frac{386}{16} = 24.125$,

$$s^2 = \frac{\sum x_i^2 - \frac{(\sum x_i)^2}{n}}{n-1} = \frac{9476 - \frac{(386)^2}{16}}{15} = 10.91667$$

and $s = \sqrt{s^2} = \sqrt{10.91667} = 3.304$.

c   Calculate $\bar{x} \pm 2s \Rightarrow 24.125 \pm 6.608$ or 17.517 to 30.733. From the original data set, 16 of the 16 measurements, or 100% fall in this interval.

2.3.31   The results of the Empirical Rule follow:

k	$\bar{x} \pm ks$	Interval	Empirical Rule
1	420 ± 5	415 to 425	approximately 0.68
2	420 ± 10	410 to 430	approximately 0.95
3	420 ± 15	405 to 435	approximately 0.997

Notice that we are assuming that attendance follows a mound-shaped distribution and hence that the Empirical Rule is appropriate.

Section 2.4

2.4.1   The ordered data are: 0, 1, 3, 4, 4, 5, 6, 6, 6, 7, 7, 8

Calculate $n = 12$, $\sum x_i = 57$ and $\sum x_i^2 = 337$. Then $\bar{x} = \frac{\sum x_i}{n} = \frac{57}{12} = 4.75$ and the sample standard deviation is

$$s = \sqrt{\frac{\sum x_i^2 - \frac{(\sum x_i)^2}{n}}{n-1}} = \sqrt{\frac{337 - \frac{(57)^2}{12}}{11}} = \sqrt{6.022727} = 2.454$$

For the smallest observation, $x = 0$, z-score $= \frac{x - \bar{x}}{s} = \frac{0 - 4.75}{2.454} = -1.94$ and for the largest observation, $x = 8$, z-score $= \frac{x - \bar{x}}{s} = \frac{8 - 4.75}{2.454} = 1.32$. Since neither z-score exceeds 2 in absolute value, none of the observations are unusually small or large.

2.4.3   The ordered data are: 2, 3, 4, 5, 6, 6, 6, 7, 8, 9, 9, 10, 25

Calculate $n = 13$, $\sum x_i = 100$ and $\sum x_i^2 = 1162$. Then $\bar{x} = \frac{\sum x_i}{n} = \frac{100}{13} = 7.692$ and the sample standard deviation is

$$s = \sqrt{\frac{\sum x_i^2 - \frac{(\sum x_i)^2}{n}}{n-1}} = \sqrt{\frac{1162 - \frac{(100)^2}{13}}{12}} = \sqrt{32.73077} = 5.721$$

For the smallest observation, $x = 2$, z-score $= \frac{x - \bar{x}}{s} = \frac{2 - 7.692}{5.721} = -.99$ and for the largest observation, $x = 25$, z-score $= \frac{x - \bar{x}}{s} = \frac{25 - 7.692}{5.721} = 3.03$. The value $x = 25$ is an outlier.

2.4.5   The ordered data are: 0, 1, 2, 3, 5, 5, 6, 6, 7. With $n = 9$, the median is in position $0.5(n+1) = 5$, the lower quartile is in position $0.25(n+1) = 2.5$ (halfway between the 2nd and 3rd ordered observations) and the upper quartile is in position $0.75(n+1) = 7.5$ (halfway between the 7th and 8th ordered observations). Hence, $m = 5$, $Q_1 = (1+2)/2 = 1.5$ and $Q_3 = (6+6)/2 = 6$.

2.4.7 The ordered data are: 0, 1, 5, 6, 7, 8, 9, 10, 12, 12, 13, 14, 16, 19, 19. With $n = 15$, the median is in position $0.5(n+1) = 8$, so that $m = 10$. The lower quartile is in position $0.25(n+1) = 4$ so that $Q_1 = 6$ and the upper quartile is in position $0.75(n+1) = 12$ so that $Q_3 = 14$. Then the five-number summary is

Min	Q₁	Median	Q₃	Max
0	6	10	14	19

and $IQR = Q_3 - Q_1 = 14 - 6 = 8$. The *lower and upper fences* are:

$$Q_1 - 1.5 IQR = 6 - 12 = -6$$
$$Q_3 + 1.5 IQR = 14 + 12 = 26$$

There are no data points that lie outside the fences. The box plot is shown next. The lower whisker connects the box to the smallest value that is not an outlier, $x = 0$. The upper whisker connects the box to the largest value that is not an outlier or $x = 19$.

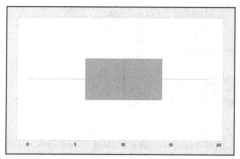

2.4.9 The ordered data are: 12, 18, 22, 23, 24, 25, 25, 26, 26, 27, 28. For $n = 11$, the position of the median is $0.5(n+1) = 0.5(11+1) = 6$ and $m = 25$. The positions of the quartiles are $0.25(n+1) = 3$ and $0.75(n+1) = 9$, so that $Q_1 = 22$, $Q_3 = 26$, and $IQR = 26 - 22 = 4$. The five-number summary is

Min	Q₁	Median	Q₃	Max
12	22	25	26	28

The *lower and upper fences* are:

$$Q_1 - 1.5 IQR = 22 - 6 = 16$$
$$Q_3 + 1.5 IQR = 26 + 6 = 32$$

The only observation falling outside the fences is $x = 12$ which is identified as an outlier. The box plot is shown next. The lower whisker connects the box to the smallest value that is not an outlier, $x = 18$. The upper whisker connects the box to the largest value that is not an outlier or $x = 28$.

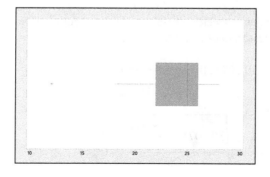

2.4.11 **Mean, standard deviation and z-scores:** From Exercise 3, the ordered data are: 2, 3, 4, 5, 6, 6, 6, 7, 8, 9, 9, 10, 25. Calculate $n = 13$, $\sum x_i = 100$ and $\sum x_i^2 = 1162$. Then $\bar{x} = \dfrac{\sum x_i}{n} = \dfrac{100}{13} = 7.692$ and the sample standard deviation is

$$s = \sqrt{\dfrac{\sum x_i^2 - \dfrac{(\sum x_i)^2}{n}}{n-1}} = \sqrt{\dfrac{1162 - \dfrac{(100)^2}{13}}{12}} = \sqrt{32.73077} = 5.721$$

For the smallest observation, $x = 2$, z-score $= \dfrac{x - \bar{x}}{s} = \dfrac{2 - 7.692}{5.721} = -.99$ and for the largest observation, $x = 25$, z-score $= \dfrac{x - \bar{x}}{s} = \dfrac{25 - 7.692}{5.721} = 3.03$. The value $x = 25$ is an outlier.

Five-number summary and box plot: For $n = 13$, the position of the median is $0.5(n+1) = 0.5(13+1) = 7$ and $m = 6$. The positions of the quartiles are $0.25(n+1) = 3.5$ and $0.75(n+1) = 10.5$, so that $Q_1 = 4.5$, $Q_3 = 9$, and $IQR = 9 - 4.5 = 4.5$. The five-number summary is

Min	Q₁	Median	Q₃	Max
2	4.5	6	9	25

The *lower and upper fences* are:

$$Q_1 - 1.5 IQR = 4.5 - 6.75 = -2.25$$
$$Q_3 + 1.5 IQR = 9 + 6.75 = 15.75$$

The only observation falling outside the fences is $x = 25$ which is identified as an outlier. The box plot is shown next. The lower whisker connects the box to the smallest value that is not an outlier, $x = 2$. The upper whisker connects the box to the largest value that is not an outlier or $x = 10$.

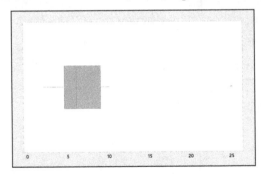

The results of the two different methods for identifying outliers are identical.

2.4.13 Since 14.5% of U.S. men are six feet or taller, $(100 - 14.5)\% = 85.5\%$ are less than six feet tall. Six feet is the 85.5th percentile.

2.4.15 Since 46% of all 19-year-old females in a certain height-weight category have a BMI greater than 21.9, $(100 - 46)\% = 54\%$ have a BMI less than or equal to 21.9. The value 21.9 is the 54th percentile.

2.4.17 Calculate $\sum x_i = 28.41$ and $\sum x_i^2 = 30.6071$, the sample mean is $\bar{x} = \dfrac{\sum x_i}{n} = \dfrac{28.41}{27} = 1.052$ and the standard deviation of the sample is

$$s = \sqrt{s^2} = \sqrt{\dfrac{\sum x_i^2 - \dfrac{(\sum x_i)^2}{n}}{n-1}} = \sqrt{\dfrac{30.6071 - \dfrac{(28.41)^2}{27}}{26}} = 0.166$$

For the smallest observation, $x = .75$, z-score $= \dfrac{x-\bar{x}}{s} = \dfrac{.75-1.052}{.166} = -1.82$ and for the largest observation, $x = 1.41$, z-score $= \dfrac{x-\bar{x}}{s} = \dfrac{1.41-1.052}{.166} = 2.16$. The value $x = 1.41$ is a suspect outlier.

2.4.19 The ordered set is: 40, 49, 52, 54, 59, 61, 67, 69, 70, 71. Since $n = 10$, the positions of m, Q_1, and Q_3 are 5.5, 2.75 and 8.25 respectively, and $m = (59+61)/2 = 60$, $Q_1 = 49+0.75(52-49) = 51.25$, $Q_3 = 69+.25(70-69) = 69.25$ and $IQR = 69.25 - 51.25 = 18$. Then the five-number summary is

Min	Q_1	Median	Q_3	Max
40	51.25	60	69.25	71

The *lower and upper fences* are:

$$Q_1 - 1.5 IQR = 51.25 - 27 = 24.25$$
$$Q_3 + 1.5 IQR = 69.25 + 27 = 96.25$$

and the box plot is shown next. There are no outliers and the data set is slightly skewed left.

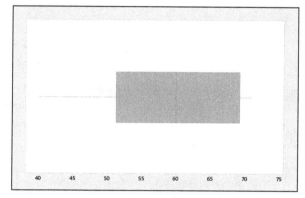

2.4.21 a For $n = 15$, the position of the median is $0.5(n+1) = 8$ and the positions of the quartiles are $0.25(n+1) = 4$ and $0.75(n+1) = 12$, while for $n = 16$, the position of the median is $0.5(n+1) = 8.5$ and the positions of the quartiles are $0.25(n+1) = 4.25$ and $0.75(n+1) = 12.75$. The sorted measurements are shown below.

Alex Smith: 14, 16, 19, 19, 20, 21, 23, 23, 25, 25, 25, 27, 27, 28, 29

Joe Flacco: 8, 9, 10, 19, 20, 20, 22, 23, 24, 25, 25, 26, 27, 29, 31, 34

For Alex Smith,

$$m = 23, Q_1 = 19 \text{ and } Q_3 = 27.$$

For Joe Flacco,

$$m = (23+24)/2 = 23.5, Q_1 = 19 + 0.25(20-19) = 19.25 \text{ and } Q_3 = 26 + 0.75(27-26) = 26.75.$$

Then the five-number summaries are

	Min	Q_1	Median	Q_3	Max
Smith	14	19	23	27	29
Flacco	8	19.25	23.5	26.75	34

b For Alex Smith, calculate $IQR = Q_3 - Q_1 = 27 - 19 = 8$. Then the *lower and upper fences* are:

$$Q_1 - 1.5IQR = 19 - 12 = 7$$
$$Q_3 + 1.5IQR = 27 + 12 = 39$$

and there are no outliers.

For Joe Flacco, calculate $IQR = Q_3 - Q_1 = 26.75 - 19.25 = 7.5$. Then the *lower and upper fences* are:

$$Q_1 - 1.5IQR = 19.25 - 11.25 = 8$$
$$Q_3 + 1.5IQR = 26.75 + 11.25 = 38$$

and there are no outliers. The box plots are shown next.

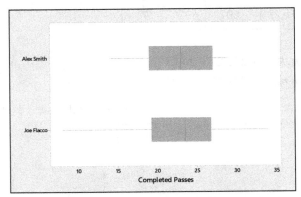

c Answers will vary. Both distributions are relatively symmetric and somewhat mound-shaped. The Flacco distribution is more variable, but Flacco has a slightly higher median number of completed passes.

2.4.23 a Just by scanning through the 20 measurements, it seems that there are a few unusually small measurements, which would indicate a distribution that is skewed to the left.

b The position of the median is $0.5(n+1) = 0.5(25+1) = 10.5$ and $m = (120+127)/2 = 123.5$. The mean is
$$\bar{x} = \frac{\sum x_i}{n} = \frac{2163}{20} = 108.15$$
which is smaller than the median, indicate a distribution skewed to the left.

c The positions of the quartiles are $0.25(n+1) = 5.25$ and $0.75(n+1) = 15.75$, so that $Q_1 = 65 - .25(87-65) = 70.5$, $Q_3 = 144 + .75(147-144) = 146.25$, and $IQR = 146.25 - 70.5 = 75.75$.

The *lower and upper fences* are:

$$Q_1 - 1.5IQR = 70.5 - 113.625 = -43.125$$
$$Q_3 + 1.5IQR = 146.25 + 113.625 = 259.875$$

The box plot is shown next. There are no outliers. The long left whisker and the median line located to the right of the center of the box indicates that the distribution that is skewed to the left.

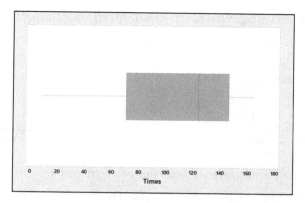

2.4.25 a Calculate $n = 50$, $\sum x_i = 418$, so that $\bar{x} = \dfrac{\sum x_i}{n} = \dfrac{418}{50} = 8.36$.

b The position of the median is $.5(n+1) = 25.5$ and $m = (4+4)/2 = 4$.

c Since the mean is larger than the median, the distribution is skewed to the right.

d Since $n = 50$, the positions of Q_1 and Q_3 are $.25(51) = 12.75$ and $.75(51) = 38.25$, respectively. Then $Q_1 = 0 + 0.75(1-0) = 12.75$, $Q_3 = 17 + .25(19-17) = 17.5$ and $IQR = 17.5 - .75 = 16.75$.

The *lower and upper fences* are:
$$Q_1 - 1.5 IQR = .75 - 25.125 = -24.375$$
$$Q_3 + 1.5 IQR = 17.5 + 25.125 = 42.625$$

and the box plot is shown next. There are no outliers and the data is skewed to the right.

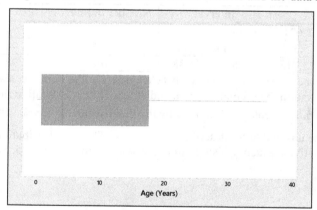

Reviewing What You've Learned

2.R.1 a Calculate $n = 14$, $\sum x_i = 367$ and $\sum x_i^2 = 9641$. Then $\bar{x} = \dfrac{\sum x_i}{n} = \dfrac{367}{14} = 26.214$ and

$$s = \sqrt{\dfrac{\sum x_i^2 - \dfrac{(\sum x_i)^2}{n}}{n-1}} = \sqrt{\dfrac{9641 - \dfrac{(367)^2}{14}}{13}} = 1.251$$

b Calculate $n = 14$, $\sum x_i = 366$ and $\sum x_i^2 = 9644$. Then $\bar{x} = \dfrac{\sum x_i}{n} = \dfrac{366}{14} = 26.143$ and

$$s = \sqrt{\dfrac{\sum x_i^2 - \dfrac{(\sum x_i)^2}{n}}{n-1}} = \sqrt{\dfrac{9644 - \dfrac{(366)^2}{14}}{13}} = 2.413$$

  c The centers are roughly the same; the Sunmaid raisins appear slightly more variable.

2.R.3 **a** Calculate $R = 499.9 - 219.9 = 280$ so that $s \approx R/4 = 280/4 = 70$.

  b Calculate $n = 25$, $\sum x_i = 7996.3$ and $\sum x_i^2 = 2634119.27$. Then $\bar{x} = \dfrac{7996.3}{25} = 319.852$

$$s^2 = \dfrac{\sum x_i^2 - \dfrac{(\sum x_i)^2}{n}}{n-1} = \dfrac{2{,}634{,}119.27 - \dfrac{(7996.3)^2}{25}}{24} = 3186.946767 \text{ and } s = \sqrt{3186.946767} = 56.453$$

which is of the same order of magnitude as the approximate value of s from part **a**.

  c Calculate $\bar{x} \pm 2s \Rightarrow 319.852 \pm 112.906$ or 206.946 to 432.758. The proportion of measurements in this interval is $24/25 = .96$, close to the proportion given by the Empirical Rule.

2.R.5 **a** Calculate $n = 50$, $\sum x_i = 418.4$ and $\sum x_i^2 = 6384.34$. Then

$$s = \sqrt{\dfrac{\sum x_i^2 - \dfrac{(\sum x_i)^2}{n}}{n-1}} = \sqrt{\dfrac{6384.34 - \dfrac{(418.4)^2}{50}}{49}} = 7.671$$

and $\bar{x} = \dfrac{\sum x_i}{n} = \dfrac{418.4}{50} = 8.368$.

The three intervals of interest is shown in the table, along with the number of observations which fall in each interval.

k	$\bar{x} \pm ks$	Interval	Number in Interval	Percentage
1	8.368±7.671	0.697 to 16.039	37	74%
2	8.368±15.342	−6.974 to 23.710	47	94%
3	8.368±23.013	−14.645 to 31.381	49	98%

  b The percentages falling in the intervals do agree with Tchebysheff's Theorem. At least 0 fall in the first interval, at least $3/4 = 0.75$ fall in the second interval, and at least $8/9 = 0.89$ fall in the third. The percentages are not too close to the percentages described by the Empirical Rule (68%, 95%, and 99.7%).

  c The Empirical Rule may be unsuitable for describing these data. The data distribution does not have a strong mound-shape (see the relative frequency histogram in the solution to Exercise 25, Section 1.4), but is skewed to the right.

2.R.7 First calculate the intervals:

   $\bar{x} \pm s = 0.17 \pm 0.01$ or 0.16 to 0.18

   $\bar{x} \pm 2s = 0.17 \pm 0.02$ or 0.15 to 0.19

   $\bar{x} \pm 3s = 0.17 \pm 0.03$ or 0.14 to 0.20

  a If no prior information as to the shape of the distribution is available, we use Tchebysheff's Theorem. We would expect at least $(1 - 1/1^2) = 0$ of the measurements to fall in the interval 0.16 to 0.18; at least $(1 - 1/2^2) = 3/4$ of the measurements to fall in the interval 0.15 to 0.19; at least $(1 - 1/3^2) = 8/9$ of the measurements to fall in the interval 0.14 to 0.20.

  b According to the Empirical Rule, approximately 68% of the measurements will fall in the interval 0.16 to 0.18; approximately 95% of the measurements will fall between 0.15 to 0.19; approximately 99.7% of the measurements will fall between 0.14 and 0.20. Since mound-shaped distributions are so frequent, if we do have a sample size of 30 or greater, we expect the sample distribution to be mound-shaped. Therefore, in this exercise, we would expect the Empirical Rule to be suitable for describing the set of data.

c If the chemist had used a sample size of four for this experiment, the distribution would not be mound-shaped. Any possible histogram we could construct would be non-mound-shaped. We can use at most 4 classes, each with frequency 1, and we will not obtain a histogram that is even close to mound-shaped. Therefore, the Empirical Rule would not be suitable for describing $n = 4$ measurements.

2.R.9 a For $n = 10$, the position of the median is $0.5(n+1) = 5.5$ and the positions of the quartiles are $0.25(n+1) = 2.75$ and $0.75(n+1) = 8.25$. The sorted data are: 5, 6, 6, 6.75, 7, 7, 7, 7.25, 8, 8.5

Then $m = (7+7)/2 = 7$, $Q_1 = 6 + 0.75(6-6) = 6$ and $Q_3 = 7.25 + 0.25(8-7.25) = 7.4375$.

Calculate $IQR = 7.4375 - 6 = 1.4375$ and the *lower and upper fences*:

$$Q_1 - 1.5 IQR = 6 - 2.15625 = 3.84$$
$$Q_3 + 1.5 IQR = 7.4375 + 2.15625 = 9.59$$

The box plot is shown next.

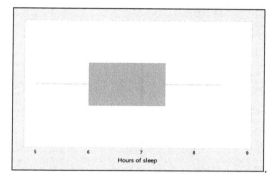

b There are no outliers, not even $x = 8.5$!

2.R.11 If the distribution is mound-shaped, then almost all of the measurements will fall in the interval $\mu \pm 3\sigma$, which is an interval 6σ in length. That is, the range of the measurements should be approximately 6σ. In this case, the range is $800 - 200 = 600$, so that $\sigma \approx 600/6 = 100$.

2.R.13 a The range is $R = 172 - 108 = 64$ and the range approximation is

$$s \approx R/4 = 64/4 = 16$$

b Calculate $n = 15$, $\sum x_i = 2041$, $\sum x_i^2 = 281{,}807$. Then $\bar{x} = \dfrac{\sum x_i}{n} = \dfrac{2041}{15} = 136.07$

$$s = \sqrt{\dfrac{\sum x_i^2 - \dfrac{(\sum x_i)^2}{n}}{n-1}} = \sqrt{\dfrac{281{,}807 - \dfrac{(2041)^2}{15}}{14}} = \sqrt{292.495238} = 17.102$$

c According to Tchebysheff's Theorem, with $k = 2$, at least 3/4 or 75% of the measurements will lie within $k = 2$ standard deviations of the mean. For this data, the two values, *a* and *b*, are calculated as

$$\bar{x} \pm 2s \Rightarrow 136.07 \pm 2(17.10) \Rightarrow 137.07 \pm 34.20 \text{ or } a = 101.87 \text{ and } b = 170.27.$$

2.R.15 a The range is $R = 19 - 4 = 15$ and the range approximation is $s \approx R/4 = 15/4 = 3.75$.

b Calculate $n = 15$, $\sum x_i = 175$, $\sum x_i^2 = 2237$. Then $\bar{x} = \dfrac{\sum x_i}{n} = \dfrac{175}{15} = 11.67$

$$s = \sqrt{\frac{\sum x_i^2 - \frac{(\sum x_i)^2}{n}}{n-1}} = \sqrt{\frac{2237 - \frac{(175)^2}{15}}{14}} = \sqrt{13.95238} = 3.735$$

 c Calculate the interval $\bar{x} \pm 2s \Rightarrow 11.67 \pm 2(3.735) \Rightarrow 11.67 \pm 7.47$ or 4.20 to 19.14. Referring to the original data set, the fraction of measurements in this interval is 14/15 = .93.

2.R.17 We must estimate s and compare with the student's value of 0.263. In this case, $n = 20$ and the range is $R = 17.4 - 16.9 = 0.5$. The estimated value for s is then $s \approx R/4 = 0.5/4 = 0.125$ which is less than 0.263. It is important to consider the magnitude of the difference between the "rule of thumb" and the calculated value. For example, if we were working with a standard deviation of 100, a difference of 0.142 would not be great. However, the student's calculation is twice as large as the estimated value. Moreover, two standard deviations, or $2(0.263) = 0.526$, already exceeds the range. Thus, the value $s = 0.263$ is probably incorrect. The correct value of s is

$$s = \sqrt{\frac{\sum x_i^2 - \frac{(\sum x_i)^2}{n}}{n-1}} = \sqrt{\frac{5851.95 - \frac{117032.41}{20}}{19}} = \sqrt{0.0173} = 0.132$$

2.R.19 **a** Calculate $n = 25$, $\sum x_i = 104.9$, $\sum x_i^2 = 454.810$. Then $\bar{x} = \frac{\sum x_i}{n} = \frac{104.9}{25} = 4.196$

$$s = \sqrt{\frac{\sum x_i^2 - \frac{(\sum x_i)^2}{n}}{n-1}} = \sqrt{\frac{454.810 - \frac{(104.9)^2}{25}}{24}} = \sqrt{.610} = .781$$

 b The ordered data set is shown below:

```
2.5   3.0   3.1   3.3   3.6
3.7   3.8   3.8   3.9   3.9
4.1   4.2   4.2   4.2   4.3
4.3   4.4   4.7   4.7   4.8
4.8   5.2   5.3   5.4   5.7
```

 c The z-scores for $x = 2.5$ and $x = 5.7$ are

$$z = \frac{x - \bar{x}}{s} = \frac{2.5 - 4.196}{.781} = -2.17 \text{ and } z = \frac{x - \bar{x}}{s} = \frac{5.7 - 4.196}{.781} = 1.93$$

Since neither of the z-scores are greater than 3 in absolute value, the measurements are not judged to be extremely unlikely; however, the minimum value, $x = -2.17$ is somewhat unlikely since it is greater than 2 standard deviations from the mean.

On Your Own

2.R.21 **a** Calculate $n = 15$, $\sum x_i = 21$ and $\sum x_i^2 = 49$. Then $\bar{x} = \frac{\sum x_i}{n} = \frac{21}{15} = 1.4$ and

$$s^2 = \frac{\sum x_i^2 - \frac{(\sum x_i)^2}{n}}{n-1} = \frac{49 - \frac{(21)^2}{15}}{14} = 1.4$$

 b Using the frequency table and the grouped formulas, calculate

$$\sum x_i f_i = 0(4) + 1(5) + 2(2) + 3(4) = 21$$

$$\sum x_i^2 f_i = 0^2(4) + 1^2(5) + 2^2(2) + 3^2(4) = 49$$

Then, as in part **a**, $\bar{x} = \dfrac{\sum x_i f_i}{n} = \dfrac{21}{15} = 1.4$

$$s^2 = \dfrac{\sum x_i^2 f_i - \dfrac{(\sum x_i f_i)^2}{n}}{n-1} = \dfrac{49 - \dfrac{(21)^2}{15}}{14} = 1.4$$

2.R.23 a The data in this exercise have been arranged in a frequency table.

x_i	0	1	2	3	4	5	6	7	8	9	10
f_i	10	5	3	2	1	1	1	0	0	1	1

Using the frequency table and the grouped formulas, calculate

$$\sum x_i f_i = 0(10) + 1(5) + \cdots + 10(1) = 51$$

$$\sum x_i^2 f_i = 0^2(10) + 1^2(5) + \cdots + 10^2(1) = 293$$

Then $\bar{x} = \dfrac{\sum x_i f_i}{n} = \dfrac{51}{25} = 2.04$

$$s^2 = \dfrac{\sum x_i^2 f_i - \dfrac{(\sum x_i f_i)^2}{n}}{n-1} = \dfrac{293 - \dfrac{(51)^2}{25}}{24} = 7.873 \text{ and } s = \sqrt{7.873} = 2.806.$$

b-c The three intervals $\bar{x} \pm ks$ for $k = 1, 2, 3$ are calculated in the table along with the actual proportion of measurements falling in the intervals. Tchebysheff's Theorem is satisfied and the approximation given by the Empirical Rule are fairly close for $k = 2$ and $k = 3$.

k	$\bar{x} \pm ks$	Interval	Fraction in Interval	Tchebysheff	Empirical Rule
1	2.04 ± 2.806	-0.766 to 4.846	$21/25 = 0.84$	at least 0	≈ 0.68
2	2.04 ± 5.612	-3.572 to 7.652	$23/25 = 0.92$	at least 0.75	≈ 0.95
3	2.04 ± 8.418	-6.378 to 10.458	$25/25 = 1.00$	at least 0.89	≈ 0.997

2.R.25 Answers will vary. The student should notice the outliers in the female group, and that the median female temperature is higher than the median male temperature.

3: Describing Bivariate Data

Section 3.1

3.1.1 The side-by-side bar measures the frequency of occurrence for each of the two categories, A and B. A separate bar is used for each of the other two categories—X and Y. The student may choose to group by X and Y first, which will change the look of the bar chart.

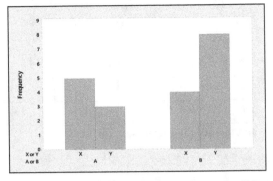

3.1.3 The side-by-side pie charts are constructed as in Chapter 1 for each of the two groups (men and women) and are displayed next using the percentages shown in the table below.

	Group 1	Group 2	Group 3	Total
Men	23%	31%	46%	100%
Women	8%	57%	35%	100%

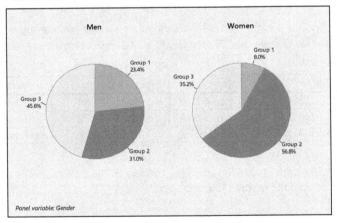

3.1.5 Use the data from Exercise 3. The stacked bar chart shows the frequency of occurrence for each of the three groups. A portion of each bar is used for men and women.

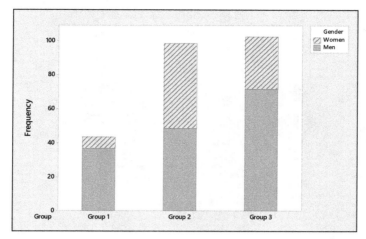

The differences in the proportions of men and women in the three groups is most graphically portrayed by the pie charts, since the unequal number of men and women tend to confuse the interpretation of the bar charts. However, the bar chart is useful in retaining the actual frequencies of occurrence in each group, which is lost in the pie chart.

3.1.7 The conditional distribution in each of the groups given that the person was male is shown in the table that follows. These are the values that were used to construct the pie charts in Exercise 3.

	Group 1	Group 2	Group 3	Total
Men	23%	31%	46%	100%

3.1.9 **a** Any of the comparative charts (side-by-side pie charts, stacked or side-by-side bar charts) can be used.

b-c The two types of comparative bar charts are shown next. The amounts spent in each of the four categories seem to be quite different for men and women, except in category C. In category C which involves the largest dollar amount of purchase, there is little difference between the genders.

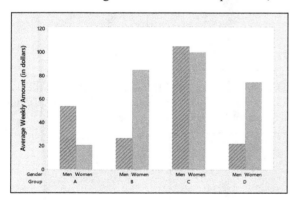 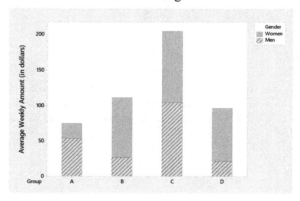

d Although it is really a matter of preference, the only advantage to the stacked chart is that the reader can easily see the total dollar amount for each category. For comparison purposes, the side-by-side chart may be better.

3.1.11 It is clear that there are more plain M&Ms than peanut M&Ms, since the peanut M&Ms are larger. We choose to ignore the differences in total numbers by using side-by-side pie charts (below). There is a distinct difference in the distribution of colors.

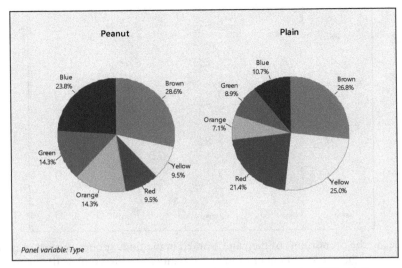

Panel variable: Type

3.1.13 **a** The side-by-side comparative bar charts that follow measure the CPIs on the vertical axis, while the clusters of bars represent the grouping into housing (stripes) and transportation (solid) over the years.

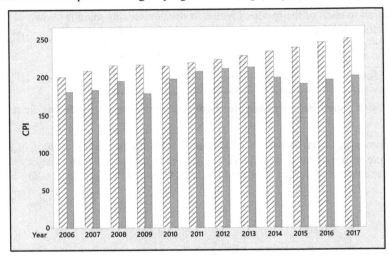

b-c The comparative line chart is shown next. The line chart is more effective, showing that the housing CPI has increased steadily, except for a small dip around 2009-2010, but that the CPI for transportation has been a lot more variable, with increases and decreases over the years.

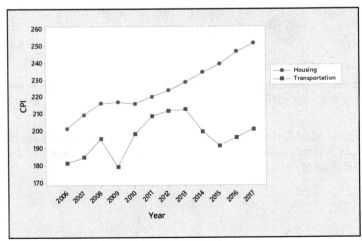

3.1.15 The side-by-side bar chart and the comparative line charts are shown next. Conclusions will vary from student to student, but it is clear that markets in Asia and the rest of the world are increasing at a faster rate than Europe, the United States and Canada.

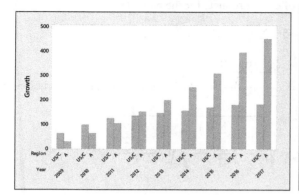 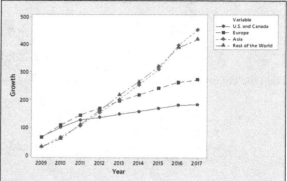

Section 3.2

3.2.1 The line is plotted next. The change in y for a one-unit change in x is defined as the *slope* of the line—the coefficient of the x variable. For this line, the slope is $b = 0.5$. The point at which the line crosses the y-axis is called the y-intercept and is the constant term in the equation of the line. For this line, the y-intercept is $a = 2.0$. When $x = 2.5$, the predicted value of y is $y = 2.0 + 0.5(2.5) = 3.25$.

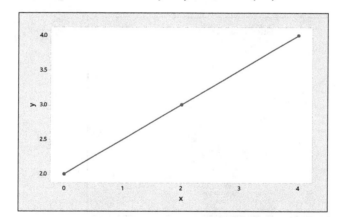

3.2.3 Similar to Exercise 1. The line is plotted next. The *slope* of the line is $b = -6$ and the y-intercept is $a = 5$. When $x = 2.5$, the predicted value of y is $y = 5 - 6(2.5) = -10$.

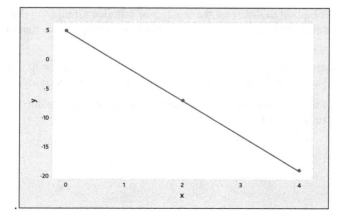

3.2.5 The pattern is strong linear, with two outliers.

3.2.7 The pattern is strong linear, with two distinct clusters.

3.2.9 Calculate the sums, sums of square sand sum of cross products for the pairs (x_i, y_i).

$$\sum x_i = 6;\ \sum y_i = 12;\ \sum x_i^2 = 14;\ \sum y_i^2 = 56;\ \sum x_i y_i = 20$$

Then the covariance is $s_{xy} = \dfrac{\sum x_i y_i - \dfrac{(\sum x_i)(\sum y_i)}{n}}{n-1} = \dfrac{20 - \dfrac{(6)(12)}{3}}{2} = -2$. The sample standard deviations are

$s_x = \sqrt{\dfrac{\sum x_i^2 - \dfrac{(\sum x_i)^2}{n}}{n-1}} = \sqrt{\dfrac{14 - \dfrac{(6)^2}{3}}{2}} = 1$ and $s_y = \sqrt{\dfrac{\sum y_i^2 - \dfrac{(\sum y_i)^2}{n}}{n-1}} = \sqrt{\dfrac{56 - \dfrac{(12)^2}{3}}{2}} = 2$

and the correlation coefficient is $r = \dfrac{s_{xy}}{s_x s_y} = \dfrac{-2}{(1)(2)} = -1$. The slope and y-intercept of the regression line

are $b = r\dfrac{s_y}{s_x} = -1\left(\dfrac{2}{1}\right) = -2$ and $a = \bar{y} - b\bar{x} = \dfrac{12}{3} - (-2)\left(\dfrac{6}{3}\right) = 8$

and the equation of the regression line is $\hat{y} = 8 - 2x$. The graph of the data points and the best fitting line is shown next. Notice that, since $r = -1$, the points all fall exactly on the straight line, which has a negative slope.

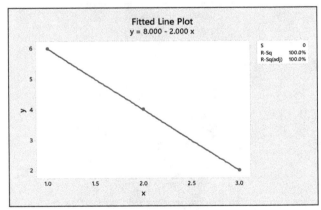

3.2.11, 13 Answers will vary, depending on the type of calculator the student is using. The results should agree with the results of Exercises 8 and 10.

3.2.15 The number of calories burned *depends on* the number of minutes running on the treadmill. Number of calories, y, is the dependent variable and number of minutes running on the treadmill, x, is the independent variable.

3.2.17 The number of ice cream cones sold *depends on* the temperature on a given day. Number of cones sold, y, is the dependent variable and daily temperature, x, is the independent variable.

3.2.19 **a-b** The scatterplot is shown next. Notice that there is a negative relationship between y and x. The measurements are decreasing over time.

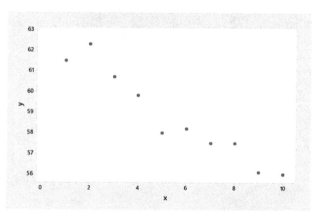

c From the Minitab printout, $s_{xy} = -6.42222$, $s_x = \sqrt{9.16667}$ and $s_y = \sqrt{4.84933}$ so that

$$r = \frac{s_{xy}}{s_x s_y} = \frac{-6.42222}{\sqrt{(9.16667)(4.84933)}} = -0.9632$$

d The slope and y-intercept of the regression line are

$$b = r\frac{s_y}{s_x} = -0.9632\left(\frac{2.2021}{3.0277}\right) = -0.700 \text{ and } a = \bar{y} - b\bar{x} = \frac{587.6}{10} - (-0.700)\left(\frac{55}{10}\right) = 62.61$$

and the equation of the regression line is $\hat{y} = 62.61 - 0.70x$.

e The graph of the data points and the best fitting line is shown next. The line fits through the points very well.

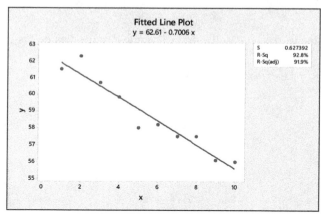

3.2.21 Since BMI is a function of income, BMI, y, will *depend on* income, x. Use your scientific calculator. You can verify that $\sum x_i = 240.5$; $\sum y_i = 162$;

$\sum x_i^2 = 12{,}370.25$; $\sum y_i^2 = 4446.22$; $\sum x_i y_i = 6090.65$. Then the covariance is

$$s_{xy} = \frac{\sum x_i y_i - \frac{(\sum x_i)(\sum y_i)}{n}}{n-1} = \frac{6090.65 - \frac{240.5(162)}{6}}{5} = -80.57$$

The sample standard deviations are $s_x = 23.3675$ and $s_y = 3.8005$ so that $r = -.9072$. Then

$$b = r\frac{s_y}{s_x} = -.14755 \text{ and } a = \bar{y} - b\bar{x} = 27 - (-.14755)(40.083333) = 32.914$$

and the equation of the regression line is $\hat{y} = 32.914 - .148x$. The graph of the data points and the best fitting line is shown next. The line provides a good description of the data.

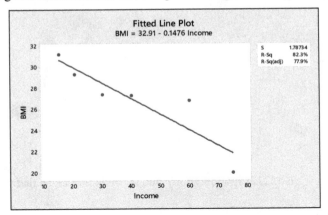

3.2.23 a Similar to previous exercises. Calculate $n = 12; \sum x_i = 20,980; \sum y_i = 4043.5; \sum x_i^2 = 37,551,600;$ $\sum y_i^2 = 1,401,773.75; \sum x_i y_i = 7,240,383$. Then the covariance is

$$s_{xy} = \frac{\sum x_i y_i - \frac{(\sum x_i)(\sum y_i)}{n}}{n-1} = 15,545.19697$$

The sample standard deviations are $s_x = 281.48416$ and $s_y = 59.75916$ so that $r = 0.9241$. Then

$$b = r\frac{s_y}{s_x} = 0.19620 \text{ and } a = \bar{y} - b\bar{x} = 336.95833 - 0.19620(1748.333) = -6.06$$

and the equation of the regression line is $y = -6.06 + 0.1962x$. The graph of the data points and the best fitting line is shown next.

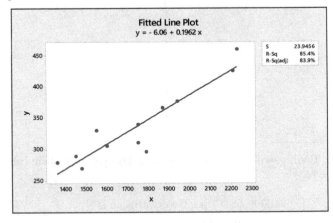

b The fitted line describes the data quite well.

3.2.25 a Since we would be interested in predicting the number of chirps based on temperature, the number of chirps is the dependent variable (y) and temperature is the independent variable (x).

b The scatterplot is shown next. There appears to be a positive linear trend, as expected.

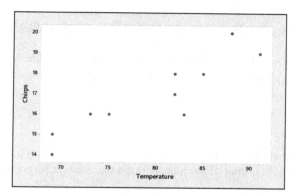

c Calculate $n = 10$; $\sum x_i = 797$; $\sum y_i = 169$; $\sum x_i^2 = 64,063$; $\sum y_i^2 = 2887$; $\sum x_i y_i = 13,586$. Then the covariance is

$$s_{xy} = \frac{\sum x_i y_i - \frac{(\sum x_i)(\sum y_i)}{n}}{n-1} = 12.966667$$

The sample standard deviations are $s_x = 7.761014$ and $s_y = 1.852926$ so that $r = 0.9017$. Then (using full accuracy), $b = r\frac{s_y}{s_x} = .21527$ and $a = \bar{y} - b\bar{x} = 16.9 - .21527(79.7) = -.257$ and the equation of the regression line is $\hat{y} = -.257 + .215x$.

d When $x = 80$, the predicted number of chirps is $y = -.257 + .215(80) = 16.943$, or approximately 17 chirps.

3.2.27 **a** The height of the building will in part *depend upon* how many floors there are. Then $y =$ height and $x =$ number of floors.

b Calculate $n = 28$; $\sum x_i = 1378$; $\sum y_i = 18,637$; $\sum x_i^2 = 70,250$; $\sum y_i^2 = 12,960,309$; $\sum x_i y_i = 952,387$. Then the covariance is

$$s_{xy} = \frac{\sum x_i y_i - \frac{(\sum x_i)(\sum y_i)}{n}}{n-1} = 1302.97619$$

The sample standard deviations are $s_x = 9.4921$ and $s_y = 143.4223$ so that $r = .9571$. There is a strong correlation between the number of floors and the height of the building.

c The scatterplot is shown next. There is a strong linear relationship with no outliers.

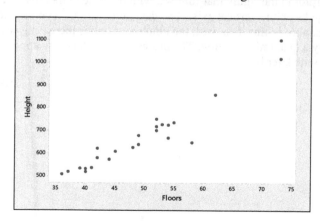

d The slope of the line is $b = r\dfrac{s_y}{s_x} = 14.4614$ and $a = \bar{y} - b\bar{x} = 665.6071 - 14.4614(49.2143) = -46.10$ and the equation of the regression line is $\hat{y} = -46.10 + 14.46x$. If $x = 48$, we predict that the height of the building will be $\hat{y} = -46.10 + 14.46(48) = 647.98$ or approximately 648 feet.

Reviewing What You've Learned

3.R.1 a-b The scatterplot is shown next. There is a fairly strong positive linear relationship between x and y, with one possible outlier in the lower right corner of the plot. Perhaps that student forgot to study for midterm 2!

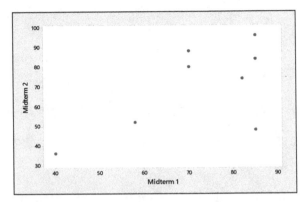

3.R.3 a-b The scatterplot is shown next. There is a strong positive linear relationship between x and y.

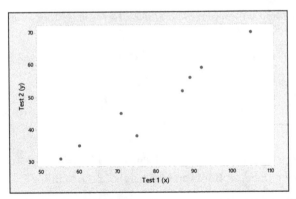

3.R.5 a-b The data is graphed as a scatterplot in the figure that follows, with the time in months plotted on the horizontal axis and the number of books on the vertical axis. The data points are then connected to form a line graph. There is a very distinct pattern in this data, with the number of books increasing with time, a response which might be modeled by a quadratic equation. The professor's productivity appears to increase, with less time required to write later books.

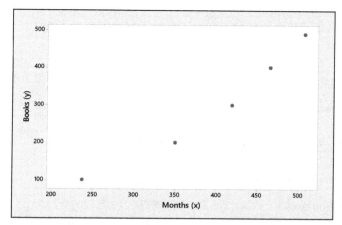

3.R.7 **a-b** The scatterplot is shown next. Four of the measurements are clustered together, and the other (Smart Beat American) is an outlier containing much less sodium and calories. There seems to be very little pattern to the plot.

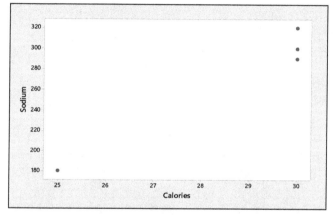

b Based only on these two variables, *Smart Beat* would be the brand of choice. However, it might taste poorly, or have some other undesirable qualities (cholesterol, price, melting ability, etc.).

3.R.9 **a** The scatterplot is shown next. The relationship is positive linear.

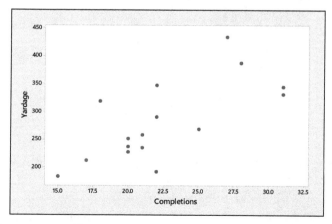

b There are no outliers to the linear trend; however, in Weeks 9 and 11, Rivers threw an unusually low number of passes.

c $n = 16$; $\sum x_i = 360$; $\sum y_i = 4515$; $\sum x_i^2 = 8432$; $\sum y_i^2 = 1,351,901$; $\sum x_i y_i = 105,161$.

Then the covariance is

$$s_{xy} = \frac{\sum x_i y_i - \frac{(\sum x_i)(\sum y_i)}{n}}{n-1} = 238.233333$$

The sample standard deviations are $s_x = 4.704608$ and $s_y = 72.0298260$ so that $r = 0.70302$.

    d   Calculate (using full accuracy)

$$b = r\frac{s_y}{s_x} = 10.76355 \text{ and } a = \bar{y} - b\bar{x} = 282.1875 - 10.76355(22.5) = 40.008$$

and the equation of the regression line is $\hat{y} = 40.01 + 10.76x$.

    e   Use the regression line from part **d**. When $x = 20$, $\hat{y} = 40.01 + 10.76(20) = 255.21$.

3.R.11 **a** The calculations for the correlation coefficients are the same as used in previous exercises. The *Minitab* printout shows the three correlation coefficients in a 2 x 2 matrix.

```
Correlations
          Al      Fe
Fe      -0.617
Mg      -0.189   0.626

Cell Contents
    Pearson correlation
```

    b   There is a strong positive relationship between iron and magnesium oxide, while there is a strong negative relationship between aluminum and iron oxide. There is very little relationship between aluminum and magnesium oxide.

3.R.13 **a** The scatterplot is shown next. There is a positive linear relationship between arm span and height.

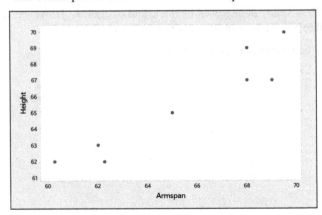

    b   Calculate $n = 8$; $\sum x_i = 524$; $\sum y_i = 525$; $\sum x_i^2 = 34413.375$; $\sum y_i^2 = 34521$; $\sum x_i y_i = 34462$.

Then the covariance is $s_{xy} = \dfrac{\sum x_i y_i - \frac{(\sum x_i)(\sum y_i)}{n}}{n-1} = 10.64286$. The sample standard deviations are $s_x = 3.61297$ and $s_y = 3.1139089$ so that $r = 0.946$.

    c   Since DaVinci indicated that a person's arm span is roughly equal to his height, the slope of the line should be approximately equal to 1.

d   Calculate (using full accuracy) $b = r\dfrac{s_y}{s_x} = .815$ and $a = \bar{y} - b\bar{x} = 65.625 - .815(65.5) = 12.22$ and the equation of the regression line is $\hat{y} = 12.22 + .815x$.

e   Use the regression line from part **d**. When $x = 62$, $\hat{y} = 12.22 + .815(62) = 62.75$ inches.

3.R.15   **a**   The scatterplot is shown next. The relationship is positive linear, but very weak.

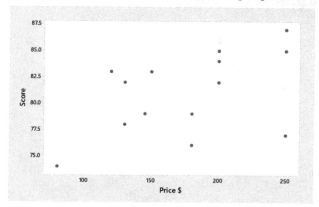

b   Calculate $n = 14$; $\sum x_i = 2465$; $\sum y_i = 1134$; $\sum x_i^2 = 470{,}425$; $\sum y_i^2 = 92{,}048$; $\sum x_i y_i = 200{,}935$. Then

the covariance is $s_{xy} = \dfrac{\sum x_i y_i - \dfrac{(\sum x_i)(\sum y_i)}{n}}{n-1} = 97.6923077$. The sample standard deviations are

$s_x = 52.921516$ and $s_y = 3.8630399$ so that $r = .47786$. Calculate (using full accuracy) $b = r\dfrac{s_y}{s_x} = .034882$

and $a = \bar{y} - b\bar{x} = 81 - .034882(176.07143) = 74.858$ and the equation of the regression line is $\hat{y} = 74.858 + .035x$.

c   Because the correlation between x and y is so weak, the fit is not good, and the regression line will not be effective.

On Your Own

3.R.17   Answers will vary. Here are the correlations for the six variables. There are strong positive correlations between total gross and number of weeks in the theatre, and between weekend gross and average gross.

Correlations					
	Weekend Gross	Theater Count	Average	Total Gross	Budget
Theater Count	0.433				
Average	0.948	0.141			
Total Gross	-0.116	0.210	-0.185		
Budget	0.087	0.397	-0.057	0.064	
No. of Weeks	-0.459	-0.047	-0.468	0.737	0.133

Cell Contents
    Pearson correlation

3.R.19   **a**   There seems to be a large cluster of points in the lower left hand corner, but the data points show no apparent relationship between the variables.

b The pattern described in parts **a** and **c** would indicate a weak correlation:

$$r = \frac{s_{xy}}{s_x s_y} = \frac{-67.770}{\sqrt{(9346.603)(705.363)}} = -0.026$$

d Number of waste sites is only slightly affected by the size of the state. Some other possible explanatory variables might be local environmental regulations, population per square mile, or geographical region in the United States.

4: Probability

Section 4.1

4.1.1, 3, 5 This experiment involves tossing a single die and observing the outcome. The sample space for this experiment consists of the following simple events:

E_1: Observe a 1 E_4: Observe a 4
E_2: Observe a 2 E_5: Observe a 5
E_3: Observe a 3 E_6: Observe a 6

Events A through F are compound events and are composed in the following manner:

1. A: {2}

3. C: {3, 4, 5, 6}

5. E: {2, 4, 6}

4.1.7 Both A and B: {E_3}

4.1.9 B or C or both: {E_1, E_2, E_3, E_4, E_5, E_7}

4.1.11 A or C or both: {E_2, E_3, E_4, E_6}

4.1.13 Both A and B: {E_2}; A or B or both: {E_1, E_2, E_3}; Not B: {E_1, E_4}

4.1.15 Both A and B: {E_1}; A or B or both: {E_1, E_3, E_4}; Not B: {E_2}

4.1.17 There are four simple events: (H, H), (H, T), (T, H), (T, T)

4.1.19 Each of the three children can be either male (M) or female (F), generating eight possible simple events:

(M, M, M)   (M, M, F)   (M, F, M)   (F, M, M)
(F, F, M)   (F, M, F)   (M, F, F)   (F, F, F)

4.1.21, 23 Tree diagrams can be used to identify simple events when the experiment is accomplished in stages. Each successive branching (column) corresponds to one of the stages necessary to generate the final outcome.

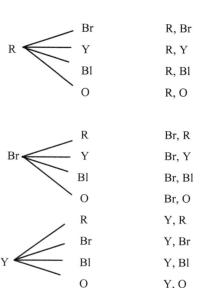
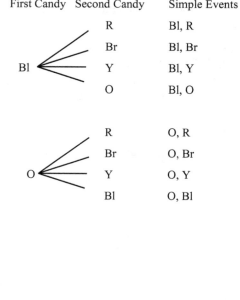

23.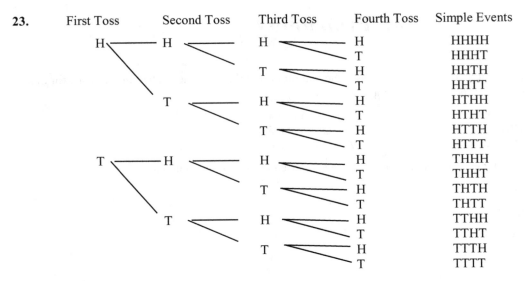

First Toss	Second Toss	Third Toss	Fourth Toss	Simple Events
H	H	H	H	HHHH
			T	HHHT
		T	H	HHTH
			T	HHTT
	T	H	H	HTHH
			T	HTHT
		T	H	HTTH
			T	HTTT
T	H	H	H	THHH
			T	THHT
		T	H	THTH
			T	THTT
	T	H	H	TTHH
			T	TTHT
		T	H	TTTH
			T	TTTT

4.1.25, 27 A table of outcomes can be used to identify simple events when the experiment is accomplished in two stages. The columns and rows correspond to the outcomes on the two stages necessary to generate the final outcome.

25.

	Homeschooled (H)	Not homeschooled (NH)
Male (M)	H, M	NH, M
Female (F)	H, F	NH, F

27.

	Blonde (BL)	Brown (BR)	Black (BLK)	Red (R)	Other (O)
Male (M)	BL, M	BR, M	BLK, M	R, M	O, M
Female (F)	BL, F	BR, F	BLK, F	R, F	O, F

4.1.29 Label the five balls as R_1, R_2, R_3, Y_1 and Y_2. The selection of two balls is accomplished in two stages to produce the simple events in the tree diagram that follows.

First Ball	Second Ball	Simple Events	First Ball	Second Ball	Simple Events
R_1	R_2	R_1R_2	Y_1	R_1	Y_1R_1
	R_3	R_1R_3		R_2	Y_1R_2
	Y_1	R_1Y_1		R_3	Y_1R_3
	Y_2	R_1Y_2		Y_2	Y_1Y_2
R_2	R_1	R_2R_1	Y_2	R_1	Y_2R_1
	R_3	R_2R_3		R_2	Y_2R_2
	Y_1	R_2Y_1		R_3	Y_2R_3
	Y_2	R_2Y_2		Y_1	Y_2Y_1
R_3	R_1	R_3R_1			
	R_2	R_3R_2			
	Y_1	R_3Y_1			
	Y_2	R_3Y_2			

When the first ball is replaced before the second ball is drawn, five additional simple events become possible:

R_1R_1, R_2R_2, R_3R_3, Y_1Y_1, and Y_2Y_2

Section 4.2

4.2.1, 3, 5 From Exercises 1-6 (Section 4.1), the simple events are:

E_1: Observe a 1 E_4: Observe a 4
E_2: Observe a 2 E_5: Observe a 5
E_3: Observe a 3 E_6: Observe a 6

Since the simple events E_i, $i = 1, 2, 3, \ldots, 6$ are equally likely, $P(E_i) = 1/6$. To find the probability of an event, we sum the probabilities assigned to the simple events in that event.

1. $P(A) = P(E_2) = \dfrac{1}{6}$

3. $P(C) = \dfrac{4}{6} = \dfrac{2}{3}$

5. $P(E) = P(E_2) + P(E_4) + P(E_6) = \dfrac{3}{6} = \dfrac{1}{2}$

4.2.7, 9 It is given that events E_1 through E_6 are equally likely, but that event E_7 is twice as likely as the others. Define $P(E)$ to be the probability of one of the first six events. Since all seven probabilities must add to 1, we can write

$$6P(E) + 2P(E) = 1 \Rightarrow 8P(E) = 1 \text{ or } P(E) = 1/8 \text{ for events } E_1 \text{ through } E_6 \text{ and } P(E_7) = 2(\tfrac{1}{8}) = \tfrac{1}{4}.$$

7. $P(A) = P(E_3) + P(E_4) + P(E_6) = \dfrac{3}{8}$

9. $P(C) = P(E_2) + P(E_4) = \dfrac{2}{8} = \dfrac{1}{4}$

4.2.11 It is given that $P(E_1) = P(E_2) = .15$ and $P(E_3) = .40$. Since $\sum_S P(E_i) = 1$, we know that

$$P(E_4) + P(E_5) = 1 - .15 - .15 - .40 = .30 \qquad (i)$$

Also, it is given that

$$P(E_4) = 2P(E_5) \qquad (ii)$$

We have two equations in two unknowns which can be solved simultaneously for $P(E_4)$ and $P(E_5)$. Substituting equation (ii) into equation (i), we have

$$2P(E_5) + P(E_5) = .3$$
$$3P(E_5) = .3 \text{ so that } P(E_5) = .1$$

Then from (i), $P(E_4) + .1 = .3$ and $P(E_4) = .2$.

4.2.13 Use the results of Exercise 11. To find the necessary probability, sum the probabilities of the simple events:

$$P(B) = P(E_2) + P(E_3) = .15 + .4 = .55$$

4.2.15 Use the results of Exercise 11. The events that are not in event A are: $\{E_2, E_5\}$ so that $P(\text{not } A) = .15 + .1 = .25$.

4.2.17 There are 52 possible outcomes when one card is drawn from a deck of 52:

Hearts: Ace, King, Queen, Jack, 10, 9, 8, 7, 6, 5, 4, 3, 2
Diamonds: Ace, King, Queen, Jack, 10, 9, 8, 7, 6, 5, 4, 3, 2
Clubs: Ace, King, Queen, Jack, 10, 9, 8, 7, 6, 5, 4, 3, 2
Spades: Ace, King, Queen, Jack, 10, 9, 8, 7, 6, 5, 4, 3, 2

Since each card is equally likely to be drawn, the probability of each simple event is 1/52 and the probability of an ace is 4/52 = 1/13.

4.2.19 The simple events are shown in Exercise 19 (Section 4.1). Each simple event is equally likely, with probability 1/8.

(M, M, M)   (M, M, F)   (M, F, M)   (F, M, M)
(F, F, M)   (F, M, F)   (M, F, F)   (F, F, F)

The probability of two boys and a girl is $P(MMF) + P(MFM) + P(FMM) = 3/8$

4.2.21 The simple events are shown in Exercise 21 (Section 4.1). Each simple event is equally likely, with probability 1/20. If we designate the first candy in the pair to be your candy, the event of interest consists of three simple events--(O, BR), (O, Y) and (O, BL)—and the probability is 3/20.

4.2.23 There are 36 pairs generated when the die is tossed twice.

(1, 1)   (1, 2)   (1, 3)   (1, 4)   (1, 5)   (1, 6)
(2, 1)   (2, 2)   (2, 3)   (2, 4)   (2, 5)   (2, 6)
(3, 1)   (3, 2)   (3, 3)   (3, 4)   (3, 5)   (3, 6)
(4, 1)   (4, 2)   (4, 3)   (4, 4)   (4, 5)   (4, 6)
(5, 1)   (5, 2)   (5, 3)   (5, 4)   (5, 5)   (5, 6)
(6, 1)   (6, 2)   (6, 3)   (6, 4)   (6, 5)   (6, 6)

Each simple event is equally likely, with probability 1/36. The event of interest consists of four simple events—(6, 3), (6, 4), (6, 5) and (6, 6)—and the probability is 4/36 = 1/9.

4.2.25 There are 38 simple events, each corresponding to a single outcome of the wheel's spin. The 38 simple events are indicated next.

E_1: Observe a 1 E_2: Observe a 2   ... E_{36}: Observe a 36
E_{37}: Observe a 0 E_{38}: Observe a 00

Since any pocket is just as likely as any other, $P(E_i) = 1/38$. The probability of observing either a 0 or a 00 contains two simple events, E_{37} and E_{38}, so that the probability is $P(E_{37}) + P(E_{38}) = 2/38 = 1/19$.

4.2.27 a It is required that $\sum_S P(E_i) = 1$. Hence, $P(E_2) = 1 - .49 - .21 - .09 = .21$

b The player will hit on at least one of the two free throws if he hits on the first, the second, or both. The associated simple events are E_1, E_2, and E_3 and $P(\text{hits on at least one}) = P(E_1) + P(E_2) + P(E_3) = .91$.

4.2.29 a The experiment is accomplished in two stages, as shown in the tree diagram shown next.

Gender	Preschool	Simple Events	Probability
Male	Yes	E_1: Male, Preschool	8/25
Male	No	E_2: Male, No preschool	6/25
Female	Yes	E_3: Female, Preschool	9/25
Female	No	E_4: Female, No preschool	2/25

b Since each of the 25 students are equally likely to be chosen, the probabilities will be proportional to the number of students in each of the four gender-preschool categories. These probabilities are shown in the last column of the tree diagram above.

c $P(\text{Male}) = P(E_1) + P(E_2) = \dfrac{8}{25} + \dfrac{6}{25} = \dfrac{14}{25}$

d $P(\text{Female and no preschool}) = P(E_4) = \dfrac{2}{25}$

4.2.31 a This experiment consists of two patients, each swallowing one of four tablets (two cold and two aspirin). There are four tablets to choose from, call them C_1, C_2, A_1 and A_2. The resulting simple events are then all possible ordered pairs which can be formed from the four choices. There are 12 simple events, each with equal probability 1/12.

$(C_1C_2) \quad (C_2C_1) \quad (A_1C_1) \quad (A_2C_1)$

$(C_1A_1) \quad (C_2A_1) \quad (A_1C_2) \quad (A_2C_2)$

$(C_1A_2) \quad (C_2A_2) \quad (A_1A_2) \quad (A_2A_1)$

Notice that it is important to consider the order in which the tablets are chosen, since it makes a difference, for example, which patient (1 or 2) swallows the cold tablet.

b $A = \{(C_1C_2),(C_1A_1),(C_1A_2),(C_2A_1),(C_2C_1),(C_2A_2)\}$ and the probability is $P(A) = 6/12 = 1/2$

c $B = \{(C_1A_1),(C_1A_2),(C_2A_1),(C_2A_2),(A_1C_1),(A_1C_2),(A_2C_1),(A_2C_2)\}$ and the probability is $P(B) = 8/12 = 2/3$.

d $C = \{(A_1A_2),(A_2A_1)\}$ and the probability is $P(C) = 2/12 = 1/6$.

4.2.33 a *Experiment*: Select three people and record their gender (M or F).

b The simple events are the same as those shown in Exercise 19 (Section 4.1):

FFF FMM MFM MMF
MFF FMF FFM MMM

c Since there are $N = 8$ equally likely simple events, each is assigned probability, $P(E_i) = 1/N = 1/8$.

d-e Sum the probabilities of the appropriate simple events:

$P(\text{only one man}) = P(MFF) + P(FMF) + P(FFM) = 3\left(\dfrac{1}{8}\right) = \dfrac{3}{8}$

$P(\text{all three are women}) = P(FFF) = \dfrac{1}{8}$

4.2.35 A taster tastes and ranks three varieties of tea A, B, and C, according to preference and the simple events in S are in triplet form, each with probability 1/6:

$E_1 : (A,B,C) \qquad E_4 : (B,C,A)$
$E_2 : (A,C,B) \qquad E_5 : (C,B,A)$
$E_3 : (B,A,C) \qquad E_6 : (C,A,B)$

Here the most desirable is in the first position, the next most desirable is in the second position, and the least desirable is in third position. Define the events

  D: variety A is ranked first   and   F: variety A is ranked third

Then $P(D) = P(E_1) + P(E_2) = 1/6 + 1/6 = 1/3$ and $P(F) = P(E_5) + P(E_6) = 1/6 + 1/6 = 1/3$

4.2.37 The four possible outcomes of the experiment, or simple events, are represented as the cells of a 2×2 table and have probabilities (when divided by 300) as given in the table.

    a P(normal eyes and normal wing size) $= 140/300 = .467$

    b P(vermillion eyes) $= (3+151)/300 = 154/300 = .513$

    c P(either vermillion eyes or miniature wings or both) $= (3+151+6)/300 = 160/300 = .533$

4.2.39   **a**   Define P: shopper prefers Pepsi and C: shopper prefers Coke. There are 16 simple events in the experiment, shown in the tree diagram shown next.

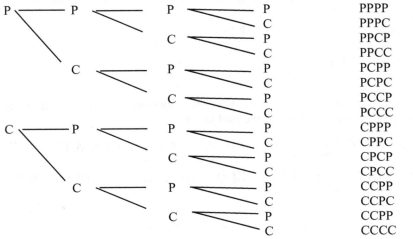

Then if there is actually no difference in the taste, each simple event is equally likely, with probability 1/16, and P(all four prefer Pepsi) $= P(PPPP) = \dfrac{1}{16} = .0625$.

    b P(exactly one prefers Pepsi) $= P(PCCC) + P(CPCC) + P(CCPC) + P(CCCP) = \dfrac{4}{16} = .25$

Section 4.3

4.3.1 Use the *mn* Rule. There are $10(8) = 80$ possible pairs.

4.3.3 Use the extended *mn* Rule. Each coin can fall in one of $n_i = 2$ ways, *and* the total number of simple events is $2(2)(2)(2) = 2^4 = 16$.

4.3.5 $P_3^5 = \dfrac{5!}{2!} = 5(4)(3) = 60$

4.3.7 $P_6^6 = \dfrac{6!}{0!} = 6! = 720$

4.3.9 $C_3^5 = \dfrac{5!}{3!2!} = \dfrac{5(4)}{2(1)} = 10$

4.3.11 $C_6^6 = \dfrac{6!}{6!0!} = 1$

4.3.13 Since order is important, you use *permutations* and $P_5^8 = \dfrac{8!}{3!} = 8(7)(6)(5)(4) = 6720$.

4.3.15 Since order is unimportant, you use *combinations* and $C_3^{10} = \dfrac{10!}{3!7!} = \dfrac{10(9)(8)}{3(2)(1)} = 120$.

4.3.17 This exercise involves the arrangement of 6 different cities in all possible orders. Each city will be visited once and only once. Hence, order is important and elements are being chosen from a single set. Permutations are used and the number of arrangements is

$$P_6^6 = \dfrac{6!}{0!} = 6! = 6(5)(4)(3)(2)(1) = 720$$

4.3.19 **a**   Each student has a choice of 52 cards, since the cards are replaced between selections. The *mn* Rule allows you to find the total number of configurations for three students as $52(52)(52) = 140{,}608$.

b   Now each student must pick a different card. That is, the first student has 52 choices, but the second and third students have only 51 and 50 choices, respectively. The total number of configurations is found using the *mn* Rule or the rule for permutations:

$$mnt = 52(51)(50) = 132{,}600 \quad \text{or} \quad P_3^{52} = \dfrac{52!}{49!} = 132{,}600.$$

c   Let A be the event of interest. Since there are 52 different cards in the deck, there are 52 configurations in which all three students pick the same card (one for each card). That is, there are $n_A = 52$ ways for the event A to occur, out of a total of $N = 140{,}608$ possible configurations from part **a**. The probability of interest is $P(A) = \dfrac{n_A}{N} = \dfrac{52}{140{,}608} = .00037$.

d   Again, let A be the event of interest. There are $n_A = 132{,}600$ ways (from part **b**) for the event A to occur, out of a total of $N = 140{,}608$ possible configurations from part **a**, and the probability of interest is

$$P(A) = \dfrac{n_A}{N} = \dfrac{132{,}600}{140{,}608} = .943$$

4.3.21 **a**   Since the order of selection for the five-card hand is unimportant, use *combinations* to find the number of possible hands as $N = C_5^{52} = \dfrac{52!}{5!47!} = \dfrac{52(51)(50)(49)(48)}{5(4)(3)(2)(1)} = 2{,}598{,}960$.

b   Since there are only four different suits, there are $n_A = 4$ ways to get a royal flush.

c   From parts **a** and **b**, $P(\text{royal flush}) = \dfrac{n_A}{N} = \dfrac{4}{2{,}598{,}960} = .000001539$

4.3.23 Notice that a sample of 10 nurses will be the same no matter in which order they were selected. Hence, order is unimportant and combinations are used. The number of samples of 10 selected from a total of 90 is

$$C_{10}^{90} = \dfrac{90!}{10!80!} = \dfrac{2.0759076(10^{19})}{3.6288(10^6)} = 5.720645(10^{12})$$

4.3.25 **a**   Use the *mn* Rule. The Western conference team can be chosen in one of $m = 6$ ways, and there are 6 ways to choose the Eastern conference team, for a total of $N = mn = 6(6) = 36$ possible pairings.

b   You must choose Los Angeles from the first group and New York from the second group, so that $n_A = (1)(1) = 1$ and the probability is $n_A/N = 1/36$.

c Since there is only one California team in the Western conference, there are five choices for the first team and six choices for the second team. Hence $n_A = 5(6) = 30$ and the probability is $n_A/N = 30/36 = 5/6$.

4.3.27 The situation presented here is analogous to drawing 5 items from a jar (the five members voting in favor of the plaintiff). If the jar contains 5 red and 3 white items (5 females and 3 males), what is the probability that all five items are red? That is, if there is no sex bias, five of the eight members are randomly chosen to be those voting for the plaintiff. What is the probability that all five are female? There are

$$N = C_5^8 = \frac{8!}{5!3!} = 56$$

simple events in the experiment, only one of which results in choosing 5 females. Hence,

$$P(\text{five females}) = \frac{1}{56}.$$

4.3.29 The monkey can place the twelve blocks in any order. Each arrangement will yield a simple event and hence the total number of simple events (arrangements) is $P_{12}^{12} = 12!$ It is necessary to determine the number of simple events in the event of interest (that he draws three of each kind, in order). First, he may draw the four different _types_ of blocks in any order. Thus we need the number of ways of arranging these four items, which is $P_4^4 = 4!$ Once this order has been chosen, the three squares can be arranged in $P_3^3 = 3!$ ways, the three triangles can be arranged in $P_3^3 = 3!$ ways, and so on. Thus, the total number of simple events in the event of interest is $P_4^4(P_3^3)^4$ and the associated probability is

$$\frac{P_4^4(P_3^3)^4}{P_{12}^{12}} = \frac{4!(3!)^4}{12!}$$

Section 4.4

4.4.1 Each simple event is equally likely, with probability $1/5$.

   a $A^C = \{E_2, E_4, E_5\}$ $P(A^C) = 3/5$

   b $A \cap B = \{E_1\}$ $P(A \cap B) = 1/5$

   c $B \cap C = \{E_4\}$ $P(B \cap C) = 1/5$

   d $A \cup B = S = \{E_1, E_2, E_3, E_4, E_5\}$ $P(A \cup B) = 1$

   e $B|C = \{E_4\}$ from among $\{E_3, E_4\}$ $P(B|C) = 1/2$

   f $A|B = \{E_1\}$ from among $\{E_1, E_2, E_4, E_5\}$ $P(A|B) = 1/4$

   g $A \cup B \cup C = S$ $P(A \cup B \cup C) = 1$

   h $(A \cap B)^C = \{E_2, E_3, E_4, E_5\}$ $P(A \cap B)^C = 4/5$

4.4.3 **a** $P(A|B) = \dfrac{P(A \cap B)}{P(B)} = \dfrac{1/5}{4/5} = \dfrac{1}{4}$ **b** $P(B|C) = \dfrac{P(B \cap C)}{P(C)} = \dfrac{1/5}{2/5} = \dfrac{1}{2}$

4.4.5 From Exercise 3, $P(A|B) = 1/4$ while $P(A) = 2/5$. Therefore, A and B are not independent.

4.4.7 $P(A \cap B) = P(B)P(A|B) = .5(.1) = .05$

4.4.9 Since $P(A \cap B) = 0$ and $P(A)P(B) = .1(.5) = .05$, $P(A \cap B) \neq P(A)P(B)$. Hence, events A and B are not independent.

4.4.11 **a** The event A will occur whether event B occurs or not (B^C). Hence,
$$P(A) = P(A \cap B) + P(A \cap B^C) = .34 + .15 = .49$$

b Similar to part **a**. $P(B) = P(A \cap B) + P(A^C \cap B) = .34 + .46 = .80$.

c The contents of the cell in the first row and first column is $P(A \cap B) = .34$.

d $P(A \cup B) = P(A) + P(B) - P(A \cap B) = .49 + .80 - .34 = .95$.

e Use the definition of conditional probability to find $P(A|B) = \dfrac{P(A \cap B)}{P(B)} = \dfrac{.34}{.80} = .425$.

f Similar to part **e**. $P(B|A) = \dfrac{P(A \cap B)}{P(A)} = \dfrac{.34}{.49} = .694$

4.4.13 From Exercise 11, $P(A|B) = .425$ and $P(A) = .49$. The two events are not independent.

4.4.15 **a** Since A and B are independent, $P(A \cap B) = P(A)P(B) = .4(.2) = .08$.

b $P(A \cup B) = P(A) + P(B) - P(A \cap B) = .4 + .2 - (.4)(.2) = .52$

4.4.17 **a** Use the definition of conditional probability to find
$$P(B|A) = \dfrac{P(A \cap B)}{P(A)} = \dfrac{.12}{.4} = .3$$

b Since $P(A \cap B) \neq 0$, A and B are not mutually exclusive.

c If $P(B) = .3$, then $P(B) = P(B|A)$ which means that A and B are independent.

4.4.19 Refer to the solution to Exercises 1-6 (Section 4.1) where the six simple events in the experiment are given, with $P(E_i) = 1/6$. Calculate

$$P(A \cap B) = P(A|B)P(B) = 1(1/3) = 1/3, \ P(A|B) = \dfrac{P(A \cap B)}{P(B)} = \dfrac{1/3}{1/3} = 1, \ P(A) = 1/2, \ P(B) = 2/6 = 1/3$$

and $P(A \cap B) = P(A|B)P(B) = 1(1/3) = 1/3$. Since $P(A|B) \neq P(A)$, A and B are not independent. Since $P(A \cap B) \neq 0$, A and B are not mutually exclusive.

4.4.21 In an experiment that consists of tossing two dice, the sample space consists of 36 ordered pairs, the first stage being the number observed on the first die, and the second being the number observed on the second die. The simple events were listed in the solution to Exercise 16 (Section 4.1).

a There are 6 pairs that have a sum of 7: (1, 6), (6, 1), (2, 5), (5, 2), (3, 4), and (4, 3). Then $P(\text{sum is } 7) = \dfrac{6}{36} = \dfrac{1}{6}$. There are 2 pairs that have a sum of 11: (5, 6) and (6,5), so that $P(\text{sum is } 11) = \dfrac{2}{36} = \dfrac{1}{18}$.

b There are 6 pairs that have the same upper face: (1, 1), (2, 2), (3, 3), (4, 4), (5, 5), and (6, 6), so that $P(\text{doubles}) = \dfrac{6}{36} = \dfrac{1}{6}$.

c   There are 9 pairs: (1, 1), (1, 3), (1, 5), (3, 1), (3, 3), (3, 5), (5, 1), (5, 3) and (5, 5), so that
$$P(\text{both odd}) = \frac{9}{36} = \frac{1}{4}.$$

4.4.23   Define the following events:

     A: project is approved for funding

     D: project is disapproved for funding

For the first panel, $P(A_1) = .2$ and $P(D_1) = .8$. For the second panel, $P[\text{same decision as first panel}] = .7$ and $P[\text{reversal}] = .3$. That is,

$$P(A_2 | A_1) = P(D_2 | D_1) = .7 \text{ and } P(A_2 | D_1) = P(D_2 | A_1) = .3.$$

a $P(A_1 \cap A_2) = P(A_1)P(A_2 | A_1) = .2(.7) = .14$

b $P(D_1 \cap D_2) = P(D_1)P(D_2 | D_1) = .8(.7) = .56$

c $P(D_1 \cap A_2) + P(A_1 \cap D_2) = P(D_1)P(A_2 | D_1) + P(A_1)P(D_2 | A_1) = .8(.3) + .2(.3) = .30$

4.4.25   The two-way table in the text gives probabilities for events A, A^C, B, B^C in the column and row marked "Totals". The interior of the table contains the four two-way intersections as shown next.

$A \cap B$	$A \cap B^C$
$A^C \cap B$	$A^C \cap B^C$

The necessary probabilities can be found using various rules of probability if not directly from the table.

a   Calculate $P(A) = .4$, $P(B | A) = P(A \cap B)/P(A) = .1/.4 = .25$ and $P(A \cap B) = .10$. Then

$$P(A \cap B) = .1 \text{ and } P(A)P(B | A) = (.4)(.25) = .1$$

b   Calculate $P(B) = .37$ and $P(A | B) = P(A \cap B)/P(B) = .1/.37 = .27$. Then

$$P(A \cap B) = .1 \text{ and } P(B)P(A | B) = (.37)(.1/.37) = .1$$

c   From the table, $P(A \cup B) = .1 + .27 + .30 = .67$ while $P(A) + P(B) - P(A \cap B) = .4 + .37 - .10 = .67$.

4.4.27   Define the following events: A: first system fails

                                     B: second system fails

A and B are independent and $P(A) = P(B) = .001$. To determine the probability that the combined missile system does not fail, we use the complement of this event; that is,

$$P[\text{system does not fail}] = 1 - P[\text{system fails}] = 1 - P(A \cap B)$$
$$= 1 - P(A)P(B) = 1 - (.001)^2$$
$$= .999999$$

4.4.29   Define the following events:

     S: student chooses Starbucks

     P: student chooses Peet's

     C: student orders a café mocha

Then $P(S) = .7$; $P(P) = .3$; $P(C | S) = P(C | P) = .60$

    a   Using the given probabilities, $P(S \cap C) = P(S)P(C|S) = .7(.6) = .42$

    b   Since $P(C) = .6$ regardless of whether the student visits Starbucks or Peet's, the two events are independent.

    c $P(P|C) = \dfrac{P(P \cap C)}{P(C)} = \dfrac{P(P)P(C|P)}{P(C)} = P(P) = .3$

    d $P(S \cup C) = P(S) + P(C) - P(S \cap C) = .7 + .6 - (.7)(.6) = .88$

4.4.31 Define the events: D: person dies   S: person smokes

It is given that $P(S) = .2$, $P(D) = .006$, and $P(D|S) = 10P(D|S^C)$. The probability of interest is $P(D|S)$. The event D, whose probability is given, can be written as the union of two mutually exclusive intersections. That is, $D = (D \cap S) \cup (D \cap S^C)$.

Then, using the Addition and Multiplication Rules,

$$P(D) = P(D \cap S) + P(D \cap S^C) = P(D|S)P(S) + P(D|S^C)P(S^C) = P(D|S)(.2) + \left[(1/10)P(D|S)\right](.8)$$

Since $P(D) = .006$, the above equation can be solved for $P(D|S)$

$$P(D|S)(.2 + .08) = .006$$
$$P(D|S) = .006/.28 = .0214$$

4.4.33 **a**   Each of the four cell events is equally likely with $P(E_i) = 1/4$. Hence, the probability of at least one dominant (R) allele is $P(rR) + P(Rr) + P(RR) = \dfrac{3}{4}$.

    b   Similar to part **a**. The probability of at least one recessive allele is $P(rR) + P(Rr) + P(rr) = \dfrac{3}{4}$.

    c   Define the events: A: plant has red flowers   B: plant has one recessive allele

Then

$$P(B|A) = \dfrac{P(A \cap B)}{P(A)} = \dfrac{P(rR) + P(Rr)}{P(rR) + P(Rr) + P(RR)} = \dfrac{2/4}{3/4} = \dfrac{2}{3}$$

4.4.35 Similar to Exercise 24-25.

    a $P(A) = \dfrac{154}{256}$ **b** $P(G) = \dfrac{155}{256}$

    c $P(A \cap G) = \dfrac{88}{256}$ **d** $P(G|A) = \dfrac{P(G \cap A)}{P(A)} = \dfrac{88/256}{154/256} = \dfrac{88}{154}$

    e $P(G|B) = \dfrac{P(G \cap B)}{P(B)} = \dfrac{44/256}{67/256} = \dfrac{44}{67}$ **f** $P(G|C) = \dfrac{P(G \cap C)}{P(C)} = \dfrac{23/256}{35/256} = \dfrac{23}{35}$

    g $P(C|P) = \dfrac{P(C \cap P)}{P(P)} = \dfrac{12/256}{101/256} = \dfrac{12}{101}$ **h** $P(B^C) = 1 - P(B) = 1 - \dfrac{67}{256} = \dfrac{189}{256}$

4.4.37 Define the following events:

        A: player makes first free throw   B: player makes second free throw

The probabilities of events A and B will depend on which player is shooting.

a $P(A \cap B) = P(A)P(B) = .81(.81) = .6561$, since the free throws are independent.

b The event that Kevin Durant makes exactly one of the two free throws will occur if he makes the first and misses the second, or vice versa. Then

$$P(\text{makes exactly one}) = P(A \cap B^C) + P(A^C \cap B)$$
$$= .62(.38) + .38(.62) = .4712$$

c This probability is the intersection of the individual probabilities for both DeAndre Jordan and Kevin Durant..

$P(\text{DeAndre makes both and Kevin makes neither}) = [.81(.81)][.38(.38)] = .0947$

Section 4.5

4.5.1 Use the Law of Total Probability, writing

$$P(A) = P(S_1)P(A|S_1) + P(S_2)P(A|S_2) = .7(.2) + .3(.3) = .23$$

4.5.3 Use the results of Exercise 1 in the form of Bayes' Rule:

$$P(S_i | A) = \frac{P(S_i)P(A|S_i)}{P(S_1)P(A|S_1) + P(S_2)P(A|S_2)}$$

For $i = 2$, $P(S_2 | A) = \frac{.3(.3)}{.7(.2) + .3(.3)} = \frac{.09}{.23} = .3913$

4.5.5 Use Bayes' Rule:

$$P(S_i | A) = \frac{P(S_i)P(A|S_i)}{P(S_1)P(A|S_1) + P(S_2)P(A|S_2) + P(S_3)P(A|S_3)}$$

For $i = 2$, $P(S_2 | A) = \frac{.5(.1)}{.2(.2) + .5(.1) + .3(.3)} = \frac{.05}{.18} = .2778$

4.5.7 Use the Law of Total Probability, writing $P(A) = P(S_1)P(A|S_1) + P(S_2)P(A|S_2) = .6(.3) + .4(.5) = .38$.

4.5.9 Define   A: machine produces a defective item   B: worker follows instructions

Then $P(A|B) = .01$, $P(B) = .90$, $P(A|B^C) = .03$, $P(B^C) = .10$. The probability of interest is

$$P(A) = P(A \cap B) + P(A \cap B^C)$$
$$= P(A|B)P(B) + P(A|B^C)P(B^C)$$
$$= .01(.90) + .03(.10) = .012$$

4.5.11 Define   L: play goes to the left
R: play goes to the right
S: right guard shifts his stance

a It is given that $P(L) = .3$, $P(R) = .7$, $P(S|R) = .8$, $P(S^C|L) = .9$, $P(S|L) = .1$, and $P(S^C|R) = .2$. Using Bayes' Rule,

$$P(L|S^C) = \frac{P(L)P(S^C|L)}{P(L)P(S^C|L) + P(R)P(S^C|R)} = \frac{.3(.9)}{.3(.9) + .7(.2)} = \frac{.27}{.41} = .6585$$

b From part a, $P(R|S^C) = 1 - P(L|S^C) = 1 - .6585 = .3415$.

c Given that the guard takes a balanced stance, it is more likely (.6585 versus .3415) that the play will go to the left.

4.5.13 The probability of interest is $P(A|H)$ which can be calculated using Bayes' Rule and the probabilities given in the exercise.

$$P(A|H) = \frac{P(A)P(H|A)}{P(A)P(H|A) + P(B)P(H|B) + P(C)P(H|C)}$$
$$= \frac{.01(.90)}{.01(.90) + .005(.95) + .02(.75)} = \frac{.009}{.02875} = .3130$$

4.5.15 a Using the probability table,

$$P(D) = .08 + .02 = .10 \qquad\qquad P(D^C) = 1 - P(D) = 1 - .10 = .90$$

$$P(N|D^C) = \frac{P(N \cap D^C)}{P(D^C)} = \frac{.85}{.90} = .94 \qquad P(N|D) = \frac{P(N \cap D)}{P(D)} = \frac{.02}{.10} = .20$$

b Using Bayes' Rule, $P(D|N) = \dfrac{P(D)P(N|D)}{P(D)P(N|D) + P(D^C)P(N|D^C)} = \dfrac{.10(.20)}{.10(.20) + .90(.94)} = .023$

c Using the definition of conditional probability, $P(D|N) = \dfrac{P(N \cap D)}{P(N)} = \dfrac{.02}{.87} = .023$

d $P(\text{false positive}) = P(P|D^C) = \dfrac{P(P \cap D^C)}{P(D^C)} = \dfrac{.05}{.90} = .056$

e $P(\text{false negative}) = P(N|D) = \dfrac{P(N \cap D)}{P(D)} = \dfrac{.02}{.10} = .20$

f The probability of a false negative is quite high, and would cause concern about the reliability of the screening method.

Reviewing What You've Learned

4.R.1 Define the following events:

A: employee fails to report fraud   B: employee suffers reprisal

It is given that $P(B|A^C) = .23$ and $P(A) = .69$. The probability of interest is

$$P(A^C \cap B) = P(B|A^C)P(A^C) = .23(.31) = .0713$$

4.R.3 Two systems are selected from seven, three of which are defective. Denote the seven systems as G_1, G_2, G_3, G_4, D_1, D_2, D_3 according to whether they are good or defective. Each simple event will represent a particular pair of systems chosen for testing, and the sample space, consisting of 21 pairs, is shown next.

G_1G_2 G_1D_1 G_2D_3 G_4D_2 G_1G_3 G_1G_2 G_3D_1 G_4D_3

G_1G_4 G_1D_3 G_3D_2 D_1D_2 G_2G_3 G_2D_1 G_3D_3 D_1D_3

G_2G_4 G_2D_2 G_4D_1 D_2D_3 G_3G_4

Note that the two systems are drawn simultaneously and that order is unimportant in identifying a simple event. Hence, the pairs G_1G_2 and G_2G_1 are not considered to represent two different simple events. The event A, "no defectives are selected", consists of the simple events $G_1G_2, G_1G_3, G_1G_4, G_2G_3, G_2G_4, G_3G_4$. Since the systems are selected at random, any pair has an equal probability of being selected. Hence, the probability assigned to each simple event is $1/21$ and $P(A) = 6/21 = 2/7$.

4.R.5 a If the fourth van tested is the last van with brake problems, then in the first three tests, we must find one van with brake problems and two without. That is, in choosing three from the six vans, we must find one faulty and two that are not faulty. Think of choosing three balls – one white and two red – from a total of six, and the probability can be calculated as

$$P(\text{one faulty and two not}) = \frac{C_1^2 C_2^4}{C_3^6} = \frac{2(6)}{20} = \frac{3}{5}$$

Once this is accomplished, the van with brake problems must be chosen on the fourth test. Using the *mn* Rule, the probability that the fourth van tested is the last with faulty brakes is $\frac{3}{5}\left(\frac{1}{3}\right) = \frac{1}{5}$.

b In order that no more than four vans must be tested, you must find one or both of the faulty vans in the first four tests. Proceed as in part **a**, this time choosing four from the six vans, and

$$P(\text{one or two faulty vans}) = \frac{C_0^2 C_4^4 + C_1^2 C_3^4}{C_4^6} = \frac{1+8}{15} = \frac{9}{15}$$

c If it is known that the first faulty van is found in the first two tests, there are four vans left from which to select those tested third and fourth. Of these four, only one is faulty. Hence,

$$P(\text{one faulty and one not} \mid \text{one faulty in first two tests}) = \frac{C_1^1 C_1^3 + C_1^2 C_3^4}{C_2^4} = \frac{3}{6} = \frac{1}{2}$$

4.R.7 a Since the Connecticut and Pennsylvania lotteries are independent, and the Connecticut lottery was fair (each ball can be chosen from a set of ten, 0, 1, …, 9).

$$P(666 \text{ in Connecticut} \mid 666 \text{ in Pennsylvania}) = P(666 \text{ in Connecticut}) = \frac{1}{(10)^3} = \frac{1}{1000}.$$

b Since $P(666 \text{ in Pennsylvania}) = 1/8$, the probability of a 666 in both lotteries is $\left(\frac{1}{8}\right)\left(\frac{1}{1000}\right) = \frac{1}{8000}$.

4.R.9 Fix the birth date of the first person entering the room. Then define the following events:

A_2: second person's birthday differs from the first

A_3: third person's birthday differs from the first and second

A_4: fourth person's birthday differs from all preceding

$\vdots$

A_n: n^{th} person's birthday differs from all preceding

Then

$$P(A) = P(A_2)P(A_3)\cdots P(A_n) = \left(\frac{364}{365}\right)\left(\frac{363}{365}\right)\cdots\left(\frac{365-n+1}{365}\right)$$

since at each step, one less birth date is available for selection. Since event B is the complement of event A,

$$P(B) = 1 - P(A)$$

a For $n = 3$, $P(A) = \frac{(364)(363)}{(365)^2} = .9918$ and $P(B) = 1 - .9918 = .0082$

b For $n=4$, $P(A) = \dfrac{(364)(363)(362)}{(365)^3} = .9836$ and $P(B) = 1 - .9836 = .0164$

4.R.11 **a** $P(T \cap P) = 27/70 = .386$

b $P(T^C \cap P) = 0$

c $P(T \cap N) = 4/70$

d $P(P|T) = \dfrac{P(T \cap N)}{P(T)} = \dfrac{27/70}{31/70} = \dfrac{27}{31}$

e $P(\text{false negative}) = P(N|T) = \dfrac{P(T \cap N)}{P(T)} = \dfrac{4/70}{31/70} = \dfrac{4}{31}$

4.R.13 Define
A: union strike fund is adequate to support a strike
C: union-management team makes a contract settlement within 2 weeks

It is given that $P(C) = .5$, $P(A) = .6$, $P(A \cap C) = .3$. Then

$$P(C|A) = \dfrac{P(A \cap C)}{P(A)} = \dfrac{.3}{.6} = .5.$$

Since $P(C) = .5$ and $P(C|A) = .5$, it appears that the settlement of the contract is independent of the ability of the union strike fund to support the strike.

4.R.15 Define the events:
A: the man waits five minutes or longer
B: the woman waits five minutes or longer

The two events are independent, and $P(A) = P(B) = .2$.

a $P(A^C) = 1 - P(A) = .8$

b $P(A^C B^C) = P(A^C)P(B^C) = (.8)(.8) = .64$

c $P[\text{at least one waits five minutes or longer}]$
$= 1 - P[\text{neither waits five minutes or longer}] = 1 - P(A^C B^C) = 1 - .64 = .36$

4.R.17 It is given that 40% of all people in a community favor the development of a mass transit system. Thus, given a person selected at random, the probability that the person will favor the system is .4. Since the pollings are independent events, when four people are selected at random,

$$P[\text{all 4 favor the system}] = (.4)^4 = .0256$$

Similarly, $P[\text{none favor the system}] = (1-.4)^4 = .1296$

4.R.19 Since the first pooled test is positive, we are interested in the probability of requiring five single tests to detect the disease in the single affected person. There are (5)(4)(3)(2)(1) ways of ordering the five tests, and there are 4(3)(2)(1) ways of ordering the tests so that the diseased person is given the final test. Hence, the desired probability is $\dfrac{4!}{5!} = \dfrac{1}{5} = .2$.

If two people are diseased, six tests are needed if the last two tests are given to the diseased people. There are 3(2)(1) ways of ordering the tests of the other three people and 2(1) ways of ordering the tests of the two diseased people. Hence, the probability that six tests will be needed is $\frac{2!3!}{5!} = \frac{1}{10} = .1$.

4.R.21 The necessary probabilities can be found by summing the necessary cells in the probability table and dividing by 200, the total number of individuals.

   a $P(Y) = \frac{96}{200} = .48$

   b $P(M \cap AE) = \frac{20}{200} = .10$

   c $P(M \mid S) = \frac{P(M \cap S)}{P(S)} = \frac{16/200}{61/200} = \frac{16}{61} = .262$

5: Discrete Probability Distributions

Section 5.1

5.1.1 $0 \le p(x) \le 1$ and $\sum p(x) = 1$

5.1.3 Shelf life is a continuous random variable since it can take on any positive real value.

5.1.5 Length is a continuous random variable, taking on any positive real value.

5.1.7 The increase in length of life achieved by a cancer patient as a result of surgery is a continuous random variable, since an increase in life (measured in units of time) can take on any of an infinite number of values in a particular interval.

5.1.9 The number of deer killed per year in a state wildlife preserve is a discrete random variable taking the values 0,1, 2, …

5.1.11 Blood pressure is a continuous random variable.

5.1.13 Since one of the requirements of a probability distribution is that $\sum_x p(x) = 1$, we need

$$p(4) = 1 - (.1 + .3 + .4 + .1 + .05) = 1 - .95 = .05$$

The probability histogram is shown in the figure that follows.

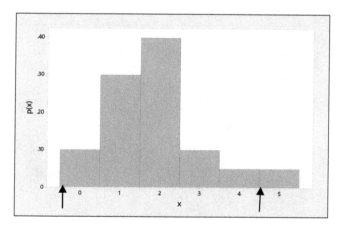

5.1.15 For the random variable x given here,

$$\mu = E(x) = \sum xp(x) = 0(.1) + 1(.3) + \cdots + 5(.05) = 1.85$$

The variance of x is defined as

$$\sigma^2 = E\left[(x-\mu)^2\right] = \sum(x-\mu)^2 p(x) = (0-1.85)^2(.1) + (1-1.85)^2(.3) + \cdots + (5-1.85)^2(.05) = 1.4275$$

and $\sigma = \sqrt{1.4275} = 1.19$.

The interval of interest is $\mu \pm 2\sigma = 1.85 \pm 2.38$ or $-.53$ to 4.23. This interval is shown on the probability histogram in Exercise 13. Then $P[-.53 \le x \le 4.23] = P[0 \le x \le 4] = .95$.

5.1.17 Since one of the requirements of a probability distribution is that $\sum_x p(x) = 1$, we need

$$p(3) = 1 - (.1 + .3 + .3 + .1) = 1 - .8 = .2$$

5.1.19 For the random variable x given here, $\mu = E(x) = \sum xp(x) = 0(.1) + 1(.3) + \cdots + 4(.1) = 1.9$

The variance of x is defined as

$$\sigma^2 = E\left[(x-\mu)^2\right] = \sum(x-\mu)^2 p(x) = (0-1.9)^2(.1) + (1-1.9)^2(.3) + \cdots + (4-1.9)^2(.1) = 1.29$$

and $\sigma = \sqrt{1.29} = 1.136$.

5.1.21 Using the table form of the probability distribution, $P(x \le 3) = 1 - P(x=4) = 1 - .1 = .9$.

5.1.23 There are $6(6) = 36$ equally likely simple events in this experiment, shown in the table that follows.

	Second Die					
First Die	1	2	3	4	5	6
1	1,1	1,2	1,3	1,4	1,5	1,6
2	2,1	2,2	2,3	2,4	2,5	2,6
3	3,1	3,2	3,3	3,4	3,5	3,6
4	4,1	4,2	4,3	4,4	4,5	4,6
5	5,1	5,2	5,3	5,4	5,5	5,6
6	6,1	6,2	6,3	6,4	6,5	6,6

Each simple event has a particular value of x associated with it, and by summing the probabilities of all simple events producing a particular value of x, the probability distribution that follows is obtained. The distribution is mound-shaped.

x	2	3	4	5	6	7	8	9	10	11	12
$p(x)$	$\frac{1}{36}$	$\frac{2}{36}$	$\frac{3}{36}$	$\frac{4}{36}$	$\frac{5}{36}$	$\frac{6}{36}$	$\frac{5}{36}$	$\frac{4}{36}$	$\frac{3}{36}$	$\frac{2}{36}$	$\frac{1}{36}$

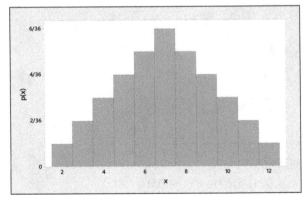

The average value of x is $\mu = E(x) = \sum xp(x) = 2\left(\frac{1}{36}\right) + 3\left(\frac{2}{36}\right) + \cdots + 12\left(\frac{1}{36}\right) = 7.0$, the variance of x is

$$\sigma^2 = E\left[(x-\mu)^2\right] = \sum(x-\mu)^2 p(x) = (2-7)^2\left(\frac{1}{36}\right) + (3-7)^2\left(\frac{2}{36}\right) + \cdots + (12-7)^2\left(\frac{1}{36}\right) = 5.83333$$

and $\sigma = \sqrt{5.83333} = 2.415$.

5.1.25 Let W_1 and W_2 be the two women while M_1, M_2 and M_3 are the three men. There are 10 ways to choose the two people to fill the positions. Let x be the number of women chosen. The 10 equally likely simple events are:

E_1: W_1W_2 $(x=2)$ $\quad$ E_6: W_2M_2 $(x=1)$
E_2: W_1M_1 $(x=1)$ $\quad$ E_7: W_2M_3 $(x=1)$

E₃: W_1M_2 $(x=1)$ E₈: M_1M_2 $(x=0)$
E₄: W_1M_3 $(x=1)$ E₉: M_1M_3 $(x=0)$
E₅: W_2M_1 $(x=1)$ E₁₀: M_2M_3 $(x=0)$

The probability distribution for x is then $p(0) = 3/10$, $p(1) = 6/10$, $p(2) = 1/10$ and the probability histogram is shown next.

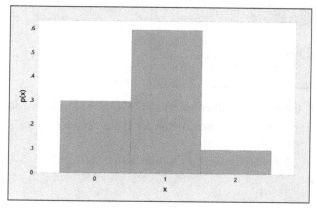

The average value of x is $\mu = E(x) = \sum xp(x) = 0\left(\frac{3}{10}\right) + 1\left(\frac{6}{10}\right) + 2\left(\frac{1}{10}\right) = .8$, the variance of x is

$$\sigma^2 = E\left[(x-\mu)^2\right] = \sum (x-\mu)^2 p(x)$$
$$= (0-.8)^2 \left(\frac{3}{10}\right) + (1-.8)^2 \left(\frac{6}{10}\right) + (2-.8)^2 \left(\frac{1}{10}\right) = .36$$

and $\sigma = \sqrt{.36} = .6$.

5.1.27 For the probability distribution given in this exercise,
$$\mu = E(x) = \sum xp(x) = 0(.1) + 1(.4) + 2(.4) + 3(.1) = 1.5.$$

5.1.29 Let x be the number of drillings until the first success (oil is struck). It is given that the probability of striking oil is $P(O) = .1$, so that the probability of no oil is $P(N) = .9$

a $p(1) = P[\text{oil struck on first drilling}] = P(O) = .1$

$p(2) = P[\text{oil struck on second drilling}]$. This is the probability that oil is not found on the first drilling, but is found on the second drilling. Using the Multiplication Law,
$$p(2) = P(NO) = (.9)(.1) = .09.$$

Finally, $p(3) = P(NNO) = (.9)(.9)(.1) = .081$.

b-c For the first success to occur on trial x, $(x-1)$ failures must occur before the first success. Thus,
$$p(x) = P(NNN\ldots NNO) = (.9)^{x-1}(.1)$$

since there are $(x-1)$ N's in the sequence. The probability histogram generated by **Minitab** follows.

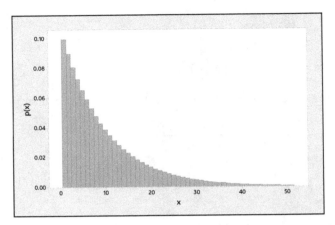

5.1.31 If a $5 bet is placed on the number 18, the gambler will either win $175 $(35 \times \$5)$ with probability 1/38 or lose $5 with probability 37/38. Hence, the probability distribution for x, the gambler's gain is

x	$p(x)$
-5	37/38
175	1/38

The expected gain is $\mu = E(x) = \sum xp(x) = -5(37/38) + 175(1/38) = \$-0.26$. The expected gain is in fact negative, a loss of $0.26.

5.1.33 The random variable G, total gain to the insurance company, will be D if there is no theft, but D – 50,000 if there is a theft during a given year. These two events will occur with probability .99 and .01, respectively. Hence, the probability distribution for G is given next.

G	$p(G)$
D	.99
D – 50,000	.01

The expected gain is

$$E(G) = \sum Gp(G) = .99D + .01(D - 50,000)$$
$$= D - 50,000$$

In order that $E(G) = 1000$, it is necessary to have $1000 = D - 500$ or $D = \$1500$.

5.1.35 a Refer to the probability distribution for x given in this exercise.

$P(\text{no coffee breaks}) = P(x = 0) = .28$

b $P(\text{more than 2 coffee breaks}) = P(x > 2) = .12 + .05 + .01 = .18$

c $\mu = E(x) = \sum xp(x) = 0(.28) + 1(.37) + \cdots + 5(.01) = 1.32$

$\sigma^2 = \sum (x - \mu)^2 p(x) = (0 - 1.32)^2(.28) + (1 - 1.32)^2(.37) + \cdots + (5 - 1.32)^2(.01) = 1.4376$ and $\sigma = \sqrt{1.4376} = 1.199$.

d Calculate $\mu \pm 2\sigma = 1.32 \pm 2.398$ or -1.078 to 3.718. Then, referring to the probability distribution of x, $P[-1.078 \le x \le 3.718] = P[0 \le x \le 3] = .28 + .37 + .17 + .12 = .94$.

5.1.37 We are asked to find the premium that the insurance company should charge in order to break even. Let c be the unknown value of the premium and x be the gain to the insurance company caused by marketing the new product. There are three possible values for x. If the product is a failure or moderately successful, x will be negative; if the product is a success, the insurance company will gain the amount of the premium and x will be positive. The probability distribution for x follows:

x	$p(x)$
c	.94
$-800{,}000 + c$	.01
$-250{,}000 + c$	.05

In order to break even, $E(x) = \Sigma x p(x) = 0$. Therefore,

$$.94(c) + .01(-800{,}000 + c) + (.05)(-250{,}000 + c) = 0$$
$$-8000 - 12{,}500 + (.01 + .05 + .94)c = 0$$
$$c = 20{,}500$$

Hence, the insurance company should charge a premium of $20,500.

Section 5.2

5.2.1 See the definition of a binomial experiment at the beginning of this section (Section 5.2)

5.2.3, 5 These probabilities can be found individually using the binomial formula, or alternatively using the cumulative binomial tables (Table 1 in Appendix I). The values in Table 1 for $n = 8$ and $p = .7$ are shown in the table that follows.

k	0	1	2	3	4	5	6	7	8
$P(x \le k)$	.000	.001	.011	.058	.194	.448	.745	.942	1.000

3. Using the formula, $P(x \ge 3) = \sum_{x=3}^{8} C_x^8 (.7)^x (.3)^{8-x} = .989$ or use Table 1, writing

$$P(x \ge 3) = 1 - P(x \le 2) = 1 - .011 = .989.$$

5. Using the formula, $P(x = 3) = C_3^8 (.7)^3 (.3)^5 = 56(.343)(.00243) = .047$ or use Table 1, writing

$$P(x = 3) = P(x \le 3) - P(x \le 2) = .058 - .011 = .047.$$

5.2.7, 9 These probabilities can be found individually using the binomial formula, or alternatively using the cumulative binomial tables (Table 1 in Appendix I). The values in Table 1 for $n = 9$ and $p = .3$ are shown in the table that follows. We choose to use the table for ease of calculation.

k	0	1	2	3	4	5	6	7	8	9
$P(x \le k)$	.040	.196	.463	.730	.901	.975	.996	1.000	1.000	1.000

7. $P(x = 2) = P(x \le 2) - P(x \le 1) = .463 - .196 = .267$

9. $P(x > 2) = 1 - P(x \le 2) = 1 - .463 = .537$

5.2.11, 13, 15 These probabilities can be found individually using the binomial formula, or alternatively using the cumulative binomial tables (Table 1 in Appendix I). The values in Table 1 for $n = 7$ and $p = .5$ are shown in the table that follows. We choose to use the table for ease of calculation.

k	0	1	2	3	4	5	6	7
$P(x \le k)$	.008	.062	.227	.500	.773	.938	.992	1.000

11. $P(x = 4) = P(x \le 4) - P(x \le 3) = .773 - .500 = .273$

13. $P(x > 1) = 1 - P(x \le 1) = 1 - .062 = .938$

15. $\sigma = \sqrt{npq} = \sqrt{7(.5)(.5)} = 1.323$

5.2.17 $C_0^4(.05)^0(.95)^4 = (.95)^4 = .8145$

5.2.19 $C_1^7(.2)^1(.8)^6 = 7(.2)(.8)^6 = .3670$

5.2.21 $C_1^8(.2)^1(.8)^7 = 8(.2)(.20971) = .3355$

5.2.23 $P(x \leq 1) = .1678 + .3355 = .5033$

5.2.25 If $p = .5$, $p(x) = C_x^n(.5)^x(.5)^{n-x} = C_x^n(.5)^n$. Since $C_k^n = \dfrac{n!}{k!(n-k)!} = C_{n-k}^n$ $C_x^n = C_{n-x}^n$, you can see that for any value of k,

$$P(x = k) = C_k^n(.5)^n = C_{n-k}^n(.5)^n = P(x = n - k)$$

This indicates that the probability distribution is exactly symmetric around the center point $\mu = np = .5n$.

5.2.27 **a** Notice that when $p = .8$, $p(x) = C_x^6(.8)^x(.2)^{6-x}$. In Exercise 26, with $p = .2$, $p(x) = C_x^6(.2)^x(.8)^{6-x}$. The probability that $x = k$ when $p = .8$ --- $C_k^6(.8)^k(.2)^{n-k}$ --- is the same as the probability that $x = n - k$ when $p = .2$ --- $C_{n-k}^6(.2)^{n-k}(.8)^k$. This follows because

$$C_k^n = \dfrac{n!}{k!(n-k)!} = C_{n-k}^n$$

Therefore, the probabilities $p(x)$ for a binomial random variable x when $n = 6$ and $p = .8$ are shown in the table. The probability distribution for x is shown followed by the probability histogram.

x	0	1	2	3	4	5	6
$p(x)$	.000	.002	.015	.082	.246	.393	.262

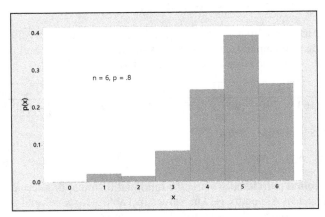

b The probabilities found in part a and the probability histogram are the mirror images of those found in Exercise 26.

5.2.29 **a** $P[x \geq 4] = 1 - P[x \leq 3] = 1 - .099 = .901$

b $P[x = 2] = P[x \leq 2] - P[x \leq 1] = .017 - .002 = .015$

c $P[x < 2] = P[x \leq 1] = .002$

d $P[x>1] = 1 - P[x \le 1] = 1 - .002 = .998$

5.2.31   **a** $\mu = 100(.01) = 1;\ \sigma = \sqrt{100(.01)(.99)} = .99$

     b $\mu = 100(.9) = 90;\ \sigma = \sqrt{100(.9)(.1)} = 3$

     c $\mu = 100(.3) = 30;\ \sigma = \sqrt{100(.3)(.7)} = 4.58$

     d $\mu = 100(.7) = 70;\ \sigma = \sqrt{100(.7)(.3)} = 4.58$

     e $\mu = 100(.5) = 50;\ \sigma = \sqrt{100(.5)(.5)} = 5$

5.2.33   **a** $p(0) = C_0^{20}(.1)^0 (.9)^{20} = .1215767$ $p(3) = C_3^{20}(.1)^3 (.9)^{17} = .1901199$

          $p(1) = C_1^{20}(.1)^1 (.9)^{19} = .2701703$ $p(4) = C_4^{20}(.1)^4 (.9)^{16} = .0897788$

          $p(2) = C_2^{20}(.1)^2 (.9)^{18} = .2851798$

so that $P[x \le 4] = p(0) + p(1) + p(2) + p(3) + p(4) = .9568255$

     b   Using Table 1, Appendix I, $P[x \le 4]$ is read directly as .957.

     c   Adding the entries for $x = 0, 1, 2, 3, 4$, we have $P[x \le 4] = .9569$.

     d $\mu = np = 20(.1) = 2$ and $\sigma = \sqrt{npq} = \sqrt{1.8} = 1.3416$

     e   For $k = 1$, $\mu \pm \sigma = 2 \pm 1.342$ or .658 to 3.342 so that

$$P[.658 \le x \le 3.342] = P[1 \le x \le 3] = .2702 + .2852 + .1901 = .7455$$

For $k = 2$, $\mu \pm 2\sigma = 2 \pm 2.683$ or $-.683$ to 4.683 so that

$$P[-.683 \le x \le 4.683] = P[0 \le x \le 4] = .9569$$

For $k = 3$, $\mu \pm 3\sigma = 2 \pm 4.025$ or -2.025 to 6.025 so that

$$P[-2.025 \le x \le 6.025] = P[0 \le x \le 6] = .9977$$

     f   The results are consistent with Tchebysheff's Theorem and the Empirical Rule.

5.2.35   If the sampling in Exercise 34 is conducted with replacement, then x is a binomial random variable with $n = 2$ independent trials, and $p = P[\text{red ball}] = 3/5$, which remains constant from trial to trial.

5.2.37   Although there are trials (telephone calls) which result in either a person who will answer (S) or a person who will not (F), the number of trials, n, is not fixed in advance. Instead of recording x, the number of *successes* in n trials, you record x, the number of *trials* until the first success. This is *not* a binomial experiment.

5.2.39   The random variable x is defined to be the number of heads observed when a coin is flipped three times. Then $p = P[\text{success}] = P[\text{head}] = 1/2$, $q = 1 - p = 1/2$ and $n = 3$. The binomial formula yields the following results.

     a $P[x=0] = p(0) = C_0^3 (1/2)^0 (1/2)^3 = 1/8$ $P[x=1] = p(1) = C_1^3 (1/2)^1 (1/2)^2 = 3/8$

          $P[x=2] = p(2) = C_2^3 (1/2)^2 (1/2)^1 = 3/8$ $P[x=3] = p(3) = C_3^3 (1/2)^3 (1/2)^0 = 1/8$

b The associated probability histogram is shown next.

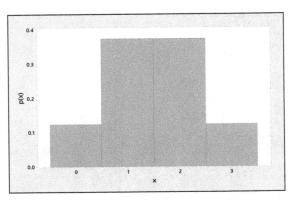

c $\mu = np = 3(1/2) = 1.5$ and $\sigma = \sqrt{npq} = \sqrt{3(1/2)(1/2)} = .866$

d The desired intervals are

$\mu \pm \sigma = 1.5 \pm .866$ or .634 to 2.366

$\mu \pm 2\sigma = 8 \pm 1.732$ or $-.232$ to 3.232

The values of x which fall in this first interval are $x = 1$ and $x = 2$, and the fraction of measurement in this interval will be $3/8 + 3/8 = 3/4$. The second interval encloses all four values of x and thus the fraction of measurements within 2 standard deviations of the mean will be 1, or 100%. These results are consistent with both Tchebysheff's Theorem and the Empirical Rule.

5.2.41 It is given that $n = 20$, $p = .5$ and x = number of patients surviving ten years.

a $P[x \geq 1] = 1 - P[x = 0] = 1 - .000 = 1$

b $P[x \geq 10] = 1 - P[x \leq 9] = 1 - .412 = .588$

c $P[x \geq 15] = 1 - P[x \leq 14] = 1 - .979 = .021$

5.2.43 Define x to be the number of failures observed among the four engines. Then $p = P[\text{engine fails}] = .01$ and $q = 1 - p = .99$, with $n = 4$.

a $P[\text{no failures}] = P[x = 0] = C_0^4 (.01)^0 (.99)^4 = .9606$

b $P[\text{no more than one failure}] = P[x \leq 1] = p(0) + p(1)$
$= C_0^4 (.01)^0 (.99)^4 + C_1^4 (.01)^1 (.99)^3 = .9606 + .0388 = .9994$

5.2.45 The random variable x, the number of California homeowners with earthquake insurance, has a binomial distribution with $n = 15$ and $p = .1$.

a $P(x \geq 1) = 1 - P(x = 0) = 1 - .206 = .794$

b $P(x \geq 4) = 1 - P(x \leq 3) = 1 - .944 = .056$

c Calculate $\mu = np = 15(.1) = 1.5$ and $\sigma = \sqrt{npq} = \sqrt{15(.1)(.9)} = 1.1619$. Then approximately 95% of the values of x should lie in the interval

$\mu \pm 2\sigma \Rightarrow 1.5 \pm 2(1.1619) \Rightarrow -.82$ to 3.82

or between 0 and 3.

5.2.47 a The random variable x, the number of plants with red petals, has a binomial distribution with $n=10$ and $p = P[\text{red petals}] = .75$.

b Since the value $p = .75$ is not given in Table 1, you must use the binomial formula to calculate

$$P(x \geq 9) = C_9^{10}(.75)^9(.25)^1 + C_{10}^{10}(.75)^{10}(.25)^0 = .1877 + .0563 = .2440$$

c $P(x \leq 1) = C_0^{10}(.75)^0(.25)^{10} + C_1^{10}(.75)^1(.25)^9 = .0000296$.

d Refer to part **c**. The probability of observing $x = 1$ or something even more unlikely $(x = 0)$ is very small – .0000296. This is a highly unlikely event if in fact $p = .75$. Perhaps there has been a nonrandom choice of seeds, or the 75% figure is not correct for this particular genetic cross.

5.2.49 The random variable x, the number of subjects who revert to their first learned method under stress, has a binomial distribution with $n = 6$ and $p = .8$. The probability that at least five of the six subjects revert to their first learned method is

$$P(x \geq 5) = 1 - P(x \leq 4) = 1 - .345 = .655$$

5.2.51 Define x to be the number of cars that are black. Then $p = P[\text{black}] = .1$ and $n = 25$. Use Table 1 in Appendix I.

a $P(x \geq 5) = 1 - P(x \leq 4) = 1 - .902 = .098$

b $P(x \leq 6) = .991$

c $P(x > 4) = 1 - P(x \leq 4) = 1 - .902 = .098$

d $P(x = 4) = P(x \leq 4) - P(x \leq 3) = .902 - .764 = .138$

e $P(3 \leq x \leq 5) = P(x \leq 5) - P(x \leq 2) = .967 - .537 = .430$

f $P(\text{more than 20 } not \text{ black}) = P(\text{less than 5 black}) = P(x \leq 4) = .902$

5.2.53 Define a success to be a patient who fails to pay his bill and is eventually forgiven. Assuming that the trials are independent and that p is constant from trial to trial, this problem satisfies the requirements for the binomial experiment with $n = 4$ and $p = .3$. You can use either the binomial formula or Table 1.

a $P[x = 4] = p(4) = C_4^4(.3)^4(.7)^0 = (.3)^4 = .0081$

b $P[x = 1] = p(1) = C_1^4(.3)^1(.7)^3 = 4(.3)(.7)^3 = .4116$

c $P[x = 0] = C_0^4(.3)^0(.7)^4 = (.7)^4 = .2401$

5.2.55 Define x to be the number of cell phone owners who indicate that they had walked into something or someone while talking on their cell phone. Then, $n = 8$ and $p = .23$.

a $P(x = 1) = C_1^8(.23)^1(.77)^7 = 8(.23)(.160485) = .2953$

b From the *Minitab* **Probability Density Function**, read $P(x = 1) = .295293$.

c Use the *Minitab* **Cumulative Distribution Function** to find
$P(x \geq 2) = 1 - P(x \leq 1) = 1 - .41887 = .58113$.

5.2.57 Define x to be the number of dog-owning households that have small dogs. Then, $n = 15$ and $p = .5$. Using the binomial tables in Appendix I,

    a $P(x = 8) = P(x \leq 8) - P(x \leq 7) = .696 - .500 = .196$

    b $P(x \leq 4) = .059$

    c $P(x > 10) = 1 - P(x \leq 10) = 1 - .941 = .059$

Section 5.3

5.3.1 The Poisson random variable can be used as an approximation when n is large and p is small so that $np < 7$.

5.3.3 Use the Poisson formula with $\mu = 2.5$.

$$P(x = 0) = \frac{2.5^0 e^{-2.5}}{0!} = .082085; \quad P(x = 1) = \frac{2.5^1 e^{-2.5}}{1!} = .205212;$$

$$P(x = 2) = \frac{2.5^2 e^{-2.5}}{2!} = .256516; \quad P(x \leq 2) = .082085 + .205212 + .256516 = .543813$$

5.3.5 Using $p(x) = \frac{\mu^x e^{-\mu}}{x!} = \frac{2^x e^{-2}}{x!}$,

$$P[x = 0] = \frac{2^0 e^{-2}}{0!} = .135335; \quad P[x = 1] = \frac{2^1 e^{-2}}{1!} = .27067$$

$$P[x > 1] = 1 - P[x \leq 1] = 1 - .135335 - .27067 = .593994$$

$$P[x = 5] = \frac{2^5 e^{-2}}{5!} = .036089$$

5.3.7 These probabilities should be found using the cumulative Poisson tables (Table 2 in Appendix I). The values in Table 2 for $\mu = 0.8$ are shown in the table that follows.

k	0	1	2	3	4	5
$P(x \leq k)$	.449	.809	.953	.991	.999	1.000

$P(x = 0) = .449$, directly from the table.

$P(x \leq 2) = .953$, directly from the table.

$P(x > 2) = 1 - P(x \leq 2) = 1 - .953 = .047$

$P(2 \leq x \leq 4) = P(x \leq 4) - P(x \leq 1) = .999 - .809 = .190$.

5.3.9 Using Table 1, Appendix I, $P[x \leq 2] = .677$. Then with $\mu = np = 20(.1) = 2$, the Poisson approximation is

$$P[x \leq 2] \approx \frac{2^0 e^{-0}}{0!} + \frac{2^1 e^{-1}}{1!} + \frac{2^2 e^{-2}}{2!}$$
$$= .135335 + .27067 + .27067 = .676675$$

The approximation is very accurate.

5.3.11 Using Table 1, Appendix I, $P[x > 6] = 1 - P(x \leq 6) = 1 - .780 = .220$. Then with $\mu = np = 25(.2) = 5$, the Poisson approximation is found from Table 2 in Appendix I to be

$$P[x>6]=1-P(x\le 6)=1-.762=.238$$

The approximation is quite accurate.

5.3.13 Let x be the number of calls to a consumer hotline in a 30-minute period. Then x has a Poisson distribution with $\mu = 5$. Use Table 2 in Appendix I.

a $P(x>8)=1-P(x\le 8)=1-.932=.068$

b Since there is an average of 5 calls per half-hour, there would be an average of 10 calls per hour. The random variable now has a Poisson distribution with $\mu = 10$.

c Refer to part b and Table 2 with $\mu = 10$. $P(x<15)=P(x\le 14)=.917$

d There is a "high probability" that x will lie within 3 standard deviations of the mean. Using Tchebysheff's Theorem with $k = 3$, we find

$$\mu \pm 3\sigma \Rightarrow 10 \pm 3\sqrt{10} \Rightarrow 10 \pm 9.49 \Rightarrow (.51, 19.49) \text{ or from } x = 1 \text{ to } x = 19.$$

5.3.15 Let x be the number of work-related accidents per week. Then x has a Poisson distribution with $\mu = 2$. Use Table 2 in Appendix I.

a $P(x=0)=.135$ **b** $P(x\ge 1)=1-P(x=0)=1-.135=.865$

c-d If we assume approximately 4 weeks per month, the number of work-related accidents per month will have a Poisson distribution with $\mu = 4(2) = 8$ and $P(x=0)=\dfrac{8^0 e^{-8}}{0!}=e^{-8}=.000335$

5.3.17 Since $np = 200(.01) = 2$, we can use the Poisson approximation with $\mu = 2$ to find $P(x\ge 5)=1-P(x\le 4)=1-.947=.053$.

5.3.19 a $p(2)=\dfrac{5^2 e^{-5}}{2!}=.08422$ and $P[x\le 2]=\dfrac{5^0 e^{-5}}{0!}+\dfrac{5^1 e^{-5}}{1!}+\dfrac{5^2 e^{-5}}{2!}=.12465$

b Recall that for the Poisson distribution, $\mu = 5$ and $\sigma = \sqrt{5} = 2.236$. Therefore the value $x = 10$ lies $(10-5)/2.236 = 2.236$ standard deviations above the mean. It is not a very likely event. Alternatively, from Table 2, $P[x>10]=1-P[x\le 10]=1-.986=.014$ which is an unlikely event.

5.3.21 Let x be the number of injuries per year, with $\mu = 2$.

a $\mu = 2$ and $\sigma = \sqrt{\mu} = \sqrt{2} = 1.414$

b The value of x should be in the interval $\mu \pm 2\sigma \Rightarrow 2 \pm 2.828 \Rightarrow -.828$ to 4.828 at least 3/4 of the time (and probably more). Hence, most value of x will lie between 0 and 4.

5.3.23 The random variable x is the number of cases per 100,000 of E. Coli. If you assume that x has an approximate Poisson distribution with $\mu = 2.85$,

a $P[x\le 2]=\dfrac{2.85^0 e^{-2.85}}{0!}+\dfrac{2.85^1 e^{-2.85}}{1!}+\dfrac{2.85^2 e^{-2.85}}{2!}=.4576$

b $P(x>3)=1-P(x\le 3)=1-\left[\dfrac{2.85^0 e^{-2.85}}{0!}+\dfrac{2.85^1 e^{-2.85}}{1!}+\dfrac{2.85^2 e^{-2.85}}{2!}+\dfrac{2.85^3 e^{-2.85}}{3!}\right]=1-.6808=.3192$

Section 5.4

5.4.1 The hypergeometric distribution is appropriate when sampling (without replacement) from a finite rather than an infinite population of successes and failures. In this case, the probability of p of a success is not constant from trial to trial and the binomial distribution is not appropriate.

5.4.5 $\dfrac{C_1^3 C_1^2}{C_2^5} = \dfrac{3(2)}{10} = .6$

5.4.7 $\dfrac{C_4^5 C_0^3}{C_4^8} = \dfrac{5(1)}{70} = .0714$

5.4.9 The formula for $p(x)$ is $p(x) = \dfrac{C_x^5 C_{4-x}^3}{C_4^8}$ for $x = 1, 2, 3, 4$. Since there are only 3 "failures" and we are selecting 4 items, we must observe at least one "success". The probability of exactly one success is

$$P(x=1) = \dfrac{C_1^5 C_3^3}{C_4^8} = \dfrac{5(1)}{70} = \dfrac{5}{70}$$

5.4.11, 13 The formula for $p(x)$ is $p(x) = \dfrac{C_x^6 C_{5-x}^4}{C_5^{10}}$ for $x = 1, 2, 3, 4, 5$. Since there are only 4 "failures" and we are selecting 5 items, we must select at least one "success".

11. Since we must select at least one "success", $P[x=0] = 0$.

13. $P[x \geq 2] = 1 - P[x \leq 1] = 1 - p(1) = 1 - \dfrac{C_1^6 C_4^4}{C_5^{10}} = 1 - \dfrac{6}{252} = .9762$

5.4.15, 17 The formula for $p(x)$ is $p(x) = \dfrac{C_x^4 C_{3-x}^{11}}{C_3^{15}}$ for $x = 0, 1, 2, 3$. Calculate

$p(0) = \dfrac{C_0^4 C_3^{11}}{C_3^{15}} = \dfrac{165}{455} = .36$ $\qquad$ $p(1) = \dfrac{C_1^4 C_2^{11}}{C_3^{15}} = \dfrac{220}{455} = .48$

$p(2) = \dfrac{C_2^4 C_1^{11}}{C_3^{15}} = \dfrac{66}{455} = .15$ $\qquad$ $p(3) = \dfrac{C_3^4 C_0^{11}}{C_3^{15}} = \dfrac{4}{455} = .01$

15. The probability histogram is shown next.

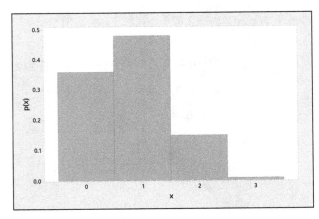

17. Calculate the intervals

$$\mu \pm 2\sigma = .8 \pm 2\sqrt{.50286} = .8 \pm 1.418 \text{ or } -.618 \text{ to } 2.218$$
$$\mu \pm 3\sigma = .8 \pm 3\sqrt{.50286} = .8 \pm 1.418 \text{ or } -1.327 \text{ to } 2.927$$

Then,
$$P[-.618 \le x \le 2.218] = p(0) + p(1) + p(2) = .99$$
$$P[-1.327 \le x \le 2.927] = p(0) + p(1) + p(2) = .99$$

These results agree with Tchebysheff's Theorem.

5.4.19 The formula for $p(x)$ is $p(x) = \dfrac{C_x^5 C_{3-x}^3}{C_3^8}$ for x = number of brown candies = $0, 1, 2, 3$. Then

$$P(x = 0) = \dfrac{C_0^5 C_3^3}{C_3^8} = \dfrac{1}{56} = .0179$$

5.4.21 The formula for $p(x)$ is $p(x) = \dfrac{C_x^3 C_{3-x}^5}{C_3^8}$ for x = number of red candies = $0, 1, 2, 3$.

5.4.23, 25 The formula for $p(x)$ is $p(x) = \dfrac{C_x^{26} C_{5-x}^{26}}{C_5^{52}}$ for x = number of red cards = $0, 1, 2, 3, 4, 5$.

23. $P(x=3) = \dfrac{C_3^{26} C_2^{26}}{C_5^{52}} = \dfrac{2600(325)}{2{,}598{,}960} = .3251.$

25. $P(x \le 1) = \dfrac{C_0^{26} C_5^{26}}{C_5^{52}} + \dfrac{C_1^{26} C_4^{26}}{C_5^{52}} = \dfrac{65780}{2{,}598{,}960} + \dfrac{26(14950)}{2{,}598{,}960} = .1749$

5.4.27 The formula for $p(x)$ is $p(x) = \dfrac{C_x^4 C_{3-x}^{48}}{C_3^{52}}$ for x = number of kings = $0, 1, 2, 3$. Then

$$P(x=0) = \dfrac{C_0^4 C_{3-0}^{48}}{C_3^{52}} = \dfrac{17296}{22100} = .782624 \qquad P(x=1) = \dfrac{C_1^4 C_{3-1}^{48}}{C_3^{52}} = \dfrac{4(1128)}{22100} = .204163$$

$$P(x=2) = \dfrac{C_2^4 C_{3-2}^{48}}{C_3^{52}} = \dfrac{6(48)}{22100} = .013032 \qquad P(x=3) = \dfrac{C_3^4 C_{3-3}^{48}}{C_3^{52}} = \dfrac{4}{22100} = .000181$$

5.4.29 a. The formula for $p(x)$ is $p(x) = \dfrac{C_x^{125} C_{10-x}^{75}}{C_{10}^{200}}$ for x = number of Republicans = $0, 1, 2, \ldots 10$.

b. The probability of selecting 10 Republicans is $p(10) = \dfrac{C_{10}^{125} C_{10-10}^{75}}{C_{10}^{200}} = .0079$.

c. The probability of selecting no Republicans is $p(0) = \dfrac{C_0^{125} C_{10-0}^{75}}{C_{10}^{200}} = .0000369$.

5.4.31 The formula for $p(x)$ is $p(x) = \dfrac{C_x^2 C_{3-x}^4}{C_3^6}$ for x = number of defectives = $0, 1, 2$.

$$p(0) = \dfrac{C_0^2 C_3^4}{C_3^6} = \dfrac{4}{20} = .2 \qquad p(1) = \dfrac{C_1^2 C_2^4}{C_3^6} = \dfrac{12}{20} = .6 \qquad p(2) = \dfrac{C_2^2 C_1^4}{C_3^6} = \dfrac{4}{20} = .2$$

These results agree with the probabilities calculated in Exercise 26 (Section 5.1).

5.4.33 **a** The random variable x has a hypergeometric distribution with $N = 8, M = 5$ and $n = 3$. Then

$$p(x) = \frac{C_x^5 C_{3-x}^3}{C_3^8} \text{ for } x = 0, 1, 2, 3$$

b $P(x = 3) = \dfrac{C_3^5 C_0^3}{C_3^8} = \dfrac{10}{56} = .1786$

c $P(x = 0) = \dfrac{C_0^5 C_3^3}{C_3^8} = \dfrac{1}{56} = .01786$

d $P(x \leq 1) = \dfrac{C_0^5 C_3^3}{C_3^8} + \dfrac{C_1^5 C_2^3}{C_3^8} = \dfrac{1+15}{56} = .2857$

5.4.35 The random variable x has a hypergeometric distribution with $N = 8, M = 2$ and $n = 4$. Then

$$P(x = 0) = \frac{C_0^2 C_4^6}{C_4^8} = \frac{15}{70} = .214 \quad \text{and} \quad P(x = 2) = \frac{C_2^2 C_2^6}{C_4^8} = \frac{15}{70} = .214$$

Reviewing What You've Learned

5.R.1 **a** Consider the event $x = 3$. This will occur only if either A or B wins three sets in a row; that is, if the event AAA or BBB occurs. Then

$$P[x = 3] = P(AAA) + P(BBB) = [P(A)]^3 + [P(B)]^3 = (.6)^3 + (.4)^3 = .28$$

Consider the event $x = 4$. This will occur if 3 A-wins are spread over 4 sets (with the last A-win in the fourth set) or if 3 B-wins are similarly spread over 4 sets. The associated simple events are

ABAA	BABB	AABA
BAAA	ABBB	BBAB

and the probability that $x = 4$ is

$$p(4) = 3(.6)^3(.4) + 3(.6)(.4)^3 = .3744$$

The event $x = 5$ will occur if 4 A-wins are spread over 5 sets (with the last A-win in the fifth set) or similarly for B. The associated simple events are

ABBAA	AABBA	BBAAA	BAABA	ABABA	BABAA
ABBAB	AABBB	BBAAB	BAABB	ABABB	BABAB

and the probability that $x = 5$ is $p(5) = 6(.6)^3(.4)^2 + 6(.6)^2(.4)^3 = .3456$

Notice that $p(3) + p(4) + p(5) = .28 + .3744 + .3456 = 1.00$.

For a general value $P(A) = p$ and $P(B) = 1 - p$, the probability distribution for x is

$$p(3) = p^3 + (1-p)^3$$
$$p(4) = 3p^3(1-p) + 3p(1-p)^3$$
$$p(5) = 6p^3(1-p)^2 + 6p^2(1-p)^3$$

For the three values of p given in this exercise, the probability distributions and values of $E(x)$ are given in the tables and parts **b**, **c**, and **d** that follow.

	$P(A)=.6$		$P(A)=.5$		$P(A)=.9$
x	p(x)	x	p(x)	x	p(x)
3	.2800	3	.25	3	.7300
4	.3744	4	.375	4	.2214
5	.3456	5	.375	5	.0486

    b $E(x) = 4.0656$ **c** $E(x) = 4.125$ **d** $E(x) = 3.3186$

    e   Notice that as the probability of winning a single set, *P(A)*, increases, the expected number of sets to win the match, *E(x)*, decreases.

5.R.3   **a** $E(x) = 2(.12) + 3(.80) + 4(.06) + 5(.02) = 2.98$

       b $E(x) = 3(.14) + 4(.80) + 5(.04) + 6(.02) = 3.94$

       c $E(x) = 4(.04) + 5(.80) + 6(.12) + 7(.04) = 5.16$

5.R.5   **a**   Define the event R: subject chooses red and N: subject does not choose red. Then $P(R) = \frac{1}{3}$ and $P(N) = \frac{2}{3}$. There are 8 simple events in the experiment:

          NNN $(x=0)$ RRN $(x=2)$

          RNN $(x=1)$ RNR $(x=2)$

          NRN $(x=1)$ NRR $(x=2)$

          NNR $(x=1)$ RRR $(x=3)$

Then

$$P(x=0) = P(NNN) = P(N)P(N)P(N) = \left(\frac{2}{3}\right)^3 = \frac{8}{27}$$

$$P(x=1) = 3P(N)P(N)P(R) = 3\left(\frac{2}{3}\right)^2\left(\frac{1}{3}\right) = \frac{12}{27}$$

$$P(x=2) = 3P(N)P(R)P(R) = 3\left(\frac{2}{3}\right)\left(\frac{1}{3}\right)^2 = \frac{6}{27}$$

$$P(x=3) = P(RRR) = P(R)P(R)P(R) = \left(\frac{1}{3}\right)^3 = \frac{1}{27}$$

The probability distribution for *x* is shown in the table.

x	0	1	2	3
p(x)	8/27	12/27	6/27	1/27

    b   The probability histogram is shown next.

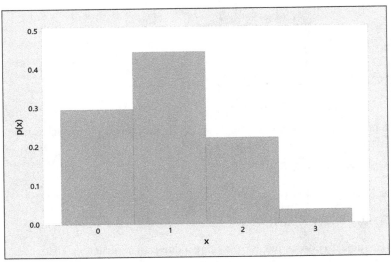

5.R.7 a Define x to be the integer between 0 and 9 chosen by a person. If the digits are equally likely to be chosen, then $p(x) = \dfrac{1}{10}$ for $x = 0,1,2,\ldots,9$.

b $P[4, 5 \text{ or } 6 \text{ is chosen}] = p(4) + p(5) + p(6) = \dfrac{3}{10}$

c $P[\text{not } 4,5 \text{ or } 6] = 1 - \dfrac{3}{10} = \dfrac{7}{10}$

5.R.9 Let x be the number of Millennials in a sample of $n = 20$ who say that they check in with the office at least once or twice a week while on vacation. Then x has a binomial distribution with $p = .6$.

a $P(x = 16) = P(x \le 16) - P(x \le 15) = .984 - .949 = .035$

b $P(16 \le x \le 18) = P(x \le 18) - P(x \le 15) = .999 - .949 = .050$

c $P(x \le 5) = .002$. This would be an unlikely occurrence since it occurs only 2 time in 1000.

5.R.11 Define x to be the number of students favoring the issue, with $n = 25$ and $p = P[\text{student favors the issue}]$ assumed to be .8. Using the binomial tables in Appendix I,

$$P[x \le 15] = .017$$

Thus, the probability of observing $x = 15$ or the more extreme values, $x = 0, 1, 2, \ldots, 14$ is quite small under the assumption that $p = .8$. We probably should conclude that p is actually smaller than .8.

5.R.13 a $p = P[\text{rain according to the forecaster}] = .3$

b With $n = 25$ and $p = .3, \mu = np = 25(.3) = 7.5$ and $\sigma = \sqrt{npq} = \sqrt{5.25} = 2.29129$

c The observed value, $x = 10$, lies $z = \dfrac{10 - 7.5}{2.29} = 1.09$ standard deviations above the mean.

d The observed event is not unlikely under the assumption that $p = .3$. We have no reason to doubt the forecaster.

5.R.15 The random variable x, the number of neighbors per square meter, has a Poisson distribution with $\mu = 4$. Use the Poisson formula or Table 2 in Appendix I.

    a $P(x=0) = .018$

    b $P(x \leq 3) = .433$

    c $P(x \geq 5) = 1 - P(x \leq 4) = 1 - .629 = .371$

    d   With $\mu = 4$ and $\sigma = \sqrt{\mu} = 2$, approximately 95% of the values of x should lie in the interval

$$\mu \pm 2\sigma \Rightarrow 4 \pm 4 \Rightarrow 0 \text{ to } 8$$

In fact, using Table 2, we can calculate the probability of observing between 0 and 8 neighbors per square meter to be $P(x \leq 8) = .979$, which is close to our approximation.

5.R.17   **a**   Since cases of adult onset diabetes is not likely to be contagious, these cases of the disorder occur independently at a rate of 12.5 per 100,000 per year. This random variable can be approximated by the Poisson random variable with $\mu = 12$.

    b   Use Table 2 to find $P(x \leq 10) = .347$.

    c $P(10 < x < 15) = P(x \leq 14) - P(x \leq 10) = .772 - .347 = .425$.

    d   The probability of observing 19 or more cases per 100,000 in a year is

$$P(x \geq 19) = 1 - P(x \leq 18) = 1 - .963 = .037$$

This is an occurrence which we would not expect to see very often, if in fact $\mu = 12$.

5.R.19 The random variable x, the number of adults who prefer milk chocolate to dark chocolate, has a binomial distribution with $n = 5$ and $p = .47$.

    a   Since $p = .47$ is not in Table 1, you must use the binomial formula to find

$$P(x=5) = C_5^5 (.47)^5 (.53)^0 = .0229$$

    b   The probability that exactly three of the adults prefer milk chocolate to dark chocolate is

$$P(x=3) = C_3^5 (.47)^3 (.53)^2 = .5833$$

    c $P(x \geq 1) = 1 - P(x=0) = 1 - C_0^5 (.47)^0 (.53)^5 = 1 - (.53)^5 = .9582$

5.R.21   **a**   The random variable x, the number of tasters who pick the correct sample, has a binomial distribution with $n = 5$ and, if there is no difference in the taste of the three samples,

$$p = P(\text{taster picks the correct sample}) = \frac{1}{3}$$

    b   The probability that exactly one of the five tasters chooses the latest batch as different from the others is

$$P(x=1) = C_1^5 \left(\frac{1}{3}\right)^1 \left(\frac{2}{3}\right)^4 = .3292$$

    c   The probability that at least one of the tasters chooses the latest batch as different from the others is

$$P(x \geq 1) = 1 - P(x=0) = 1 - C_0^5 \left(\frac{1}{3}\right)^0 \left(\frac{2}{3}\right)^5 = .8683$$

5.R.23 Refer to Exercise 22. The random variable x, the number of questionnaires that are filled out and returned, has a binomial distribution with $n = 20$ and $p = .7$.

    a   The average value of x is $\mu = np = 20(.7) = 14$.

    b   The standard deviation of x is $\sigma = \sqrt{npq} = \sqrt{20(.7)(.3)} = \sqrt{4.2} = 2.049$.

    c   The z-score corresponding to $x = 10$ is $z = \dfrac{x - \mu}{\sigma} = \dfrac{10 - 14}{2.049} = -1.95$. Since this z-score does not exceed 3 in absolute value, we would not consider the value $x = 10$ to be an unusual observation.

5.R.25 The random variable x has a Poisson distribution with $\mu = 2$. Use Table 2 in Appendix I or the Poisson formula to find the following probabilities.

    a $P(x = 0) = \dfrac{2^0 e^{-2}}{0!} = e^{-2} = .135335$

    b $P(x \leq 2) = \dfrac{2^0 e^{-2}}{0!} + \dfrac{2^1 e^{-2}}{1!} + \dfrac{2^2 e^{-2}}{2!}$
$= .135335 + .270671 + .270671 = .676676$

5.R.27 The random variable x, the number of applicants who will actually enroll in the freshman class, has a binomial distribution with $n = 1360$ and $p = .9$. Calculate $\mu = np = 1360(.9) = 1224$ and $\sigma = \sqrt{npq} = \sqrt{1360(.9)(.1)} = 11.06$. Then approximately 95% of the values of x should lie in the interval
$$\mu \pm 2\sigma \Rightarrow 1224 \pm 2(11.06) \Rightarrow 1201.87 \text{ to } 1246.12.$$
or between 1202 and 1246.

5.R.29 **a**   The random variable x, the number of Californians who believe that college is not important, has a binomial distribution with $n = 25$ and $p = .5$.

    b $P(x \geq 17) = 1 - P(x \leq 16) = 1 - .946 = .054$.

    c   From Table 1 in Appendix I, the largest value of c (labeled "k" in Table 1) for which $P(x \leq c) \leq .05$ is the value $c = 7$ for which $P(x \leq 7) = .022$.

    d $P(x \leq 6) = 007$. This is a very unlikely event if $p = .5$.

On Your Own

5.R.31 Define x to be the number of times the mouse chooses the red door. Then, if the mouse actually has no preference for color, $p = P[\text{red door}] = .5$ and $n = 10$. Since $\mu = np = 5$ and $\sigma = \sqrt{npq} = 1.58$, you would expect that, if there is no color preference, the mouse should choose the red door
$$\mu \pm 2\sigma \Rightarrow 5 \pm 3.16 \Rightarrow 1.84 \text{ to } 8.16$$
or between 2 and 8 times. If the mouse chooses the red door more than 8 or less than 2 times, the unusual results might suggest a color preference.

6: The Normal Probability Distribution

Section 6.1

6.1.1-7 For a uniform random variable x over the interval from a to b, the probability density function is given as $f(x) = 1/(b-a)$ for $a \le x \le b$. The probability that x lies in any particular interval is calculated as the area under the probability density function (a rectangle) over that interval.

1. $P(x < 5) = \dfrac{5-0}{10} = \dfrac{1}{2}$
3. $P(x > 8) = \dfrac{10-8}{10} = \dfrac{1}{5}$
5. $P(x < 0) = \dfrac{0-(-1)}{1-(-1)} = \dfrac{1}{2}$
7. $P(-.5 < x < .5) = \dfrac{.5-(-.5)}{1-(-1)} = \dfrac{1}{2}$

6.1.9-15 For an exponential random variable x with parameter λ, the probability density function is given as $f(x) = \lambda e^{-\lambda x}$ for $x \ge 0$ and $\lambda > 0$. The probability that x lies in any particular interval is calculated as the area under the probability density function over that interval. Use the fact that $P(x > a) = e^{-\lambda a}$ and $P(x \le a) = 1 - e^{-\lambda a}$.

9. $P(x > 1) = e^{-1} = .3679$

11. $P(x < 1.5) = 1 - e^{-1.5} = .7769$

13. $P(x > 6) = e^{-.2(6)} = .3012$

15. $P(x < 5) = 1 - e^{-.2(5)} = .6321$

6.1.17 It is given that x has a uniform distribution from $a = 0$ to $b = 30$, so that $f(x) = \dfrac{1}{30}$ for $0 \le x \le 30$.

a. $P(x > 20) = \dfrac{30-20}{30} = \dfrac{1}{3}$
b. $P(x < 10) = \dfrac{10-0}{30} = \dfrac{1}{3}$
c. $P(10 < x < 20) = \dfrac{20-10}{30} = \dfrac{1}{3}$

6.1.19 It is given that x has an exponential distribution with $\lambda = 1/365$.

a. $P(x < 180) = 1 - e^{-\frac{180}{365}} = .3893$
b. $P(x > 365) = e^{-\frac{365}{365}} = .3679$

c. If there are two batteries required and the batteries operate independently,

$P(\text{both batteries last more than one year}) = (.3679)(.3679) = .1353$

Section 6.2

6.2.1-23 The first exercises are designed to provide practice for the student in evaluating areas under the normal curve. The following notes may be of some assistance.

I Table 3, Appendix I tabulates the cumulative area under a standard normal curve to the left of a specified value of z.

II Since the total area under the curve is one, the total area lying to the right of a specified value of z and the total area to its left must add to 1. Thus, in order to calculate a "tail area", such as the one shown in Figure 6.1, the value of $z = c$ will be indexed in Table 3, and the area that is obtained will be subtracted from 1. Denote the area obtained by indexing $z = c$ in Table 3 by $A(c)$ and the desired area by A. Then, in the above example, $A = 1 - A(c)$.

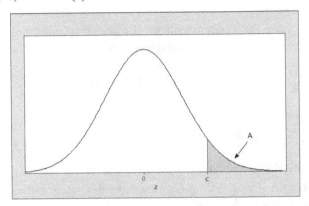

III To find the area under the standard normal curve between two values, c_1 and c_2, calculate the difference in their cumulative areas, $A = A(c_2) - A(c_1)$.

IV Note that z, similar to x, is actually a random variable which may take on an infinite number of values, both positive and negative. Negative values of z lie to the left of the mean, $z = 0$, and positive values lie to the right.

1. $P(z < 2) = P(z < 2.00) = A(2.00) = .9772$

3. $P(-2.33 < z < 2.33) = A(2.33) - A(-2.33) = .9901 - .0099 = .9802$

5. $P(z > 5) < P(z > 3.49) = 1 - .9998 = .0002$ or "approximately zero". Notice that the values in Table 3 approach 1 as the value of z increases. When the value of z is larger than $z = 3.49$ (the largest value in the table), we can assume that the area to its left is approximately 1, and the area to its right is approximately zero.

7. $P(z < 2.81) = A(2.81) = .9975$

9. $P(z < 2.33) = A(2.33) = .9901$

11. $P(z > 1.96) = 1 - A(1.96) = 1 - .9750 = .0250$

13. It is necessary to find the area to the left of $z = 1.6$. That is, $A = A(1.6) = .9452$.

15. This probability is the mirror image of the probability in Exercise 14. The area to the right of $z = -1.83$ is $1 - A(-1.83) = 1 - .0336 = .9664$

17. The area to the right of -1.96 is $1 - A(-1.96) = 1 - .0250 = .9750$

19. $P(-1.43 < z < .68) = A(.68) - A(-1.43) = .7517 - .0764 = .6753$

21. $P(z > 1.34) = 1 - A(1.34) = 1 - .9099 = .0901$

23. $P(.58 < z < 1.74) = A(1.74) - A(.58) = .9591 - .7190 = .2401$

6.2.25-27 The pth percentile of the standard normal distribution is a value of z which has area $p/100$ to its left. Since the four percentiles in these exercises are greater than the 50th percentile, the value of z will all lie to the right of $z = 0$, as shown for the 90th percentile in the figure that follows.

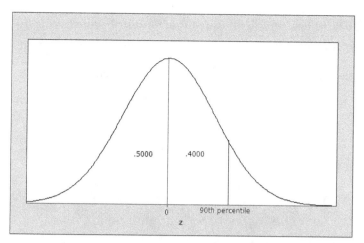

25. From the figure, the area to the left of the 90th percentile is .9000. From Table 3, the appropriate value of z is closest to $z = 1.28$ with area .8997. Hence the 90th percentile is approximately $z = 1.28$.

27. The area to the left of the 98th percentile is .9800. From Table 3, the appropriate value of z is closest to $z = 2.05$ with area .9798. Hence the 98th percentile is approximately $z = 2.05$.

6.2.29 We need to find a c such that $P(z > c) = .025$. This is equivalent to finding an indexed area of $1 - .025 = .975$. Search the interior of Table 3 until you find the four-digit number **.9750**. The corresponding z-value is **1.96**; that is, $A(1.96) = .9750$. Therefore, $c = 1.96$ is the desired z-value (see the figure that follows).

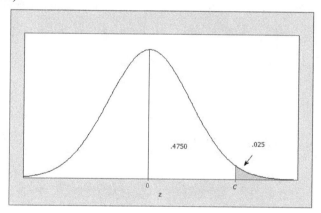

6.2.31 We want to find a z-value such that $P(-c < z < c) = .8262$ (see the next figure.)

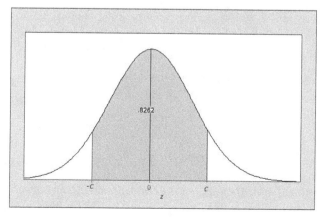

Since $A(c) - A(-c) = .8262$, the total area in the two tails of the distribution must be $1 - .8262 = .1738$ so that the lower tail area must be $A(-c) = .1738/2 = .0869$. From Table 3, $-c = -1.36$ and $c = 1.36$.

6.2.33 It is given that the area to the left of c is .05, shown as A_1 in the figure that follows.

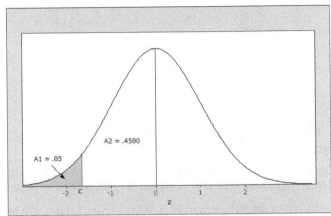

The desired area is not tabulated in Table 3 but falls between two tabulated values, .0505 and .0495. One solution is to choose a value of c which lies halfway between the two corresponding z-values, $z = -1.65$ and $z = -1.64$, or $c = -1.645$.

6.2.35 Refer to the figure above and consider $P(-c < z < c) = A_1 + A_2 = .9900$. Then

$A(-c) = \frac{1}{2}(1 - .9900) = .0050$. Linear interpolation must now be used to determine the value of $-c$, which will lie between $z_1 = -2.57$ and $z_2 = -2.58$. Hence, using a method similar to that in Exercise 10, we find

$$c = z_1 + \frac{P_c - P_1}{P_2 - P_1}(z_2 - z_1) = -2.57 + \left(\frac{.4950 - .4949}{.4951 - .4949}\right)(-2.58 + 2.57)$$

$$= -2.57 - \left(\frac{1}{2}\right)(.01) = -2.575$$

If Table 3 were correct to more than 4 decimal places, you would find that the actual value of c is $c = 2.576$; many texts chose to round up and use the value $c = 2.58$.

6.2.37 Calculate $z = \frac{x - \mu}{\sigma} = \frac{8.2 - 10}{2} = -0.9$. Then $P(x < 8.2) = P(z < -0.9) = A(-.9) = .1841$. This probability is the shaded area in the left tail of the normal distribution above.

6.2.39 Calculate $z_1 = \frac{1.00 - 1.20}{.15} = -1.33$ and $z_2 = \frac{1.10 - 1.20}{.15} = -.67$. Then

$$P(1.00 < x < 1.10) = P(-1.33 < z < -.67) = .2514 - .0918 = .1596$$

6.2.41 Calculate $z_1 = \frac{1.35 - 1.20}{.15} = 1$ and $z_2 = \frac{1.50 - 1.20}{.15} = 2$. Then

$$P(1.35 < x < 1.50) = P(1 < z < 2) = .9772 - .8413 = .1359.$$

6.2.43 The z-value corresponding to $x = 0$ is $z = \frac{x - \mu}{\sigma} = \frac{0 - 50}{15} = -3.33$. Since the value $x = 0$ lies more than three standard deviations away from the mean, it is considered an unusual observation. The probability of observing a value of z as large or larger than $z = -3.33$ is $A(-3.33) = .0004$.

6.2.45 The random variable x is normal with unknown μ and σ. However, it is given that

$P(x>4) = P\left(z > \dfrac{4-\mu}{\sigma}\right) = .9772$ and $P(x>5) = P\left(z > \dfrac{5-\mu}{\sigma}\right) = .9332$. These probabilities are shown in the figure that follows.

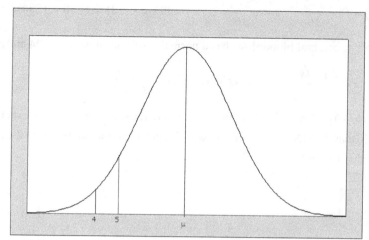

The value $\dfrac{4-\mu}{\sigma}$ is negative, with $A\left(\dfrac{4-\mu}{\sigma}\right) = 1 - .9772 = .0228$ or $\dfrac{4-\mu}{\sigma} = -2$ (i)

The value $\dfrac{5-\mu}{\sigma}$ is also negative, with $A\left(\dfrac{5-\mu}{\sigma}\right) = 1 - .9332 = .0668$ or $\dfrac{5-\mu}{\sigma} = -1.5$ (ii)

Equations (i) and (ii) provide two equations in two unknowns which can be solved simultaneously for μ and σ. From (i), $\sigma = \dfrac{\mu - 4}{2}$ which, when substituted into (ii) yields

$$5 - \mu = -1.5\left(\dfrac{\mu - 4}{2}\right)$$
$$10 - 2\mu = -1.5\mu + 6$$
$$\mu = 8$$

and from (i), $\sigma = \dfrac{8-4}{2} = 2$.

6.2.47 Calculate $z = \dfrac{x - \mu}{\sigma} = \dfrac{16 - 14.1}{1.6} = 1.19$. Then $P(x > 16) = P(z > 1.19) = 1 - .8830 = .1170$.

6.2.49 The random variable x, the height of a male human, has a normal distribution with $\mu = 69.5$ and $\sigma = 3.5$. A height of 6'0" represents $6(12) = 72$ inches, so that

$$P(x > 72) = P\left(z > \dfrac{72 - 69.5}{3.5}\right) = P(z > .71) = 1 - .7611 = .2389$$

6.2.51 The random variable x, the height of a male human, has a normal distribution with $\mu = 69.5$ and $\sigma = 3.5$. A height of 6'2" represents $6(12) + 2 = 74$ inches, which has a z-value of

$$z = \dfrac{74 - 69.5}{3.5} = 1.29$$

The probability of observing a value of x as large or larger than $x = 74$ is $1 - A(1.29) = 1 - .9015 = .0985$. This is not an unlikely event.

6.2.53 The random variable x, cerebral blood flow, has a normal distribution with $\mu = 74$ and $\sigma = 16$.

$$P(60 < x < 80) = P\left(\frac{60-74}{16} < z < \frac{80-74}{16}\right) = P(-.88 < z < .38) = .6480 - .1894 = .4586$$

6.2.55 The random variable x, cerebral blood flow, has a normal distribution with $\mu = 74$ and $\sigma = 16$.

$$P(x < 40) = P\left(z < \frac{40-74}{16}\right) = P(z < -2.12) = .0170$$

6.2.57 For this exercise $\mu = 70$ and $\sigma = 12$. The object is to determine a value, c, for the random variable x so that $P(x < c) = .90$ (that is, 90% of the students will finish the examination before the set time limit). Refer to the figure that follows.

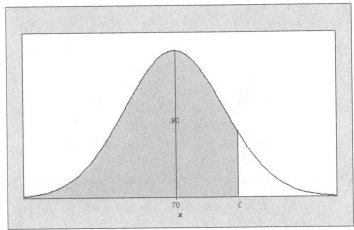

We must have

$$P(x < c) = P\left(z \leq \frac{c-70}{12}\right) = .90$$

$$A\left(\frac{c-70}{12}\right) = .90$$

Consider $z = \frac{c-70}{12}$. Without interpolating, the approximate value for z is

$$z = \frac{c-70}{12} = 1.28 \quad \text{or} \quad c = 85.36$$

6.2.59 The random variable x, the gestation time for a human baby is normally distributed with $\mu = 278$ and $\sigma = 12$.

a. To find the upper and lower quartiles, first find these quartiles for the standard normal distribution. We need a value c such that $P(z > c) = 1 - A(c) = .2500$. Hence, $A(c) = .7500$. The desired value, c, will be between $z_1 = .67$ and $z_2 = .68$ with associated probabilities $P_1 = .2514$ and $P_2 = .2483$. Since the desired tail area, .2500, is closer to $P_1 = .2514$, we approximate c as $c = .67$. The values $z = -.67$ and $z = .67$ represent the 25th and 75th percentiles of the standard normal distribution. Converting these values to their equivalents for the general random variable x using the relationship $x = \mu + z\sigma$, you have:

The lower quartile: $x = -.67(12) + 278 = 269.96$ and
The upper quartile: $x = .67(12) + 278 = 286.04$

b If you consider a month to be approximately 30 days, the value $x = 6(30) = 180$ is unusual, since it lies
$z = \dfrac{x-\mu}{\sigma} = \dfrac{180-278}{12} = -8.167$ standard deviations below the mean gestation time.

6.2.61 The random variable x, the weight of a package of ground beef, has a normal distribution with $\mu = 1$ and $\sigma = .15$.

a $P(x > 1) = P\left(z > \dfrac{1-1}{.15}\right) = P(z > 0) = 1 - .5 = .5$

b $P(.95 < x < 1.05) = P\left(\dfrac{.95-1}{.15} < z < \dfrac{1.05-1}{.15}\right) = P(-.33 < z < .33) = .6293 - .3707 = .2586$

c $P(x < .80) = P\left(z < \dfrac{.80-1}{.15}\right) = P(z < -1.33) = .0918$

d The z-value corresponding to $x = 1.45$ is $z = \dfrac{x-\mu}{\sigma} = \dfrac{1.45-1}{.15} = 3$, which would be considered an unusual observation. Perhaps the setting on the scale was accidentally changed to 1.5 pounds!

6.2.63 It is given that x is normally distributed with $\mu = 140$ and $\sigma = 16$.

a $P(x \leq 120) = P\left(z \leq \dfrac{120-140}{16}\right) = P(z \leq -1.25) = .1056$

$P(x \leq 140) = P\left(z \leq \dfrac{140-140}{16}\right) = P(z \leq 0) = .5$

b In order to avoid a collision, you must brake within 160 feet or less. Hence,

$P(x \leq 160) = P\left(z \leq \dfrac{160-140}{16}\right) = P(z \leq 1.25) = .8944$

6.2.65 It is given that x, the unsupported stem diameter of a sunflower plant, is normally distributed with $\mu = 35$ and $\sigma = 3$.

a $P(x > 40) = P\left(z > \dfrac{40-35}{3}\right) = P(z > 1.67) = 1 - .9525 = .0475$

b From part **a**, the probability that one plant has stem diameter of more than 40 mm is .0475. Since the two plants are independent, the probability that two plants both have diameters of more than 40 mm is

$(.0475)(.0475) = .00226$

c Since 95% of all measurements for a normal random variable lie within 1.96 standard deviations of the mean, the necessary interval is

$\mu \pm 1.96\sigma \Rightarrow 35 \pm 1.96(3) \Rightarrow 35 \pm 5.88$

or in the interval 29.12 to 40.88.

d The 90th percentile of the standard normal distribution was found in Exercise 25 to be $z = 1.28$. Since the relationship between the general normal random variable x and the standard normal z is $z = \dfrac{x-\mu}{\sigma}$, the corresponding percentile for this general normal random variable is found by solving for $x = \mu + z\sigma$.

$x = 35 + 1.28(3)$ or $x = 38.84$

6.2.67 It is given that the counts of the number of bacteria are normally distributed with $\mu = 85$ and $\sigma = 9$. The z-value corresponding to $x = 100$ is $z = \dfrac{x-\mu}{\sigma} = \dfrac{100-85}{9} = 1.67$ and

$$P(x > 100) = P(z > 1.67) = 1 - .9525 = .0475$$

6.2.69 The pulse rates are normally distributed with $\mu = 78$ and $\sigma = 12$.

    a The z-values corresponding to $x = 60$ and $x = 100$ are $z = \dfrac{x-\mu}{\sigma} = \dfrac{60-78}{12} = -1.5$. and

$z = \dfrac{x-\mu}{\sigma} = \dfrac{100-78}{12} = 1.83$ Then $P(60 < x < 100) = P(-1.5 < z < 1.83) = .9664 - .0668 = .8996$.

    b From Exercise 26, we found that the 95th percentile of the *standard* normal (z) distribution is $z = 1.645$. Since $z = \dfrac{x-\mu}{\sigma} = \dfrac{x-78}{12}$, solve for x to find the 95th percentile for the pulse rates:

$$1.645 = \dfrac{x-78}{12} \Rightarrow x = 78 + 1.645(12) = 97.74$$

    c The z-score for $x = 110$ is $z = \dfrac{x-\mu}{\sigma} = \dfrac{110-78}{12} = 2.67$ and

$$P(x > 110) = P(z > 2.67) = 1 - .9962 = .0038.$$

The z-score is between 2 and 3; the probability of observing a value this large or larger is quite small. This pulse rate would be somewhat unusual.

Section 6.3

6.3.1-3 The normal distribution can be used to approximate binomial probabilities when both *np* and *nq* are greater than 5. If *n* is large and *np* is small (preferably less than 7), the Poisson distribution can be used to approximate binomial probabilities.

    1. Calculate $np = 25(.6) = 15$ and $nq = 25(.4) = 10$. The normal approximation is appropriate.

    3. Calculate $np = 25(.3) = 7.5$ and $nq = 25(.7) = 17.5$. The normal approximation is appropriate.

6.3.5 Calculate $\mu = np = 15$ and $\sigma = \sqrt{npq} = \sqrt{25(.6)(.4)} = 2.449$. To find the probability of more than 9 successes, we need to include the values $x = 10, 11, \ldots 25$. To include the entire block of probability for the first value of $x = 10$, we need to start at 9.5. Then the probability of more than 9 successes is approximated as

$$P(x > 9.5) = P\left(z > \dfrac{9.5-15}{2.449}\right) = P(z > -2.25) = 1 - .0122 = .9878.$$

6.3.7 For this binomial random variable, $\mu = np = 100(.2) = 20$ and $\sigma = \sqrt{npq} = 100\sqrt{25(.2)(.8)} = 4$. To include the entire rectangles for $x = 21$ and $x = 24$, the approximating probability is

$$P(20.5 < x < 24.5) = P(.12 < z < 1.12) = .8686 - .5478 = .3208$$

6.3.9 For this binomial random variable, $\mu = np = 100(.2) = 20$ and $\sigma = \sqrt{npq} = 100\sqrt{25(.2)(.8)} = 4$. The approximating probability is $P(x > 21.5)$ since the entire rectangle corresponding to $x = 22$ must be included.

$$P(x > 21.5) = P\left(z > \frac{21.5 - 20}{4}\right) = P(z > .38) = 1 - .6480 = .3520$$

6.3.11 The normal approximation with "correction for continuity" is $P(354.5 < x < 360.5)$ where x is normally distributed with mean and standard deviation given by

$$\mu = np = 400(.9) = 360 \text{ and } \sigma = \sqrt{npq} = \sqrt{400(.9)(.1)} = 6$$

Then
$$P(354.5 < x < 360.5) = P\left(\frac{354.5 - 360}{6} < z < \frac{360.5 - 360}{6}\right)$$
$$= P(-.92 < z < .08) = .5319 - .1788 = .3531$$

6.3.13 Given a binomial random variable x with $n = 25$ and $p = .2$, use Table 1 to calculate

$$P(4 \le x \le 6) = P(x \le 6) - P(x \le 3) = .780 - .234 = .546$$

To approximate this probability, we find the area under the binomial distribution for $x = 4, 5, 6$, using the "correction for continuity" and find the area under the normal curve between 3.5 and 6.5. This is done in order to include the entire area under the rectangles associated with the different values of x. First find the mean and standard deviation of the binomial random variable x:

$\mu = np = 25(.2) = 5$ and $\sigma = \sqrt{npq} = \sqrt{25(.2)(.8)} = 2$. Then

$$P(3.5 < x < 6.5) = P\left(\frac{3.5 - 5}{2} < z < \frac{6.5 - 5}{2}\right) = P(-.75 < z < .75) = .7734 - .2266 = .5468$$

The actual and approximated probabilities are very close.

6.3.15 Similar to previous exercises. With $n = 20$ and $p = .4$, use Table 1 to find
$P(x \ge 10) = 1 - P(x \le 9) = 1 - .755 = .245$.

To use the normal approximation, find the mean and standard deviation of this binomial random variable:

$$\mu = np = 20(.4) = 8 \text{ and } \sigma = \sqrt{npq} = \sqrt{20(.4)(.6)} = \sqrt{4.2} = 2.191.$$

Using the continuity correction, it is necessary to find the area to the right of 9.5. The z-value corresponding to $x = 9.5$ is
$$z = \frac{9.5 - 8}{2.191} = .68 \text{ and}$$

$$P(x \ge 10) \approx P(z > .68) = 1 - .7517 = .2483.$$

Note that the normal approximation is very close to the exact binomial probability.

6.3.17 The random variable x has a binomial distribution with $n = 100$ and $p = .51$. Calculate
$\mu = np = 100(.51) = 51$ and $\sigma = \sqrt{npq} = \sqrt{100(.51).49} = 4.999$. The probability that x is 60 or more is approximated as

$$P(x \ge 59.5) = P\left(z \ge \frac{59.5 - 51}{4.999}\right) = P(z \ge 1.70) = 1 - .9554 = .0446$$

6.3.19 The random variable of interest is x, the number of persons not showing up for a given flight. This is a binomial random variable with $n = 160$ and $p = P[\text{person does not show up}] = .05$. If there is to be a seat available for every person planning to fly, then there must be at least five persons not showing up. Hence, the probability of interest is $P(x \ge 5)$. Calculate

$$\mu = np = 160(.05) = 8 \text{ and } \sigma = \sqrt{npq} = \sqrt{160(.05)(.95)} = \sqrt{7.6} = 2.7$$

Referring to the figure that follows, a correction for continuity is made to include the entire area under the rectangle associated with the value $x = 5$, and the approximation becomes $P(x \geq 4.5)$. The z-value corresponding to $x = 4.5$ is $z = \dfrac{x - \mu}{\sigma} = \dfrac{4.5 - 8}{\sqrt{7.6}} = -1.27$ so that

$$P(x \geq 4.5) = P(z \geq -1.27) = 1 - .1020 = .8980$$

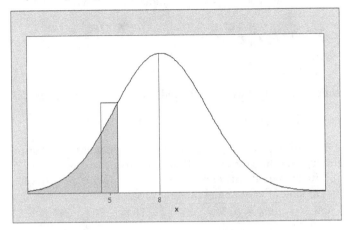

6.3.21 Define x to be the number of guests claiming a reservation at the motel. Then $p = P[\text{guest claims reservation}] = 1 - .1 = .9$ and $n = 215$. The motel has only 200 rooms. Hence, if $x > 200$, a guest will not receive a room. The probability of interest is then $P(x \leq 200)$. Using the normal approximation, calculate

$$\mu = np = 215(.9) = 193.5 \text{ and } \sigma = \sqrt{215(.9)(.1)} = \sqrt{19.35} = 4.399$$

The probability $P(x \leq 200)$ is approximated by the area under the appropriate normal curve to the left of 200.5. The z-value corresponding to $x = 200.5$ is $z = \dfrac{200.5 - 193.5}{\sqrt{19.35}} = 1.59$ and

$$P(x \leq 200) \approx P(z < 1.59) = .9441$$

6.3.23 Define x to be the number of elections in which the taller candidate won. If Americans are not biased by height, then the random variable x has a binomial distribution with $n = 53$ and $p = .5$. Calculate

$$\mu = np = 53(.5) = 26.5 \text{ and } \sigma = \sqrt{53(.5)(.5)} = \sqrt{13.25} = 3.640$$

a Using the normal approximation with correction for continuity, we find the area to the right of $x = 27.5$:

$$P(x > 27.5) = P\left(z > \dfrac{27.5 - 26.5}{3.640}\right) = P(z > 0.27) = 1 - .6064 = .3936$$

b Since the occurrence of 28 out of 53 taller choices is not unusual, based on the results of part **a**, there is insufficient evidence to conclude that Americans consider height when casting a vote for a candidate.

6.3.25 Define x to be the number of consumers who preferred a *Coke* product. Then the random variable x has a binomial distribution with $n = 500$ and $p = .36$, if *Coke's* market share is indeed 36%. Calculate

$$\mu = np = 500(.36) = 180 \text{ and } \sigma = \sqrt{500(.36)(.64)} = \sqrt{115.2} = 10.733$$

a Using the normal approximation with correction for continuity, we find the area between $x = 199.5$ and $x = 200.5$:

$$P(199.5 < x < 200.5) = P\left(\frac{199.5 - 180}{10.733} < z < \frac{200.5 - 180}{10.733}\right) = P(1.82 < z < 1.91) = .9719 - .9656 = .0063$$

b Find the area between $x = 174.5$ and $x = 200.5$:

$$P(174.5 < x < 200.5) = P\left(\frac{174.5 - 180}{10.733} < z < \frac{200.5 - 180}{10.733}\right) = P(-.51 < z < 1.91) = .9719 - .3050 = .6669$$

c Find the area to the left of $x = 199.5$:

$$P(x < 199.5) = P\left(z < \frac{199.5 - 180}{10.733}\right) = P(z < 1.82) = .9656$$

d The value $x = 232$ lies $z = \frac{232 - 180}{10.733} = 4.84$ standard deviations above the mean, if *Coke's* market share is indeed 36%. This is such an unusual occurrence that we would conclude that *Coke's* market share is higher than claimed.

6.3.27 It is given that the probability of a successful single transplant from the early gastrula stage is .65. In a sample of 100 transplants, the mean and standard deviation of the binomial distribution are

$$\mu = np = 100(.65) = 65 \quad \text{and} \quad \sigma = \sqrt{npq} = \sqrt{100(.65)(.35)} = 4.770$$

It is necessary to find the probability that more than 70 transplants will be successful. This is approximated by the area under a normal curve with $\mu = 65$ and $\sigma = 4.77$ to the right of 70.5. The z-value corresponding to $x = 70.5$ is $\quad z = \frac{x - \mu}{\sigma} = \frac{70.5 - 65}{4.77} = 1.15 \quad \text{and} \quad P(x > 70) \approx P(z > 1.15) = 1 - .8749 = .1251$

6.3.29 Define x to be the number of American shoppers who use their smartphones to search for coupons. Then the random variable x has a binomial distribution with $n = 50$ and $p = .49$.

a-b Calculate $\mu = np = 50(.49) = 24.5$ and $\sigma = \sqrt{50(.49)(.51)} = \sqrt{12.495} = 3.535$

c Using the normal approximation with correction for continuity to approximate the probability that x is less than or equal to 15, we find the area to the right of $x = 15.5$:

$$P(x < 15.5) = P\left(z < \frac{15.5 - 24.5}{3.535}\right) = P(z < -2.55) = .0054$$

This would be considered a very unusual occurrence.

Reviewing What You've Learned

6.R.1 It is given that x has an exponential distribution with $\lambda = 1/5 = .2$.

a $P(x < 4) = 1 - e^{-.2(4)} = .5507$ **b** $P(x > 6) = e^{-.2(6)} = .3012$

c If there are two such switches that operate independently,

$P(\text{neither switch fails before year 8}) = [P(x > 8)]^2 = [e^{-.2(8)}]^2 = .0408$

6.R.3 It is given that x is approximately normally distributed with $\mu = 75$ and $\sigma = 12$.

a Calculate $\quad z = \frac{x - \mu}{\sigma} = \frac{60 - 75}{12} = -1.25$. Then $P(x < 60) = P(z < -1.25) = .1056$

b $P(x > 60) = 1 - P(x < 60) = 1 - .1056 = .8944$

c If the bit is replaced after more than 90 hours, then $x > 90$. Calculate $z = \dfrac{x - \mu}{\sigma} = \dfrac{90 - 75}{12} = 1.25$. Then
$P(x > 90) = P(z > 1.25) = 1 - .8944 = .1056$.

6.R.5 It is given that x is normally distributed with $\mu = 30$ and $\sigma = 11$. The probability of interest is

$$P(x > 50) = P\left(z > \dfrac{50 - 30}{11}\right) = P(z > 1.82) = 1 - .9656 = .0344$$

6.R.7 For the binomial random variable x, the mean and standard deviation are calculated under the assumption that the advertiser's claim is correct and $p = .2$. Then

$$\mu = np = 1000(.2) = 200 \text{ and } \sigma = \sqrt{npq} = \sqrt{1000(.2)(.8)} = 12.649$$

If the advertiser's claim is correct, the z-score for the observed value of x, $x = 184$, is

$$z = \dfrac{x - \mu}{\sigma} = \dfrac{184 - 200}{12.649} = -1.26$$

That is, the observed value lies 1.26 standard deviations below the mean. This is not an unlikely occurrence. Hence, we would have no reason to doubt the advertiser's claim.

6.R.9 Define x to be the percentage of returns audited for a particular state. It is given that x is normally distributed with $\mu = 1.55$ and $\sigma = .45$.

a $P(x > 2.5) = P\left(z > \dfrac{2.5 - 1.55}{.45}\right) = P(z > 2.11) = 1 - .9826 = .0174$

b $P(x < 1) = P\left(z < \dfrac{1 - 1.55}{.45}\right) = P(z < -1.22) = .1112$

6.R.11 Define x to be the number of men who have fished in the last year. Then x has a binomial distribution with $n = 180$ and $p = .41$. Calculate

$$\mu = np = 180(.41) = 73.8 \quad \text{and} \quad \sigma = \sqrt{npq} = \sqrt{180(.41)(.59)} = 6.5986$$

a Using the normal approximation with correction for continuity,

$$P(x < 50) \approx P\left(z < \dfrac{49.5 - 73.8}{6.5986}\right) = P(z < -3.68) \approx 0$$

b $P(50 \le x \le 75) \approx P\left(\dfrac{49.5 - 73.8}{6.5986} \le z \le \dfrac{75.5 - 73.8}{6.5986}\right) = P(-3.68 \le z \le .26) = .6026 - .0000 = .6026$

c The sample is not random, since mailing lists for a sporting goods company will probably contain more fishermen than the population in general. Since the sampling was not random, the survey results are not reliable.

6.R.13 a It is given that the scores on a national achievement test were approximately normally distributed with a mean of 540 and standard deviation of 110. It is necessary to determine how far, in standard deviations, a score of 680 departs from the mean of 540. Calculate

$$z = \dfrac{x - \mu}{\sigma} = \dfrac{680 - 540}{110} = 1.27.$$

b To find the percentage of people who scored higher than 680, we find the area under the standardized normal curve greater than 1.27. Using Table 3, this area is equal to

$$P(x > 680) = P(z > 1.27) = 1 - .8980 = .1020$$

Thus, approximately 10.2% of the people who took the test scored higher than 680.

On Your Own

6.R.15 The random variable x, total weight of 8 people, has a mean of $\mu = 1200$ and a standard deviation $\sigma = 99$. It is necessary to find $P(x > 1300)$ and $P(x > 1500)$ if the distribution of x is approximately normal. Refer to the figure that follows.

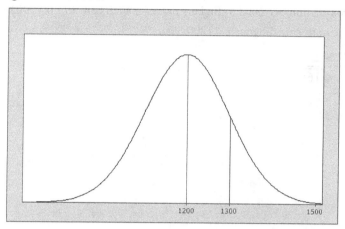

The z-value corresponding to $x_1 = 1300$ is $z_1 = \dfrac{x_1 - \mu}{\sigma} = \dfrac{1300 - 1200}{99} = 1.01$. Hence,

$$P(x > 1300) = P(z > 1.01) = 1 - A(1.01) = 1 - .8438 = .1562.$$

Similarly, the z-value corresponding to $x_2 = 1500$ is $z_2 = \dfrac{x_2 - \mu}{\sigma} = \dfrac{1500 - 1200}{99} = 3.03$.

and $\quad P(x > 1500) = P(z > 3.03) = 1 - A(3.03) = 1 - .9988 = .0012.$

6.R.17 Let w be the number of words specified in the contract. Then x, the number of words in the manuscript, is normally distributed with $\mu = w + 20,000$ and $\sigma = 10,000$. The publisher would like to specify w so that

$$P(x < 100,000) = .95.$$

As in Exercise 16, calculate $\quad z = \dfrac{100,0000 - (w + 20,000)}{10,000} = \dfrac{80,000 - w}{10,000}.$

Then $\quad P(x < 100,000) = P\left(z < \dfrac{80,000 - w}{10,000}\right) = .95$. It is necessary that $z_0 = (80,000 - w)/10,000$ be such that

$$P(z < z_0) = .95 \implies A(z_0) = .9500 \text{ or } z_0 = 1.645.$$

Hence,

$$\dfrac{80,000 - w}{10,000} = 1.645 \text{ or } w = 63,550.$$

6.R.19 The random variable x is the size of the freshman class. That is, the admissions office will send letters of acceptance to (or accept deposits from) a certain number of qualified students. Of these students, a certain number will actually enter the freshman class. Since the experiment results in one of two outcomes (enter or not enter), the random variable x, the number of students entering the freshman class, has a binomial distribution with

n = number of deposits accepted and

$p = P$[student, having been accepted, enters freshman class] $= .8$

a It is necessary to find a value for n such $P(x \leq 1200) = .95$. Note that,

$$\mu = np = .8n \text{ and } \sigma = \sqrt{npq} = \sqrt{.16n}$$

Using the normal approximation, we need to find a value of n such that $P(x \leq 1200) = .95$, as shown in the next figure. The z-value corresponding to $x = 1200.5$ is

$$z = \frac{x - \mu}{\sigma} = \frac{1200.5 - .8n}{\sqrt{.16n}}$$

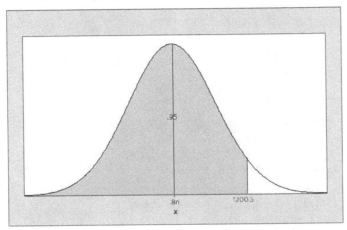

From Table 3, the z-value corresponding to an area of .05 in the right tail of the normal distribution is 1.645. Then,

$$\frac{1200.5 - .8n}{\sqrt{.16n}} = 1.645$$

Solving for n in the above equation, we obtain the following quadratic equation:

$$.8n + .658\sqrt{n} - 1200.5 = 0$$

Let $x = \sqrt{n}$. Then the equation takes the form $ax^2 + bx + c = 0$ which can be solved using the quadratic formula, $x = \frac{-b \pm \sqrt{b^2 - 4ac}}{2a}$, or

$$x = \frac{-.658 \pm \sqrt{.432964 + 4(960.4)}}{1.6} = \frac{-.658 \pm 61.9841}{2}$$

Since x must be positive, the desired root is

$$x = \sqrt{n} = \frac{61.3261}{1.6} = 38.329 \text{ or } n = (38.329)^2 = 1469.1.$$

Thus, 1470 deposits should be accepted.

b Once $n = 1470$ has been determined, the mean and standard deviation of the distribution are

$$\mu = np = 1470(.8) = 1176 \text{ and } \sigma = \sqrt{npq} = \sqrt{235.2} = 15.3362.$$

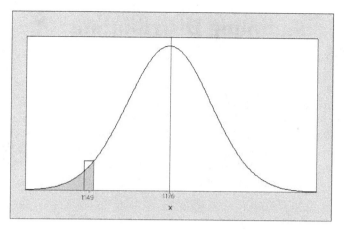

Then the approximation for $P(x < 1150) = P(x \leq 1149)$, shown above, is

$$P(x \leq 1149.5) = P\left(z \leq \frac{1149.5 - 1176}{15.3362}\right) = P(z \leq -1.73) = .0418$$

6.R.21 The measurements are approximately normal with mean $\mu = 39.83$ and standard deviation $\sigma = 2.05$.

a

$$P(36.5 < x < 43.5) = P\left(\frac{36.5 - 39.83}{2.05} < z < \frac{43.5 - 39.83}{2.05}\right) = P(-1.62 < z < 1.79) = .9633 - .0526 = .9107$$

b From Exercise 11 (Section 6.2), we know that the area to the right of $z = 1.96$ is .025. The area to the left of $z = -1.96$ is also .025 so that $P(-1.96 < z < 1.96) = .95$ for the standard normal z. Substituting for z in the probability statement, we write

$$P\left(-1.96 < \frac{x - \mu}{\sigma} < 1.96\right) \Rightarrow P(\mu - 1.96\sigma < x < \mu + 1.96\sigma) = .95$$

That is, 95% of the chest measurements, x, will lie between $\mu - 1.96\sigma$ and $\mu + 1.96\sigma$, or

$$39.83 \pm 1.96(2.05) \text{ which is } 35.812 \text{ to } 43.848.$$

c Refer to the data distribution given in Exercise 20, and count the number of measurements falling in each of the two intervals.

- Between 36.5 and 43.5: $\frac{420 + 749 + ... + 370}{5738} = \frac{5283}{5738} = .921$ (compared to .9107)

- Between 35.812 and 43.848: $\frac{185 + 420 + ... + 370}{5738} = \frac{5468}{5738} = .953$ (compared to .95)

7: Sampling Distributions

Section 7.1

7.1.1 The experimental units are each assigned two random numbers (001 and 501 to unit 1, 002 and 502 to unit 2, ..., 499 and 999 to unit 499, and 000 and 500 to unit 500). You can select a simple random sample of size $n = 20$ using Table 10 in Appendix I. First choose a starting point and consider the first three digits in each number. The three digits OR the (three digits – 500) will identify the proper experimental unit. For example, if the three digits are 742, you should select the experimental unit numbered $742 - 500 = 242$. The probability that any three-digit number is selected is $2/1000 = 1/500$. One possible selection for the sample size $n = 20$ is

242	134	173	128	399
056	412	188	255	388
469	244	332	439	101
399	156	028	238	231

7.1.3 Students may choose different sampling schemes, and will obtain different samples, using Table 10 in Appendix I.

7.1.5 Each home represents a cluster within which each occupant is surveyed. The ten homes represent a *cluster sample*.

7.1.7 A random sample is selected from each of six departments. This is a *stratified sample* with the departments representing the strata.

7.1.9 Since the student procrastinated until the last moment to complete the survey, he chooses the most convenient people (his fraternity brothers) to participate. This is a *convenience sample*.

7.1.11 The professor would like to be certain that he has a high return rate on his questionnaire, so he chooses his twenty most reliable students. This is a *convenience sample*. Some students may decide that this is a *judgement sample*, since the professor is using his best judgement in the selection of the students to receive the questionnaire.

7.1.13 This is a 1-in-10 systematic sample.

7.1.15 This is a 1-in-10 systematic sample.

7.1.17 The survey responses do not represent a representative sample from the total number of surveys that were mailed out. Since only 30% of the surveys were returned, there is a problem of *nonresponse* and the 90% figure may indicate that only those people who were in favor of the proposed zoning changes bothered to return their survey.

7.1.19 Only readers of your Facebook page will respond to the survey. There is a problem with *undercoverage* and the responses will not be representative of the entire population.

7.1.21 The questionnaires that were returned do not constitute a representative sample from the 1000 questionnaires that were randomly sent out. It may be that the voters who chose to return the questionnaire were particularly adamant about the Parks and Recreation surcharge, while the others had no strong feelings one way or the other. The nonresponse of half the voters in the sample will undoubtedly bias the resulting statistics.

7.1.23 The wording of the question is biased to suggest that a "yes" response is the correct one. A more unbiased way to phrase the question might be: "Is there too much sex and violence during prime TV viewing hours?"

7.1.25 **a** Since the question is particularly sensitive to people of different ethnic origins, you may find that the answers may not always be truthful, depending on the ethnicity of the interviewer and the person being interviewed.

b Notice that the percentage in favor of affirmative action increases as the ethnic origin of the interviewer changes from Caucasian to Asian to African-American. The people being interviewed may be changing their response to match what they perceive to be the response which the interviewer wants to hear.

7.1.27 **a** Since each subject must be randomly assigned to either a tai chi class or a wellness education class with equal probability, assign the digits 0-4 to the tai chi treatment, and the digits 5-9 to the wellness treatment. As each subject enters the study, choose a random digit using Table 10 and assign the appropriate treatment.

b The randomization scheme in part **a** does not guarantee an equal number of subjects in each group.

7.1.29 Answers will vary from student to student. Paying cash for opinions will not necessarily produce a random sample of opinions of all Pepsi and Coke drinkers.

Section 7.2

7.2.1 $C_4^6 = \dfrac{6!}{4!2!} = 15$ samples are possible. The 15 samples along with the sample means and medians for each are shown in the table that follows.

Sample	Observations	$\bar{x}$	m	Sample	Observations	$\bar{x}$	m
1	1, 3, 4, 7	3.75	3.5	9	1, 4, 10, 11	6.5	7
2	1, 3, 4, 10	4.5	3.5	10	3, 4, 7, 10	6	5.5
3	1, 3, 4, 11	4.75	3.5	11	3, 4, 7, 11	6.25	5.5
4	1, 3, 10, 11	6.25	6.5	12	3, 4, 10, 11	7	7
5	1, 3, 7, 10	5.25	5	13	3, 7, 10, 11	7.75	8.5
6	1, 3, 7, 11	5.5	5	14	4, 7, 10, 11	8	8.5
7	1, 4, 7, 10	5.5	5.5	15	1, 7, 10, 11	7.25	8.5
8	1, 4, 7, 11	5.75	5.5				

There are 15 possible samples which are equally likely (due to random sampling), each with probability 1/15. However, there are only 13 distinct values of $\bar{x}$, since two values of $\bar{x}$ (6.25 and 5.5) occur twice. Therefore, the sampling distribution of $\bar{x}$ is given as

$$p(\bar{x}) = \dfrac{1}{15} \text{ for } \bar{x} = 3.75, 4.5, 4.75, 5.25, 5.75, 6, 6.5, 7, 7.25, 7.75, 8$$

$$p(\bar{x}) = \dfrac{2}{15} \text{ for } \bar{x} = 5.5, 6.25$$

The sampling distribution is shown next.

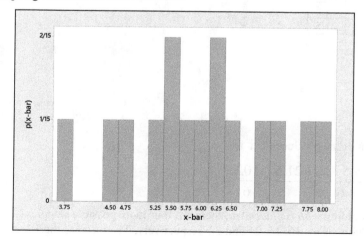

7.2.3-5 $C_3^5 = \dfrac{5!}{3!2!} = 10$ samples are possible. The 10 samples along with the sample mean, median, range and sample variance for each are shown next.

Sample	Observations	$\bar{x}$	m	R	s^2
1	11, 12, 15	12.67	12	4	4.33
2	11, 12, 18	13.67	12	7	14.33
3	11, 12, 20	14.33	12	9	24.33
4	11, 15, 18	14.67	15	7	12.33
5	11, 15, 20	15.33	15	9	20.33
6	11, 18, 20	16.33	18	9	22.33
7	12, 15, 18	15.00	15	6	9.00
8	12, 15, 20	15.67	15	8	16.33
9	12, 18, 20	16.67	18	8	17.33
10	15, 18, 20	17.67	18	5	6.33

3. There are 10 unique values for $\bar{x}$, each with probability 1/10 because of the random sampling. The sampling distribution of $\bar{x}$ is:

$$p(\bar{x}) = \frac{1}{10} \text{ for } \bar{x} = 12.67, 13.67, 14.33, 14.67, 15, 15.33, 15.67, 16.33, 16.67, 17.67$$

5. There are 10 values for the range, each with probability 1/10, but only 6 unique values. The sampling distribution of R is

R	4	5	6	7	8	9
p(R)	1/10	1/10	1/10	2/10	2/10	3/10

7.2.7 There are $C_2^4 = \dfrac{4!}{2!2!} = 6$ possible samples without replacement, shown as follows along with the value of $\bar{x}$ for each. Since each value of $\bar{x}$ is unique, each has probability 1/6.

Sample	Observations	$\bar{x}$
1	10, 15	12.5
2	10, 21	15.5
3	10, 22	16
4	15, 21	18
5	15, 22	18.5
6	21, 22	21.5

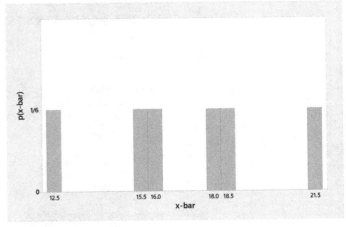

7.2.9 For Exercise 7, the mean of the 6 values of $\bar{x}$ is

$$\mu_{\bar{x}} = \frac{12.5 + 15.5 + \ldots + 21.5}{6} = \frac{102}{6} = 17.$$

For Exercise 8, when sampling *with replacement*, there are four more possible samples—(10, 10), (15, 15), (21, 21) and (22, 22)—with means 10, 15, 21 and 22, respectively. The mean of the 10 values of $\bar{x}$ for Exercise 8 is

$$\mu_{\bar{x}} = \frac{12.5+15.5+\ldots+21.5+10+15+21+22}{10} = \frac{170}{10} = 17.$$

In both cases, the mean of the sampling distribution of $\bar{x}$ is the same as the mean $\mu = 17$ of the original distribution.

7.2.11 The distribution is an example of a hypergeometric distribution with $N = 5$ and $n = 2$. When sampling without replacement, there are $C_2^5 = 10$ possible samples, listed next along with the proportion of successes in the sample.

Sample	Outcome	Proportion of Successes	Sample	Outcome	Proportion of Successes
1	$S_1 S_2$	1	6	$S_2 F_2$	1/2
2	$S_1 F_1$	1/2	7	$S_2 F_3$	1/2
3	$S_1 F_2$	1/2	8	$F_1 F_2$	0
4	$S_1 F_3$	1/2	9	$F_1 F_3$	0
5	$S_2 F_1$	1/2	10	$F_2 F_3$	0

7.2.13 Both of the sampling distributions are centered at the mean $\mu = 5.5$, but distribution 2 is much less variable. Distribution 2 would be the better plan to use in estimating the population mean.

Section 7.3

7.3.1-3 If the sampled populations are normal, the distribution of $\bar{x}$ is also normal *for all values of n*. The sampling distribution of the sample mean will have a mean μ equal to the mean of the population from which we are sampling, and a standard deviation equal to $\sigma/\sqrt{n}$.

1. $\mu = 10;\ \sigma/\sqrt{n} = 3/\sqrt{36} = .5$

3. $\mu = 120;\ \sigma/\sqrt{n} = 1/\sqrt{8} = .3536$

7.3.5 The Central Limit Theorem states that for sample sizes as small as $n = 30$, the sampling distribution of $\bar{x}$ will be approximately normal. But regardless of the shape of the population from which we are sampling, the sampling distribution of the sample mean will have a mean μ equal to the mean of the population from which we are sampling, and a standard deviation equal to $\sigma/\sqrt{n}$. The sample size, $n = 10$, is too small to assume that the distribution of $\bar{x}$ is approximately normal. The mean and standard error are
$\mu = 15;\ \sigma/\sqrt{n} = \sqrt{4}/\sqrt{10} = .6325$.

7.3.7-13 For a population with $\sigma = 1$, the standard error of the mean is

$$\sigma/\sqrt{n} = 1/\sqrt{n}$$

The values of $\sigma/\sqrt{n}$ for various values of n are tabulated in the following table.

Exercise Number	7	9	11	13
n	1	4	16	100
$SE(\bar{x}) = \sigma/\sqrt{n}$	1.00	.500	.250	.100

7.3.15 Since $n = 49$, the Central Limit Theorem is used. The sampling distribution of $\bar{x}$ will be approximately normal. The mean of the sampling distribution of $\bar{x}$ is $\mu = 53$ and the standard deviation (or standard error) is $\sigma/\sqrt{n} = 21/\sqrt{49} = 3$.

7.3.17 If the sample population is normal, the sampling distribution of $\bar{x}$ will also be normal (regardless of the sample size) with mean $\mu = 106$ and standard deviation (or *standard error*) given as

$$\sigma/\sqrt{n} = 12/\sqrt{25} = 2.4$$

7.3.19 From Exercise 16, the sampling distribution of $\bar{x}$ will be approximately normal, with mean $\mu = 100$ and standard error $\sigma/\sqrt{n} = 20/\sqrt{40} = 3.16$. The probability of interest is $P(105 < \bar{x} < 110)$. When $\bar{x} = 105$ and $\bar{x} = 110$,

$$z = \frac{\bar{x} - \mu}{\sigma/\sqrt{n}} = \frac{105 - 100}{3.16} = 1.58 \text{ and } z = \frac{\bar{x} - \mu}{\sigma/\sqrt{n}} = \frac{110 - 100}{3.16} = 3.16$$

Then, $P(105 < \bar{x} < 110) = P(1.58 < z < 3.16) = .9992 - .9429 = .0563$.

7.3.21 Since the weights are normally distributed, the sampling distribution for the average weight for a sample of $n = 16$ baby girls will also be normally distributed with mean $\mu = 12.9$ and standard error $\sigma/\sqrt{n} = 1.6/\sqrt{16} = 0.4$.

a Calculate $z = \frac{\bar{x} - \mu}{\sigma/\sqrt{n}} = \frac{14 - 12.9}{0.4} = 2.75$, so that $P(\bar{x} > 14) = P(z > 2.75) = 1 - .9970 = .0030$.

b Calculate $z = \frac{\bar{x} - \mu}{\sigma/\sqrt{n}} = \frac{11 - 12.9}{0.4} = -4.75$, so that $P(\bar{x} < 11) = P(z < -4.75) \approx 0$.

c Since the probability in part b is close to zero, this would be a very unlikely event, assuming that the average weight of babies at this facility is 12.9 as it is in the general population. Perhaps babies in this area are underweight.

7.3.23 Since $n = 400$, the Central Limit Theorem ensures that the sampling distribution of $\bar{x}$ will be approximately normal with mean $\mu = 1110$ and standard deviation $\sigma/\sqrt{n} = 80/\sqrt{400} = 4$.

a $P(1100 < \bar{x} < 1110) = P(\frac{1100 - 1110}{4} < z < \frac{1110 - 1110}{4}) = P(-2.5 < z < 0) = .5000 - .0062 = .4938$.

b $P(\bar{x} > 1120) = P(z > \frac{1120 - 1110}{4}) = P(z > 2.5) = 1 - .9938 = .0062$.

c $P(\bar{x} < 900) = P(z < \frac{900 - 1110}{4}) = P(z < -52.5) = .0000$.

7.3.25 a Since the sample size is large, the sampling distribution of $\bar{x}$ will be approximately normal with mean $\mu = 75,878$ and standard deviation $\sigma/\sqrt{n} = 4000/\sqrt{60} = 516.3978$.

b From the Empirical Rule (and the general properties of the normal distribution), approximately 95% of the measurements will lie within 2 standard deviations of the mean:

$\mu \pm 2SE \Rightarrow 75,878 \pm 2(516.3978)$

$75,878 \pm 1032.80$ or $74,845.20$ to $76,910.80$

c Use the mean and standard deviation for the distribution of $\bar{x}$ given in part **a**.

$$P(\bar{x} > 78,000) = P\left(z > \frac{78,000 - 75,878}{516.3978}\right)$$
$$= P(z > 4.11) \approx 0$$

d Refer to part **c**. You have observed a very unlikely occurrence (one that almost never happens), assuming that $\mu = 75,878$. Perhaps your sample was not a random sample, or perhaps the average salary of $75,878 is no longer correct.

7.3.27 The weight of a package of 12 tomatoes is the sum of the 12 individual tomato weights. Hence, since weights and heights are in general normally distributed, so will be the sum of the 12 weights, according to the Central Limit Theorem.

7.3.29 **a** Since the original population is normally distributed, the sample mean $\bar{x}$ is also normally distributed (for any sample size) with mean μ and standard deviation

$$\sigma/\sqrt{n} \approx 2/\sqrt{10}$$

b If $\mu = 21$ and $n = 10$, the z-value corresponding to $\bar{x} = 20$ is

$$z = \frac{\bar{x} - \mu}{\sigma/\sqrt{n}} = \frac{20 - 21}{2/\sqrt{10}} = -1.58 \text{ and } P(\bar{x} < 20) = P(z < -1.58) = .0571$$

c Refer to part **b** and assume that μ is unknown. It is necessary to find a value c so that $P(\bar{x} < 20) = .001$. Recall that if $\mu = c$, the z-value corresponding to $\bar{x} = 20$ is

$$z = \frac{\bar{x} - c}{\sigma/\sqrt{n}} = \frac{20 - c}{2/\sqrt{10}} = \frac{20 - c}{.632}.$$

It is necessary then to have $P(\bar{x} < 20) = P\left(z < \frac{20-c}{.632}\right) = A\left(\frac{20-c}{.632}\right) = .001$.

Refer to the figure that follows.

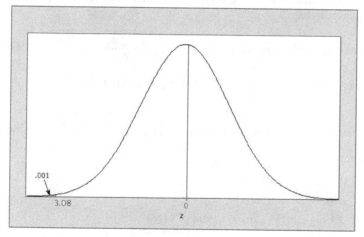

From Table 3, the value of z that cuts off .001 in the left-hand tail of the normal distribution is $z = -3.08$. Hence, the two values, $(20-c)/.632$ and $z = -3.08$, must be the same. Solving for c,

$$\frac{20-c}{.632} = -3.08 \text{ or } c = 21.948.$$

7.3.31 **a** Since the total daily sales is the sum of the sales made by a fixed number of customers on a given day, it is a sum of random variables, which, according to the Central Limit Theorem, will have an approximate normal distribution.

b Let x_i be the total sales for a single customer, with i = 1, 2, ..., 30. Then x_i has a probability distribution with $\mu = 8.50$ and $\sigma = 2.5$. The total daily sales can now be written as $x = \Sigma x_i$. If $n = 30$, the mean and standard deviation of the sampling distribution of x are given as

$$n\mu = 30(8.5) = 255 \text{ and } \sigma\sqrt{n} = 2.5\sqrt{30} = 13.693$$

7.3.33 The sampled population has a mean of 5.97 with a standard deviation of 1.95.

a With $n = 31$, calculate $z = \frac{\bar{x} - \mu}{\sigma/\sqrt{n}} = \frac{6.5 - 5.97}{1.95/\sqrt{31}} = 1.51$, so that $P(\bar{x} \leq 6.5) = P(z \leq 1.51) = .9345$.

b Calculate $z = \dfrac{\bar{x} - \mu}{\sigma/\sqrt{n}} = \dfrac{9.80 - 5.97}{1.95/\sqrt{31}} = 10.94$, so that $P(\bar{x} \geq 9.80) = P(z \geq 10.94) \approx 1 - 1 = 0$.

c The probability of observing an average diameter of 9.80 or higher is extremely unlikely, if indeed the average diameter in the population of injured tendons was no different from that of healthy tendons (5.97). We would conclude that the average diameter in the population of patients with injured tendons is higher than 5.97.

Section 7.5

7.5.1 $p = .3;\ SE(\hat{p}) = \sqrt{\dfrac{pq}{n}} = \sqrt{\dfrac{.3(.7)}{100}} = .0458$

7.5.3 $p = .6;\ SE(\hat{p}) = \sqrt{\dfrac{pq}{n}} = \sqrt{\dfrac{.6(.4)}{250}} = .0310$

7.5.5 For this binomial distribution, calculate $np = 30$ and $nq = 45$. Since both values are greater than 5, the normal approximation is appropriate.

7.5.7 For this binomial distribution, calculate $np = 50$ and $nq = 450$. Since both values are greater than 5, the normal approximation is appropriate.

7.5.9 Since $\hat{p}$ is approximately normal, with standard deviation $SE(\hat{p}) = \sqrt{\dfrac{pq}{n}} = \sqrt{\dfrac{.4(.6)}{75}} = .0566$, the probability of interest is $P(\hat{p} \leq .43) = P\left(z \leq \dfrac{.43 - .4}{.0566}\right) = P(z \leq .53) = .7019$.

7.5.11 Since $\hat{p}$ is approximately normal, with standard deviation $SE(\hat{p}) = \sqrt{\dfrac{pq}{n}} = \sqrt{\dfrac{.4(.6)}{75}} = .0566$, the probability of interest is approximated as

$$P(.35 \leq \hat{p} \leq .43) = P\left[\dfrac{.35 - .4}{.0566} \leq z \leq \dfrac{.43 - .4}{.0566}\right]$$
$$= P(-.88 \leq z \leq .53) = .7019 - .1894 = .5125$$

7.5.13-15 For $n = 500$ and $p = .1$, $np = 50$ and $nq = 450$ are both greater than 5. Therefore, the normal approximation will be appropriate.

13. $P(\hat{p} > .12) = P\left(z > \dfrac{.12 - .1}{.0134}\right) = P(z > 1.49) = 1 - .9319 = .0681$

15. $P(-.02 \leq (\hat{p} - p) \leq .02) = P(-1.49 \leq z \leq 1.49) = .9319 - .0681 = .8638$

7.5.17-21 The values $SE = \sqrt{pq/n}$ for $n = 100$ and various values of p are tabulated next.

Exercise Number	17	19	21
p	.10	.50	.90
$SE(\hat{p})$	.03	.05	.03

7.5.23 Use the values of *SE* calculated in Exercises 16-22 to produce a graph similar to the one shown as follows. Notice that *SE* is a maximum for $p = .5$ and becomes very small for p near zero and one.

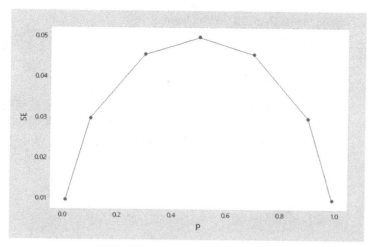

7.5.25 Since $np = 20$ and $nq = 60$, the normal approximation is appropriate. The sampling distribution of $\hat{p}$ will be approximately normal. The mean of the sampling distribution of $\hat{p}$ is $p = .25$ and the standard deviation (or standard error) is $\sqrt{\dfrac{pq}{n}} = \sqrt{\dfrac{.25(.75)}{80}} = .04841$. The probability of interest is $P(.18 < \hat{p} < .44)$.
Calculate

$$z = \dfrac{\hat{p} - p}{\sqrt{\dfrac{pq}{n}}} = \dfrac{.18 - .25}{.04841} = -1.45 \text{ and } z = \dfrac{\hat{p} - p}{\sqrt{\dfrac{pq}{n}}} = \dfrac{.44 - .25}{.04841} = 3.92$$

Then $P(.18 < \hat{p} < .44) = P(-1.45 < z < 3.92) = 1 - .0735 = .9265$.

7.5.27 Since the coin is fair, we can assume that $p = P(\text{head}) = .5$. Then for $n = 80$ and $p = .5$, $np = 40$ and $nq = 40$ are both greater than 5. Therefore, the normal approximation will be appropriate, with mean $p = .5$ and $SE = \sqrt{\dfrac{pq}{n}} = \sqrt{\dfrac{.5(.5)}{80}} = .0559$. The required probability is

$$P(.44 < \hat{p} < .61) = P\left(\dfrac{.44 - .5}{.0559} < z < \dfrac{.61 - .5}{.0559}\right) = P(-1.07 < z < 1.97) = .9756 - .1423 = .8333$$

7.5.29 Let x be the number of cell phone owners who have walked into someone or something while they were on the phone. Then x has a binomial distribution with $n = 200$ and $p = .23$. Since $np = 46$ and $nq = 154$, the normal approximation is appropriate, and

$$P(\hat{p} < .15) = P\left(z < \dfrac{.15 - .23}{\sqrt{\dfrac{.23(.77)}{200}}}\right) = P(z < -2.69) = .0036$$

7.5.31 **a** Let x be the number of Americans who associate "recycling" with Earth Day. Then x has a binomial distribution with $n = 100$ and $p = .47$. Since $np = 47$ and $nq = 53$, the normal approximation is appropriate, with mean $p = .47$ and $SE = \sqrt{\dfrac{pq}{n}} = \sqrt{\dfrac{.47(.53)}{100}} = .0499$.

b $P(\hat{p} < .45) = P\left(z < \dfrac{.45 - .47}{.0499}\right) = P(z < -.40) = .3446$

c $P(.42 < \hat{p} < .45) = P\left(\dfrac{.42 - .47}{.0499} < z < \dfrac{.45 - .47}{.0499}\right) = P(-1.00 < z < -.40) = .3446 - .1587 = .1859$.

d The probability that the proportion is less than .30 is calculated as

$$P(\hat{p} < .30) = P\left(z < \frac{.30 - .47}{.0499}\right) = P(z < -3.41) = .0003$$

Since this is such an unlikely occurrence, perhaps the 47% figure is too high.

7.5.33 **a** The random variable $n\hat{p}$, where $\hat{p}$ is the sample proportion of brown M&Ms in a package of $n = 55$, has a binomial distribution with $n = 55$ and $p = .13$. Since $np = 7.15$ and $nq = 47.85$ are both greater than 5, the distribution of $\hat{p}$ can be approximated by a normal distribution with mean $p = .13$ and

$$SE = \sqrt{\frac{.13(.87)}{55}} = .045347.$$

b $P(\hat{p} < .2) = P\left(z < \frac{.2 - .13}{.045347}\right) = P(z < 1.54) = .9382$

c $P(\hat{p} > .35) = P\left(z > \frac{.35 - .13}{.045347}\right) = P(z > 4.85) \approx 1 - 1 = 0$

d From the Empirical Rule (and the general properties of the normal distribution), approximately 95% of the measurements will lie within 2 (or 1.96) standard deviations of the mean:

$$p \pm 2SE \;\Rightarrow\; .13 \pm 2(.045347)$$

$.13 \pm .09$ or .04 to .22

7.5.35 **a** The random variable $n\hat{p}$, where $\hat{p}$ is the sample proportion of consumers who like nuts or caramel in their chocolate, has a binomial distribution with $n = 200$ and $p = .75$. Since $np = 150$ and $nq = 50$ are both greater than 5, the distribution of $\hat{p}$ can be approximated by a normal distribution with mean $p = .75$ and

$$SE = \sqrt{\frac{.75(.25)}{200}} = .03062.$$

b $P(\hat{p} > .80) = P\left(z > \frac{.80 - .75}{.03062}\right) = P(z > 1.63) = 1 - .9484 = .0516$

c From the Empirical Rule (and the general properties of the normal distribution), approximately 95% of the measurements will lie within 2 (or 1.96) standard deviations of the mean:

$$p \pm 2SE \;\Rightarrow\; .75 \pm 2(.03062)$$

$.75 \pm .06$ or .69 to .81

Section 7.6

7.6.1 The $\bar{x}$ chart is used to monitor the process variable—the average value of a sample of quantitative data—detecting shifts that might indicate control problems.

7.6.3 The $\bar{x}$ chart is used to monitor the average value of a sample of quantitative data, while the p chart is used to monitor qualitative data by counting the number of defective items and tracking the percentage defective.

7.6.5 The upper and lower control limits are

$$UCL = \bar{\bar{x}} + 3\frac{s}{\sqrt{n}} = 155.9 + 3\frac{4.3}{\sqrt{5}} = 155.9 + 5.77 = 161.67$$

$$LCL = \bar{\bar{x}} - 3\frac{s}{\sqrt{n}} = 155.9 - 3\frac{4.3}{\sqrt{5}} = 155.9 - 5.77 = 150.13$$

The control chart is constructed by plotting two horizontal lines, one the upper control limit and one the lower control limit (see Figure 7.17 in the text). Values of $\bar{x}$ are plotted and should remain within the control limits. If not, the process should be checked.

7.6.7 The upper and lower control limits for the p chart are

$$UCL = \bar{p} + 3\sqrt{\frac{\bar{p}(1-\bar{p})}{n}} = .041 + 3\sqrt{\frac{.041(.959)}{200}} = .041 + .042 = .083$$

$$LCL = \bar{p} - 3\sqrt{\frac{\bar{p}(1-\bar{p})}{n}} = .041 - 3\sqrt{\frac{.041(.959)}{200}} = .041 - .042 = -.001$$

or LCL = 0 (since p cannot be negative). The control chart is constructed by plotting two horizontal lines, one the upper control limit and one the lower control limit (see Figure 7.18 in the text). Values of $\hat{p}$ are plotted and should remain within the control limits. If not, the process should be checked.

7.6.9 a The upper and lower control limits for the p chart are

$$UCL = \bar{p} + 3\sqrt{\frac{\bar{p}(1-\bar{p})}{n}} = .021 + 3\sqrt{\frac{.021(.979)}{400}} = .021 + .022 = .043$$

$$LCL = \bar{p} - 3\sqrt{\frac{\bar{p}(1-\bar{p})}{n}} = .021 - 3\sqrt{\frac{.021(.979)}{400}} = .021 - .022 = -.001$$

or LCL = 0 (since p cannot be negative).

b The manager can use the control chart to detect changes in the production process which might produce an unusually large number of defectives.

7.6.11 a The upper and lower control limits are

$$UCL = \bar{\bar{x}} + 3\frac{s}{\sqrt{n}} = 7.24 + 3\frac{.07}{\sqrt{3}} = 7.24 + .12 = 7.36$$

$$LCL = \bar{\bar{x}} - 3\frac{s}{\sqrt{n}} = 7.24 - 3\frac{.07}{\sqrt{3}} = 7.24 - .12 = 7.12$$

b The $\bar{x}$ chart will allow the manager to monitor the percentage of ash in the coal to see whether there is a problem on any particular day.

7.6.13 The upper and lower control limits are

$$UCL = \bar{\bar{x}} + 3\frac{s}{\sqrt{n}} = 31.7 + 3\frac{.2064}{\sqrt{5}} = 31.7 + .277 = 31.977$$

$$LCL = \bar{\bar{x}} - 3\frac{s}{\sqrt{n}} = 31.7 - 3\frac{.2064}{\sqrt{5}} = 31.7 - .277 = 31.423$$

The control chart is constructed by plotting two horizontal lines, one the upper control limit and one the lower control limit (see Figure 7.17 in the text). Values of $\bar{x}$ are plotted and should remain within the control limits. If not, the process should be checked.

7.6.15 a If the process is in control, $\sigma = 1.20$. Calculate $\bar{\bar{x}} = \frac{23.1 + 21.3 + \cdots + 21.3}{30} = \frac{636.1}{30} = 21.2033$.

With $n = 5$, the upper and lower control limits are

$$\bar{\bar{x}} \pm 3\frac{s}{\sqrt{n}} = 21.2033 \pm 3\frac{1.2}{\sqrt{5}} = 21.2033 \pm 1.6100$$

or $LCL = 19.5933$ and $UCL = 22.8133$.

b The centerline is $\bar{\bar{x}} = 21.2033$ and the graph is omitted. Only one sample exceeds the UCL; the process is probably in control.

Reviewing What You've Learned

7.R.1 **a** $C_2^4 = \dfrac{4!}{2!2!} = 6$ samples are possible.

b-c The 6 samples along with the sample means for each are shown on the next page.

Sample	Observations	$\bar{x}$
1	6, 1	3.5
2	6, 3	4.5
3	6, 2	4.0
4	1, 3	2.0
5	1, 2	1.5
6	3, 2	2.5

d Since each of the 6 distinct values of $\bar{x}$ are equally likely (due to random sampling), the sampling distribution of $\bar{x}$ is given as

$$p(\bar{x}) = \frac{1}{6} \quad \text{for } \bar{x} = 1.5, 2, 2.5, 3.5, 4, 4.5$$

The sampling distribution follows.

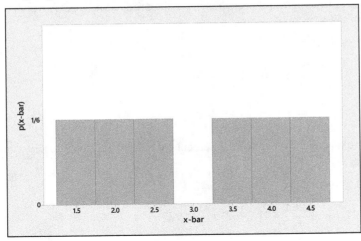

e The population mean is $\mu = (6+1+3+2)/4 = 3$. Notice that none of the samples of size $n = 2$ produce a value of $\bar{x}$ exactly equal to the population mean.

7.R.3 **a** Regardless of the shape of the original sampled distribution, the sampling distribution of $\bar{x}$ will have a mean of $\mu = 1$ and a standard deviation of $\sigma/\sqrt{n} = .36/\sqrt{5} = .160997 \approx .161$.

b Since the sampled distribution is normal, the distribution of $\bar{x}$ will also be normal, even for a small value of n. Therefore,

$$P(\bar{x} > 1.3) = P\left(z > \frac{1.3 - 1}{.161}\right) = P(z > 1.86) = 1 - .9686 = .0314.$$

c $P(\bar{x} < .5) = P\left(z < \dfrac{.5 - 1}{.161}\right) = P(z < -3.11) = .0009.$

d $P(|\bar{x}-\mu|>.4) = P(|z|>\frac{.4}{.161}) = P(|z|>2.48) = P(z>2.48) + P(z<-2.48) = 2(.0066) = .0132$.

7.R.5 a Using the range approximation, the standard deviation σ can be approximated as

$$\sigma \approx \frac{R}{4} = \frac{55-5}{4} = 12.5$$

b The sampling distribution of $\bar{x}$ is approximately normal with mean μ and standard error

$$\sigma/\sqrt{n} \approx 12.5/\sqrt{400} = .625$$

Then $P(|\bar{x}-\mu|\le 2) = P\left(\frac{-2}{.625} \le z \le \frac{2}{.625}\right) = P(-3.2 \le z \le 3.2) = .9993 - .0007 = .9986$.

c If the scientists are worried that the estimate of $\mu = 35$ is too high, and if your estimate is $\bar{x} = 31.75$, then your estimate lies

$$z = \frac{\bar{x}-\mu}{\sigma/\sqrt{n}} = \frac{31.75-35}{.625} = -5.2$$

standard deviations below the mean. This is a very unlikely event if in fact $\mu = 35$. It is more likely that the scientists are correct in assuming that the mean is an overestimate of the mean biomass for tropical woodlands.

7.R.7 a From the 50 lettuce seeds, the researcher must choose a group of 26 and a group of 13 for the experiment. Identify each seed with a number from 01 to 50 and then select random numbers from Table 10. The first 26 numbers chosen will identify the seeds in the first Petri dish, and the next 13 will identify the seeds in the third Petri dish. If the same number is picked twice, simply ignore it and go on to the next number. Use a similar procedure to choose the groups of 26 and 13 radish seeds.

b The seeds in these two packages must be representative of all seeds in the general population of lettuce and radish seeds.

7.R.9 Referring to Table 10 in Appendix I, we will select 20 numbers. First choose a starting point and consider the first four digits in each number. If the four digits are a number greater than 7000, discard it. Continue until 20 numbers have been chosen. The customers have already been numbered from 0001 to 7000. One possible selection for the sample size $n = 20$ is

1048	2891	5108	4866
2236	6355	0236	5416
2413	0942	0101	3263
4216	1036	5216	2933
3757	0711	0705	0248

7.R.11 a The number of packages which can be assembled in 8 hours is the sum of 8 observations on the random variable described here. Hence, its mean is $n\mu = 8(16.4) = 131.2$ and its standard deviation is $\sigma\sqrt{n} = 1.3\sqrt{8} = 3.677$.

b If the original population is approximately normal, the sampling distribution of a sum of 8 normal random variables will also be approximately normal. Since the original population is exactly normal, so will be the sampling distribution of the sum.

c $P(x>135) = P\left(z > \frac{135-131.2}{3.677}\right) = P(z>1.03) = 1-.8485 = .1515$

7.R.13 a The average proportion of defectives is

$$\bar{p} = \frac{.04+.02+\cdots+.03}{25} = .032$$

and the control limits are

$$UCL = \bar{p} + 3\sqrt{\frac{\bar{p}(1-\bar{p})}{n}} = .032 + 3\sqrt{\frac{.032(.968)}{100}} = .0848$$

and $$LCL = \bar{p} - 3\sqrt{\frac{\bar{p}(1-\bar{p})}{n}} = .032 - 3\sqrt{\frac{.032(.968)}{100}} = -.0208$$

If subsequent samples do not stay within the limits, $UCL = .0848$ and $LCL = 0$, the process should be checked.

b From part **a**, we must have $\hat{p} > .0848$.

c An erroneous conclusion will have occurred if in fact $p < .0848$ and the sample has produced $\hat{p} = .15$ by chance. One can obtain an upper bound on the probability of this particular type of error by calculating $P(\hat{p} \geq .15$ when $p = .0848)$.

7.R.15 Refer to Exercise 13, in which $UCL = .0848$ and $LCL = 0$. For the next 5 samples, the values of $\hat{p}$ are .02, .04, .09, .07, .11. Hence, samples 3 and 5 are producing excess defectives. The process should be checked.

On Your Own

7.R.17 Answers will vary. There is an overrepresentation of Christian schools, which may bias the answers (depending on what questions are asked).

7.R.19 a For the data given in the text, the mean and standard deviation are calculated as

$$\bar{x} = \frac{\sum x_i}{n} = \frac{224.3}{50} = 4.486 \text{ and } s = \sqrt{\frac{\sum x_i^2 - \frac{(\sum x_i)^2}{n}}{n-1}} = \sqrt{\frac{1025.23 - \frac{224.3^2}{50}}{49}} = .623$$

b The values calculated in part **a** are close to the theoretical values of the mean and standard deviation for the sampling distribution of $\bar{x}$, given as $\mu_{\bar{x}} = \mu = 4.4$ and $\sigma_{\bar{x}} = \frac{\sigma}{\sqrt{n}} = \frac{2.15}{\sqrt{10}} = .680$.

7.R.21 Define x_i to be the weight of a particular man or woman using the elevator. It is given that x_i is approximately normally distributed with $\mu = 150$ and $\sigma = 35$. Then according to the Central Limit Theorem, the sum or total weight, $\sum x_i$, will be normally distributed with mean $n\mu = 150n$ and standard deviation $\sigma\sqrt{n} = 35\sqrt{n}$. It is necessary to find a value of n such that

$$P(\sum x_i > 2000) = .01$$

The z-value corresponding to $\sum x_i = 2000$ is $z = \frac{\sum x_i - n\mu}{\sigma\sqrt{n}} = \frac{2000 - 150n}{35\sqrt{n}}$

so that we need $P\left(z > \frac{2000 - 150n}{35\sqrt{n}}\right) = .01$

From Table 3, we conclude that $\frac{2000 - 150n}{35\sqrt{n}} = 2.33$. Manipulating the above equation, we obtain a quadratic equation in n.

$$(2000 - 150n) = 81.55\sqrt{n}$$
$$4,000,000 - 606,650.4025n + 22,500n^2 = 0$$

Using the quadratic formula, the necessary value of n is found.

$$n = \frac{-b \pm \sqrt{b^2 - 4ac}}{2a} = \frac{606,650.4025 \pm 89,580.734}{45,000}$$

Choosing the smaller root of the equation,

$$n = \frac{517,069.67}{45,000} = 11.49.$$

Hence, $n = 11$ or $n = 12$ will provide a sample size with $P(\sum x_i > 2000) \approx .01$. In fact, for $n = 11$,

$$P(\sum x_i > 2000) = P(z > 3.02) = 1 - .9987 = .0013$$

while for $n = 12$,

$$P(\sum x_i > 2000) = P(z > 1.65) = 1 - .9505 = .0495$$

8: Large-Sample Estimation

Section 8.2

8.2.1 The margin of error in estimation provides a practical upper bound to the difference between a particular estimate and the parameter which it estimates. In this chapter, the margin of error is $1.96 \times$ (standard error of the estimator).

8.2.3-5 For the estimate of μ given as $\bar{x}$, the margin of error is $1.96\ SE = 1.96 \dfrac{\sigma}{\sqrt{n}}$. Notice that as the population variance σ^2 increases, the margin of error also increases.

3. $1.96\sqrt{\dfrac{0.2}{30}} = .160$ **5.** $1.96\sqrt{\dfrac{1.5}{30}} = .438$

8.2.7-9 The margin of error is $1.96\ SE = 1.96 \dfrac{\sigma}{\sqrt{n}}$, where σ can be estimated by the sample standard deviation s for large values of n. Notice that as the sample size n increases, the margin of error decreases.

7. $1.96\sqrt{\dfrac{4}{50}} = .554$ **9.** $1.96\sqrt{\dfrac{4}{500}} = .175$

8.2.11-13 For the estimate of p given as $\hat{p} = x/n$, the margin of error is $1.96\ SE = 1.96\sqrt{\dfrac{pq}{n}}$. Notice that, as with the margin of error for $\bar{x}$, when the sample size n increases, the margin of error for $\hat{p}$ decreases.

11. $1.96\sqrt{\dfrac{(.5)(.5)}{30}} = .179$ **13.** $1.96\sqrt{\dfrac{(.5)(.5)}{400}} = .049$

8.2.15-19 For the estimate of p given as $\hat{p} = x/n$, the margin of error is $1.96\ SE = 1.96\sqrt{\dfrac{pq}{n}}$. Use the value given in the exercise for p. The largest margin of error occurs when $p = .5$.

15. $1.96\sqrt{\dfrac{(.1)(.9)}{100}} = .0588$ **17.** $1.96\sqrt{\dfrac{(.5)(.5)}{100}} = .098$

19. $1.96\sqrt{\dfrac{(.9)(.1)}{100}} = .0588$

8.2.21 The point estimate of μ is $\bar{x} = 29.7$ and the margin of error in estimation with $s^2 = 10.8$ and $n = 75$ is

$$1.96\ SE = 1.96\dfrac{\sigma}{\sqrt{n}} \approx 1.96\dfrac{s}{\sqrt{n}} = 1.96\sqrt{\dfrac{10.8}{75}} = .744$$

8.2.23 The point estimate for p is given as $\hat{p} = \dfrac{x}{n} = \dfrac{450}{500} = .90$ and the margin of error is approximately

$$1.96\sqrt{\dfrac{\hat{p}\hat{q}}{n}} = 1.96\sqrt{\dfrac{.90(.10)}{500}} = .0263$$

8.2.25 The point estimate of μ is $\bar{x} = 34$ and the margin of error in estimation with $s = 3$ and $n = 100$ is

$$1.96 \, SE = 1.96 \frac{\sigma}{\sqrt{n}} \approx 1.96 \frac{s}{\sqrt{n}} = 1.96 \frac{3}{\sqrt{100}} = .588$$

8.2.27 The point estimate of μ is $\bar{x} = 2.705$ and the margin of error in estimation with $s = .028$ and $n = 36$ is

$$1.96 \, SE = 1.96 \frac{\sigma}{\sqrt{n}} \approx 1.96 \frac{s}{\sqrt{n}} = 1.96 \frac{.028}{\sqrt{36}} = .009$$

8.2.29 The point estimate of μ is $\bar{x} = 4.2$ and the margin of error with $s = 1.5$ and $n = 75$ is

$$1.96 \, SE = 1.96 \frac{\sigma}{\sqrt{n}} \approx 1.96 \frac{s}{\sqrt{n}} = 1.96 \frac{1.5}{\sqrt{75}} = .339$$

8.2.31 a The point estimate for p is given as $\hat{p} = \frac{x}{n} = \frac{170}{250} = .68$ and the margin of error is approximately

$$1.96 \sqrt{\frac{\hat{p}\hat{q}}{n}} = 1.96 \sqrt{\frac{.68(.32)}{250}} = .0578$$

b The point estimate for p is given as $\hat{p} = \frac{x}{n} = \frac{120}{250} = .48$ and the margin of error is approximately

$$1.96 \sqrt{\frac{\hat{p}\hat{q}}{n}} = 1.96 \sqrt{\frac{.48(.52)}{250}} = .0619$$

8.2.33 a This method of sampling would not be random, since only interested viewers (those who were adamant in their approval or disapproval) would reply.

b The results of such a survey will not be valid, and a margin or error would be useless, since its accuracy is based on the assumption that the sample was random.

8.2.35 A point estimate for the mean length of time is $\bar{x} = 19.3$, with margin of error

$$1.96 \, SE = 1.96 \frac{\sigma}{\sqrt{n}} \approx 1.96 \frac{s}{\sqrt{n}} = 1.96 \frac{5.2}{\sqrt{30}} = 1.86$$

Section 8.3

8.3.1-7 Use the information given in one of this section's "Need a Tip?", reproduced below:

Right Tail Area	.05	.025	.01	.005
z-value	1.645	1.96	2.33	2.58

1. The z-value for a 90% confidence interval will have $(1-\alpha)100\% = 90\%$ of the area in the center of the standard normal distribution, leaving $\alpha/2 = .05$ for each of the two tails. From the table above, $z_{.05} = 1.645$.

3. The z-value for a 98% confidence interval will have $(1-\alpha)100\% = 98\%$ of the area in the center of the standard normal distribution, leaving $\alpha/2 = .01$ for each of the two tails. From the table above, $z_{.01} = 2.33$.

5. A confidence coefficient of $\alpha = .02$ corresponds to a 98% confidence interval with area $\alpha/2 = .01$ in each of the two tails. From the table above, $z_{.01} = 2.33$.

7. A confidence coefficient of $\alpha = .01$ corresponds to a 99% confidence interval with area $\alpha/2 = .005$ in each of the two tails. From the table above, $z_{.005} = 2.58$.

8.3.9 A 95% confidence interval for the population mean μ is given by $\bar{x} \pm 1.96 \frac{\sigma}{\sqrt{n}}$ where σ can be estimated by the sample standard deviation s for large values of n.

$$13.1 \pm 1.96 \sqrt{\frac{3.42}{36}} = 13.1 \pm .604 \quad \text{or} \quad 12.496 < \mu < 13.704$$

Intervals constructed in this manner will enclose the true value of μ 95% of the time in repeated sampling. Hence, we are fairly confident that this particular interval will enclose μ.

8.3.11 A 90% confidence interval for μ is given as $\bar{x} \pm 1.645 \frac{\sigma}{\sqrt{n}}$ where σ can be estimated by the sample standard deviation s for large values of n.

$$.84 \pm 1.645 \sqrt{\frac{.086}{125}} = .84 \pm .043 \quad \text{or} \quad .797 < \mu < .883$$

Intervals constructed in this manner will enclose the true value of μ 90% of the time in repeated sampling. Hence, we are fairly confident that this particular interval will enclose μ.

8.3.13 $\bar{x} \pm z_{.005} \frac{\sigma}{\sqrt{n}} = \bar{x} \pm 2.58 \frac{\sigma}{\sqrt{n}} \approx 34 \pm 2.58 \sqrt{\frac{12}{38}} = 34 \pm 1.450$ or $32.550 < \mu < 35.450$.

8.3.15 $\bar{x} \pm z_{.025} \frac{\sigma}{\sqrt{n}} = \bar{x} \pm 1.96 \frac{\sigma}{\sqrt{n}} \approx 66.3 \pm 1.96 \sqrt{\frac{2.48}{89}} = 66.3 \pm .327$ or $65.973 < \mu < 66.627$.

8.3.17 Calculate $\hat{p} = \frac{x}{n} = \frac{27}{500} = .054$. Then an approximate 95% confidence interval for p is

$$\hat{p} \pm 1.96 \sqrt{\frac{\hat{p}\hat{q}}{n}} = .054 \pm 1.96 \sqrt{\frac{.054(.946)}{500}} = .054 \pm .020$$

or $.034 < p < .074$. Notice that the interval is narrower than the one calculated in Exercise 17, even though the confidence coefficient is larger and n is larger. This is because the value of p (estimated by $\hat{p}$) is quite close to zero, causing $\sigma_{\hat{p}}$ to be small.

8.3.19 The width of a 95% confidence interval for μ is given as $1.96 \frac{\sigma}{\sqrt{n}}$. When $n = 200$, the width is

$$2 \left(1.96 \frac{10}{\sqrt{200}} \right) = 2(1.386) = 2.772.$$ If you compare this result to the results when $n = 100$ and $n = 400$, you will find that when the sample size is doubled, the width is decreased by $1/\sqrt{2}$; when the sample size is quadrupled, the width is decreased by $1/\sqrt{4} = 1/2$.

8.3.21-23 The width of a $100(1-\alpha)\%$ confidence interval for μ is given as $z_{\alpha/2} \frac{\sigma}{\sqrt{n}}$. Notice that as the confidence coefficient increases, so does the width of the confidence interval. If we wish to be more confident of enclosing the unknown parameter, we must make the interval wider.

21. For a 99% confidence interval for μ the width is $2 \left(2.58 \frac{\sigma}{\sqrt{n}} \right) = 2 \left(2.58 \frac{10}{\sqrt{100}} \right) = 2(2.58) = 5.16$

23. For a 90% confidence interval for μ the width is $2\left(1.645\dfrac{\sigma}{\sqrt{n}}\right) = 2\left(1.645\dfrac{10}{\sqrt{100}}\right) = 2(1.645) = 3.29$

8.3.25 The percentages given in the table (when divided by 100) are the sample proportion of women ($n = 500$) who answered "yes" to each of the reasons for dieting.

a Let p be the proportion of all women who would consider dieting to improve their health. The 95% confidence interval for p is

$$\hat{p} \pm 1.96\sqrt{\dfrac{\hat{p}\hat{q}}{n}} = .68 \pm 1.96\sqrt{\dfrac{.68(.32)}{500}} = .68 \pm .041 \quad \text{or} \quad .639 < p < .721.$$

b Let p be the proportion of all women who would consider dieting to have more energy. The 90% confidence interval for p is

$$\hat{p} \pm 1.645\sqrt{\dfrac{\hat{p}\hat{q}}{n}} = .39 \pm 1.645\sqrt{\dfrac{.39(.61)}{500}} = .39 \pm .036 \quad \text{or} \quad .354 < p < .426.$$

8.3.27 The approximate 95% confidence interval for μ is

$$\bar{x} \pm 1.96\dfrac{s}{\sqrt{n}} = 21.51 \pm 1.96\dfrac{2.84}{\sqrt{36}} = 21.51 \pm .928$$

or $20.582 < \mu < 22.438$. Since $\mu = 24.98$ is not in this interval, it is unlikely that the average paid to grounds persons in Auburn, Washington is the same as the average paid by other school districts. Since the possible values given in the confidence interval are all lower than $24.98, we can conclude that Auburn, Washington pays significantly more per hour than the second school district.

8.3.29 With $n = 30$, $\bar{x} = 145$ and $s = .0051$, a 90% confidence interval for μ is approximated by

$$\bar{x} \pm 1.645\dfrac{s}{\sqrt{n}} = .145 \pm 1.645\dfrac{.0051}{\sqrt{30}} = .145 \pm .0015 \text{ or } .1435 < \mu < .1465$$

8.3.31 a An approximate 95% confidence interval for μ is

$$\bar{x} \pm 1.96\dfrac{s}{\sqrt{n}} = 54 \pm 1.96\dfrac{12.7}{\sqrt{1136}} = 54 \pm .739 \text{ or } 53.261 < \mu < 54.739$$

b An approximate 95% confidence interval for μ is

$$\bar{x} \pm 1.96\dfrac{s}{\sqrt{n}} = 72 \pm 1.96\dfrac{10.4}{\sqrt{795}} = 72 \pm .723 \text{ or } 71.277 < \mu < 72.723$$

8.3.33 a The 90% confidence interval for p is

$$\hat{p} \pm 1.645\sqrt{\dfrac{\hat{p}\hat{q}}{n}} = .64 \pm 1.645\sqrt{\dfrac{.64(.36)}{1011}} = .64 \pm .025 \quad \text{or } .615 < p < .665.$$

b The 90% confidence interval for p is

$$\hat{p} \pm 1.645\sqrt{\dfrac{\hat{p}\hat{q}}{n}} = .34 \pm 1.645\sqrt{\dfrac{.34(.66)}{1011}} = .34 \pm .025 \quad \text{or } .315 < p < .365.$$

c The margin of error is calculated by using the maximum margin of error using $p = .5$, and by rounding off to the nearest percent:

$$1.96\sqrt{\dfrac{\hat{p}\hat{q}}{n}} = 1.96\sqrt{\dfrac{.5(.5)}{1011}} = .0308 \text{ or } \pm 3.1\%$$

It appears that the Gallup poll is rounding the margin of error up to $\pm 4\%$.

8.3.35 **a** Since n is large, the Central Limit Theorem ensures that the sample mean $\bar{x}$ is approximately normal, and the standard normal distribution can be used to construct a confidence interval for μ. The 95% confidence interval for μ is

$$\bar{x} \pm 1.96 \frac{s}{\sqrt{n}} = 232 \pm 1.96 \frac{20.2}{\sqrt{35}} = 232 \pm 6.692 \text{ or } 225.308 < \mu < 238.692$$

b Since the value $\mu = 238$ lies in the confidence interval, it is likely that the claim is correct.

8.3.37 **a** The point estimate of p is $\hat{p} = \frac{x}{n} = .58$, and the approximate 98% confidence interval for p is

$$\hat{p} \pm 2.33 \sqrt{\frac{\hat{p}\hat{q}}{n}} = .58 \pm 2.33 \sqrt{\frac{.58(.42)}{100}} = .58 \pm .115$$

or $.465 < p < .695$.

b Since the possible values for p given in the confidence interval includes the value $p = .67$, there is no reason to doubt the claim of the *Pew Research* survey.

Section 8.4

8.4.1 When estimating the difference $\mu_1 - \mu_2$, the $(1-\alpha)100\%$ confidence interval is $(\bar{x}_1 - \bar{x}_2) \pm z_{\alpha/2} \sqrt{\frac{\sigma_1^2}{n_1} + \frac{\sigma_2^2}{n_2}}$.

Estimating σ_1^2 and σ_2^2 with s_1^2 and s_2^2, the approximate 90% confidence interval is

$$(9.7 - 7.4) \pm 1.645 \sqrt{\frac{10.78}{35} + \frac{16.44}{49}} = 2.3 \pm 1.320 \text{ or } .980 < \mu_1 - \mu_2 < 3.620.$$

The approximate 99% confidence interval is

$$(9.7 - 7.4) \pm 2.58 \sqrt{\frac{10.78}{35} + \frac{16.44}{49}} = 2.3 \pm 2.070 \text{ or } .230 < \mu_1 - \mu_2 < 4.370$$

In repeated sampling, 90% (or 99%) of all intervals constructed in this manner will enclose $\mu_1 - \mu_2$. Hence, we are fairly certain that this particular interval contains $\mu_1 - \mu_2$.

8.4.3 Refer to Exercise 1. Since the value $\mu_1 - \mu_2 = 0$ is not in either confidence interval, it is unlikely that $\mu_1 = \mu_2$. You should conclude that there is a difference in the two population means.

8.4.5 The 95% confidence interval for $\mu_1 - \mu_2$ is approximately

$$(\bar{x}_1 - \bar{x}_2) \pm 1.96 \sqrt{\frac{s_1^2}{n_1} + \frac{s_2^2}{n_2}}$$

$$(125.2 - 123.7) \pm 1.96 \sqrt{\frac{5.6^2}{50} + \frac{6.8^2}{50}}$$

$$1.5 \pm 2.442 \quad \text{or} \quad -.942 < (\mu_1 - \mu_2) < 3.942$$

The point estimate for $\mu_1 - \mu_2$ is $\bar{x}_1 - \bar{x}_2 = 125.2 - 123.7 = 1.5$ and the margin of error is approximately

$$MOE = 1.96\sqrt{\frac{s_1^2}{n_1} + \frac{s_2^2}{n_2}} = 1.96\sqrt{\frac{5.6^2}{50} + \frac{6.8^2}{50}} = 2.442$$

The results are similar. Since the value $\mu_1 - \mu_2 = 0$ is in the confidence interval, it is possible that $\mu_1 = \mu_2$. You should not conclude that there is a difference in the two population means.

8.4.7 Since the value $\mu_1 - \mu_2 = 0$ is in the confidence interval, it is possible that $\mu_1 = \mu_2$. You should not conclude that there is a difference in the two population means.

8.4.9 Since the value $\mu_1 - \mu_2 = 0$ is not in the confidence interval, it is unlikely that $\mu_1 = \mu_2$. You can conclude that there is a difference in the two population means.

8.4.11 The following information is available:

$$n_1 = n_2 = 30 \quad \bar{x}_1 = 167.1 \quad \bar{x}_2 = 140.9$$
$$s_1 = 24.3 \quad s_2 = 17.6$$

The 95% confidence interval for $\mu_1 - \mu_2$ is approximately

$$(\bar{x}_1 - \bar{x}_2) \pm 1.96\sqrt{\frac{s_1^2}{n_1} + \frac{s_2^2}{n_2}}$$

$$(167.1 - 140.9) \pm 1.96\sqrt{\frac{(24.3)^2}{30} + \frac{(17.6)^2}{30}}$$

$$26.2 \pm 10.737 \quad \text{or} \quad 15.463 < (\mu_1 - \mu_2) < 36.937$$

In repeated sampling, 95% of all intervals constructed in this manner will enclose $\mu_1 - \mu_2$. Hence, we are fairly certain that this particular interval contains $\mu_1 - \mu_2$.

8.4.13 a The 95% confidence interval for $\mu_1 - \mu_2$ is approximately

$$(\bar{x}_1 - \bar{x}_2) \pm 1.96\sqrt{\frac{s_1^2}{n_1} + \frac{s_2^2}{n_2}}$$

$$(18.5 - 16.5) \pm 1.96\sqrt{\frac{8.03^2}{365} + \frac{6.96^2}{298}}$$

$$2.0 \pm 1.142 \quad \text{or} \quad .858 < (\mu_1 - \mu_2) < 3.142$$

b Since the confidence interval in part **a** has two positive endpoints, it does not contain the value $\mu_1 - \mu_2 = 0$. Hence, it is not likely that the means are equal. It appears that there is a real difference in the mean scores.

8.4.15 a The point estimate of the difference $\mu_1 - \mu_2$ is

$$\bar{x}_1 - \bar{x}_2 = 62,428 - 57,762 = 4666$$

and the margin of error is

$$1.96\sqrt{\frac{\sigma_1^2}{n_1} + \frac{\sigma_2^2}{n_2}} \approx 1.96\sqrt{\frac{12,500^2}{50} + \frac{13,330^2}{50}} = 5065.293$$

b Since the margin of error allows the estimate of the difference $\mu_1 - \mu_2$ to be negative—the lower limit is $4666 - 5065.293 = -399.293$—it is possible that the mean for engineering majors the mean could be the same as for computer science majors. We would not conclude that there is a difference.

8.4.17 a The 99% confidence interval for $\mu_1 - \mu_2$ is approximately

$$(\bar{x}_1 - \bar{x}_2) \pm 2.58\sqrt{\frac{s_1^2}{n_1} + \frac{s_2^2}{n_2}}$$

$$(15-23) \pm 2.58\sqrt{\frac{4^2}{30} + \frac{10^2}{40}}$$

$$-8 \pm 4.49 \quad \text{or} \quad -12.49 < (\mu_1 - \mu_2) < -3.51$$

b Since the confidence interval in part **a** has two negative endpoints, it does not contain the value $\mu_1 - \mu_2 = 0$. Hence, it is not likely that the means are equal. It appears that there is a real difference in the mean times to completion for the two groups.

8.4.19 a The point estimate of the difference $\mu_1 - \mu_2$ is $\bar{x}_1 - \bar{x}_2 = 1122 - 1048 = 74$

and the margin of error is

$$1.96\sqrt{\frac{\sigma_1^2}{n_1} + \frac{\sigma_2^2}{n_2}} \approx 1.96\sqrt{\frac{194^2}{100} + \frac{165^2}{100}} = 49.917$$

b The 95% confidence interval for $\mu_1 - \mu_2$ is approximately

$$(\bar{x}_1 - \bar{x}_2) \pm 1.96\sqrt{\frac{s_1^2}{n_1} + \frac{s_2^2}{n_2}}$$

$$(1122 - 1048) \pm 1.96\sqrt{\frac{194^2}{100} + \frac{165^2}{100}}$$

$$74 \pm 49.917 \quad \text{or} \quad 24.083 < (\mu_1 - \mu_2) < 123.917$$

Since the value $\mu_1 - \mu_2 = 0$ is not in the confidence interval, it is unlikely that $\mu_1 = \mu_2$. You can conclude that there is a difference in the two population means.

8.4.21 a The point estimate of the difference $\mu_1 - \mu_2$ is $\bar{x}_1 - \bar{x}_2 = 42.7 - 37.5 = 5.2$

and the margin of error is

$$1.96\sqrt{\frac{\sigma_1^2}{n_1} + \frac{\sigma_2^2}{n_2}} \approx 1.96\sqrt{\frac{5.3^2}{50} + \frac{4.2^2}{50}} = 1.874$$

b The 98% confidence interval for $\mu_1 - \mu_2$ is approximately

$$(\bar{x}_1 - \bar{x}_2) \pm 2.33\sqrt{\frac{s_1^2}{n_1} + \frac{s_2^2}{n_2}}$$

$$(42.7 - 37.5) \pm 2.33\sqrt{\frac{5.3^2}{50} + \frac{4.2^2}{50}}$$

$$5.2 \pm 2.228 \quad \text{or} \quad 2.972 < (\mu_1 - \mu_2) < 7.428$$

Since the value $\mu_1 - \mu_2 = 0$ is not in the confidence interval, it is unlikely that $\mu_1 = \mu_2$. You can conclude that there is a difference in the two population means.

Section 8.5

8.5.1 Calculate $\hat{p}_1 = \frac{x_1}{n_1} = \frac{120}{500} = .24$ and $\hat{p}_2 = \frac{x_2}{n_2} = \frac{147}{500} = .294$. The approximate 95% confidence interval is

$$(\hat{p}_1 - \hat{p}_2) \pm 1.96\sqrt{\frac{\hat{p}_1\hat{q}_1}{n_1} + \frac{\hat{p}_2\hat{q}_2}{n_2}}$$

$$(.240 - .294) \pm 1.96\sqrt{\frac{.24(.76)}{500} + \frac{.294(.706)}{500}}$$

$$-.054 \pm .0547 \quad \text{or} \quad -.1087 < (p_1 - p_2) < .0007$$

The approximate 99% confidence interval is

$$(\hat{p}_1 - \hat{p}_2) \pm 2.58\sqrt{\frac{\hat{p}_1\hat{q}_1}{n_1} + \frac{\hat{p}_2\hat{q}_2}{n_2}}$$

$$(.240 - .294) \pm 2.58\sqrt{\frac{.24(.76)}{500} + \frac{.294(.706)}{500}}$$

$$-.054 \pm .0721 \quad \text{or} \quad -.1261 < (p_1 - p_2) < .0181$$

Intervals constructed in this manner will enclose $(p_1 - p_2)$ 95% (or 99%) of the time in repeated sampling. Hence, we are fairly certain that this particular interval encloses $(p_1 - p_2)$.

8.5.3 Refer to Exercise 1. Since the value $p_1 - p_2 = 0$ is in both confidence intervals, it is possible that $p_1 = p_2$. You should not conclude that there is a difference in the two population proportions.

8.5.5 The best estimate of $p_1 - p_2$ is $\hat{p}_1 - \hat{p}_2 = \frac{565}{1250} - \frac{621}{1100} = .452 - .565 = -.113$. The standard error is calculated by estimating p_1 and p_2 with $\hat{p}_1$ and $\hat{p}_2$ in the formula:

$$SE = \sqrt{\frac{p_1q_1}{n_1} + \frac{p_2q_2}{n_2}} \approx \sqrt{\frac{\hat{p}_1\hat{q}_1}{n_1} + \frac{\hat{p}_2\hat{q}_2}{n_2}} = \sqrt{\frac{.452(.548)}{1250} + \frac{.565(.435)}{1100}} = .02053$$

and the approximate margin of error is

$$1.96\sqrt{\frac{.452(.548)}{1250} + \frac{.565(.435)}{1100}} = 1.96(.02053) = .040$$

8.5.7 Since the value $p_1 - p_2 = 0$ is not in the confidence interval, it is unlikely that $p_1 = p_2$. You can conclude that there is a difference in the two population proportions.

8.5.9 Since the value $p_1 - p_2 = 0$ is not in the confidence interval, it is unlikely that $p_1 = p_2$. You can conclude that there is a difference in the two population proportions.

8.5.11 a Calculate $\hat{p}_1 = \frac{x_1}{n_1} = \frac{849}{1265} = .671$ and $\hat{p}_2 = \frac{x_2}{n_2} = \frac{910}{1688} = .539$. The approximate 99% confidence interval is

$$(\hat{p}_1 - \hat{p}_2) \pm 2.58\sqrt{\frac{\hat{p}_1\hat{q}_1}{n_1} + \frac{\hat{p}_2\hat{q}_2}{n_2}}$$

$$(.671 - .539) \pm 2.58\sqrt{\frac{.671(.329)}{1265} + \frac{.539(.461)}{1688}}$$

$$.132 \pm .046 \quad \text{or} \quad .086 < (p_1 - p_2) < .178$$

In repeated sampling, 99% of all intervals constructed in this manner will enclose $p_1 - p_2$. Hence, we are fairly certain that this particular interval contains $p_1 - p_2$.

b Since the value $p_1 - p_2 = 0$ is not in the confidence interval, it is not likely that $p_1 = p_2$. You should conclude that there is a difference in the two population proportions.

8.5.13 **a** Calculate $\hat{p}_1 = \dfrac{x_1}{n_1} = \dfrac{12}{56} = .214$ and $\hat{p}_2 = \dfrac{x_2}{n_2} = \dfrac{8}{32} = .25$. The approximate 95% confidence interval is

$$(\hat{p}_1 - \hat{p}_2) \pm 1.96 \sqrt{\dfrac{\hat{p}_1 \hat{q}_1}{n_1} + \dfrac{\hat{p}_2 \hat{q}_2}{n_2}}$$

$$(.214 - .25) \pm 1.96 \sqrt{\dfrac{.214(.786)}{56} + \dfrac{.25(.75)}{32}}$$

$$-.036 \pm .185 \quad \text{or} \quad -.221 < (p_1 - p_2) < .149$$

b Since the value $p_1 - p_2 = 0$ is in the confidence interval, it is possible that $p_1 = p_2$. You should not conclude that there is a difference in the proportion of red candies in plain and peanut M&Ms.

8.5.15 **a** With $\hat{p}_1 = \dfrac{x_1}{1001} = .45$ and $\hat{p}_2 = \dfrac{x_2}{1001} = .51$. The approximate 99% confidence interval is

$$(\hat{p}_1 - \hat{p}_2) \pm 2.58 \sqrt{\dfrac{\hat{p}_1 \hat{q}_1}{n_1} + \dfrac{\hat{p}_2 \hat{q}_2}{n_2}}$$

$$(.45 - .51) \pm 2.58 \sqrt{\dfrac{.45(.55)}{1001} + \dfrac{.51(.49)}{1001}}$$

$$-.06 \pm .058 \quad \text{or} \quad -.118 < (p_1 - p_2) < -.002$$

b Since the interval in part **a** contains only negative values of $p_1 - p_2$, it is likely that $p_1 - p_2 < 0 \Rightarrow p_1 < p_2$. This would indicate that the proportion of adults who claim to be fans is higher in November than in March.

8.5.17 **a** Define sample #1 as the responses of the 72 people under the age of 34 and sample #2 as the responses of the 55 people who are 65 or older. Then $\hat{p}_1 = .37$ and $\hat{p}_2 = .13$.

b The approximate 95% confidence interval is

$$(\hat{p}_1 - \hat{p}_2) \pm 1.96 \sqrt{\dfrac{\hat{p}_1 \hat{q}_1}{n_1} + \dfrac{\hat{p}_2 \hat{q}_2}{n_2}}$$

$$(.37 - .13) \pm 1.96 \sqrt{\dfrac{.37(.63)}{72} + \dfrac{.13(.87)}{55}}$$

$$.24 \pm .143 \quad \text{or} \quad .097 < (p_1 - p_2) < .383$$

c Since the value $p_1 - p_2 = 0$ is not in the confidence interval, it is unlikely that $p_1 = p_2$. You should conclude that there is a difference in the proportion of people in the two groups who more likely to "haggle". In fact, since all the probable values of $p_1 - p_2$ are positive, the proportion of young people who "haggle" appears to be larger than the proportion of older people.

8.5.19 The following sample information is available:

$$n_1 = n_2 = 200 \qquad \hat{p}_1 = \dfrac{142}{200} = .71 \qquad \hat{p}_2 = \dfrac{120}{200} = .60.$$

The approximate 95% confidence interval is

$$(\hat{p}_1 - \hat{p}_2) \pm 1.96 \sqrt{\dfrac{\hat{p}_1 \hat{q}_1}{n_1} + \dfrac{\hat{p}_2 \hat{q}_2}{n_2}}$$

$$(.71 - .60) \pm 1.96 \sqrt{\dfrac{.71(.29)}{200} + \dfrac{.60(.40)}{200}}$$

$$.11 \pm .093 \quad \text{or} \quad .017 < (p_1 - p_2) < .203$$

Intervals constructed in this manner will enclose the true value of $p_1 - p_2$ 95% of the time in repeated sampling. Hence, we are fairly certain that this particular interval encloses $p_1 - p_2$.

8.5.21 **a** The approximate 90% confidence interval is

$$(\hat{p}_1 - \hat{p}_2) \pm 1.645 \sqrt{\frac{\hat{p}_1 \hat{q}_1}{n_1} + \frac{\hat{p}_2 \hat{q}_2}{n_2}}$$

$$(.44 - .23) \pm 1.645 \sqrt{\frac{.44(.56)}{491} + \frac{.23(.77)}{398}}$$

$$.21 \pm .051 \quad \text{or} \quad .159 < (p_1 - p_2) < .261$$

b The approximate 99% confidence interval is

$$(\hat{p}_1 - \hat{p}_2) \pm 2.58 \sqrt{\frac{\hat{p}_1 \hat{q}_1}{n_1} + \frac{\hat{p}_2 \hat{q}_2}{n_2}}$$

$$(.44 - .41) \pm 2.58 \sqrt{\frac{.44(.56)}{491} + \frac{.41(.59)}{398}}$$

$$.03 \pm .086 \quad \text{or} \quad -.056 < (p_1 - p_2) < .116$$

c For part a, the value $p_1 - p_2 = 0$ is not in the confidence interval, so it is unlikely that $p_1 = p_2$. You should conclude that there is a difference in the proportion of men and women who feel very comfortable driving in winter conditions. For part b, the value $p_1 - p_2 = 0$ is in the confidence interval, so it is possible that $p_1 = p_2$. You should not conclude that there is a difference in the proportion of men and women who feel kind of comfortable driving in winter conditions.

8.5.23 **a** The point estimate for p is given as $\hat{p} = \frac{x}{n} = \frac{23}{41} = .561$ and the margin of error is approximately

$$1.96 \sqrt{\frac{\hat{p}\hat{q}}{n}} = 1.96 \sqrt{\frac{.56(.44)}{41}} = .152$$

b Calculate $\hat{p}_1 = \frac{10}{32} = .3125$ and $\hat{p}_2 = \frac{23}{41} = .561$. The approximate 95% confidence interval is

$$(\hat{p}_1 - \hat{p}_2) \pm 1.96 \sqrt{\frac{\hat{p}_1 \hat{q}_1}{n_1} + \frac{\hat{p}_2 \hat{q}_2}{n_2}}$$

$$(.3125 - .561) \pm 1.96 \sqrt{\frac{.3125(.6875)}{32} + \frac{.561(.439)}{41}}$$

$$-.2485 \pm .2211 \quad \text{or} \quad -.4696 < (p_1 - p_2) < -.0274$$

Section 8.6

8.6.1-3 The z-value for a $100(1 - \alpha)\%$ one-sided confidence bound is a value with area α to its right.

1. For $100(1 - \alpha)\% = 90\%$, $\alpha = .10$ so that $z_\alpha = z_{.10} = 1.28$.

3. For $100(1 - \alpha)\% = 98\%$, $\alpha = .02$ so that $z_\alpha = z_{.02} = 2.05$.

8.6.5 The parameter to be estimated is the population mean μ and the 90% upper confidence bound is calculated using a value $z_\alpha = z_{.10} = 1.28$. The upper bound is approximately

$$\bar{x} + 1.28 \frac{s}{\sqrt{n}} = 75 + 1.28 \sqrt{\frac{65}{40}} = 75 + 1.632 \text{ or } \mu < 76.632$$

8.6.7 Similar to Exercise 5. The 99% lower bound is approximately

$$\bar{x} - 2.33 \frac{s}{\sqrt{n}} = 101.4 - 2.33 \sqrt{\frac{25.8}{55}} = 101.4 - 1.596 \text{ or } \mu > 99.804$$

8.6.9 The 95% lower confidence bound is calculated using $\hat{p} = x/n = 54/60 = 0.9$ and a value $z_\alpha = z_{.05} = 1.645$. The lower confidence bound for the binomial parameter p is approximately

$$\hat{p} - 1.645 \sqrt{\frac{\hat{p}\hat{q}}{n}} = .9 - 1.645 \sqrt{\frac{.9(.1)}{60}} = .9 - .064 \text{ or } p > .836$$

8.6.11 Calculate $\hat{p}_1 = \frac{x_1}{n_1} = \frac{120}{500} = .24$ and $\hat{p}_2 = \frac{x_2}{n_2} = \frac{147}{500} = .294$. The approximate 98% lower confidence bound is

$$(\hat{p}_1 - \hat{p}_2) - 2.05 \sqrt{\frac{\hat{p}_1 \hat{q}_1}{n_1} + \frac{\hat{p}_2 \hat{q}_2}{n_2}}$$

$$(.24 - .294) - 2.05 \sqrt{\frac{.24(.76)}{500} + \frac{.294(.706)}{500}}$$

$$-.054 - .057 \text{ or } (p_1 - p_2) > -.111$$

8.6.13 For the difference $\mu_1 - \mu_2$ in the population means for two quantitative populations, the 95% upper confidence bound uses $z_{.05} = 1.645$ and is calculated as

$$(\bar{x}_1 - \bar{x}_2) + 1.645 \sqrt{\frac{s_1^2}{n_1} + \frac{s_2^2}{n_2}} = (9.7 - 7.4) + 1.645 \sqrt{\frac{10.78}{35} + \frac{16.44}{49}}$$

$$2.3 + 1.320 \text{ or } (\mu_1 - \mu_2) < 3.620$$

Since the value $\mu_1 - \mu_2 = 0$ is in the confidence interval, it is possible that $\mu_1 = \mu_2$. You cannot conclude that one population mean is larger than the other.

8.6.15 The 99% lower confidence bound is calculated using $\hat{p} = x/n = 196/400 = .49$ and a value $z_\alpha = z_{.01} = 2.33$. The lower confidence bound for the binomial parameter p is approximately

$$\hat{p} - 1.645 \sqrt{\frac{\hat{p}\hat{q}}{n}} = .49 - 1.645 \sqrt{\frac{.49(.51)}{400}} = .49 - .058 \text{ or } p > .432$$

8.6.17 a The point estimate of the difference $\mu_1 - \mu_2$ is $\bar{x}_1 - \bar{x}_2 = 62,428 - 57,762 = \$4666$.

b The 95% lower confidence bound uses $z_{.05} = 1.645$ and is calculated as

$$(\bar{x}_1 - \bar{x}_2) - 1.645 \sqrt{\frac{s_1^2}{n_1} + \frac{s_2^2}{n_2}} = (4666) - 1.645 \sqrt{\frac{13,330^2}{50} + \frac{12,500^2}{50}}$$

$$4666 - 4251.228 \text{ or } (\mu_1 - \mu_2) > 414.772$$

Since the value $\mu_1 - \mu_2 = 0$ is not in the confidence interval, it is unlikely that $\mu_1 = \mu_2$. You can conclude that electrical engineers have a higher average starting salary than computer scientists.

8.6.19 The approximate 95% lower confidence bound is

$$(\hat{p}_1 - \hat{p}_2) - 1.645\sqrt{\frac{\hat{p}_1\hat{q}_1}{n_1} + \frac{\hat{p}_2\hat{q}_2}{n_2}}$$

$$(.62 - .35) - 1.645\sqrt{\frac{.62(.38)}{96} + \frac{.35(.65)}{105}}$$

$$.27 - .112 \quad \text{or} \quad (p_1 - p_2) > .158$$

Since the value $p_1 - p_2 = 0$ is not in the confidence interval, it is unlikely that $p_1 = p_2$. You can conclude that p_1, the proportion with fewer social contacts who catch cold is greater than the proportion with many contacts. Perhaps those with many social contacts build up an immunity to cold viruses.

8.6.21   **a**   For the difference $\mu_1 - \mu_2$ in the population means this year and ten years ago, the 99% lower confidence bound uses $z_{.01} = 2.33$ and is calculated as

$$(\bar{x}_1 - \bar{x}_2) - 2.33\sqrt{\frac{s_1^2}{n_1} + \frac{s_2^2}{n_2}} = (73 - 63) - 2.33\sqrt{\frac{25^2}{400} + \frac{28^2}{400}}$$

$$10 - 4.37 \quad \text{or} \quad (\mu_1 - \mu_2) > 5.63$$

b   Since the difference in the means is positive, you can conclude that there has been a decrease in the average per-capita beef consumption over the last ten years.

Section 8.7

8.7.1   It is necessary to find the sample size required to estimate a certain parameter to within a given bound with confidence $(1-\alpha)$. Recall from Section 8.5 that we may estimate a parameter with $(1-\alpha)$ confidence within the interval (estimator) $\pm z_{\alpha/2} \times$ (std error of estimator). Thus, $z_{\alpha/2} \times$ (std error of estimator) provides the margin of error with $(1-\alpha)$ confidence. The experimenter will specify a given bound B. If we let $z_{\alpha/2} \times$ (std error of estimator) $\leq B$, we will be $(1-\alpha)$ confident that the estimator will lie within B units of the parameter of interest.

For this exercise, the parameter of interest is μ, $B = 1.6$ and $1 - \alpha = .95$. Hence, we must have

$$1.96\frac{\sigma}{\sqrt{n}} \leq 1.6 \Rightarrow 1.96\frac{12.7}{\sqrt{n}} \leq 1.6$$

$$\sqrt{n} \geq \frac{1.96(12.7)}{1.6} = 15.5575$$

$$n \geq 242.04 \quad \text{or} \quad n \geq 243$$

8.7.3   In this exercise, the parameter of interest is $\mu_1 - \mu_2$, $n_1 = n_2 = n$, and $\sigma_1^2 \approx \sigma_2^2 \approx 27.8$. Then we must have

$$z_{\alpha/2} \times (\text{std error of } \bar{x}_1 - \bar{x}_2) \leq B$$

$$1.645\sqrt{\frac{\sigma_1^2}{n_1} + \frac{\sigma_2^2}{n_2}} \leq .17 \Rightarrow 1.645\sqrt{\frac{27.8}{n} + \frac{27.8}{n}} \leq .17$$

$$\sqrt{n} \geq \frac{1.645\sqrt{55.6}}{.17} \Rightarrow n \geq 5206.06 \quad \text{or} \quad n_1 = n_2 = 5207$$

8.7.5   For this exercise, $B = .04$ for the binomial estimator $\hat{p}$, where $SE(\hat{p}) = \sqrt{\frac{pq}{n}}$. Assuming maximum variation, which occurs if $p = .3$ (since we suspect that $.1 < p < .3$) and $z_{.025} = 1.96$, we have

$$1.96\sigma_{\hat{p}} \le B \Rightarrow 1.96\sqrt{\frac{pq}{n}} \le B$$

$$1.96\sqrt{\frac{.3(.7)}{n}} \le .04 \Rightarrow \sqrt{n} \ge \frac{1.96\sqrt{.3(.7)}}{.04} \Rightarrow n \ge 504.21 \text{ or } n \ge 505$$

8.7.7 For this exercise, the parameter of interest is μ, $B = 0.5$ and $1-\alpha = .95$. Hence, we must have

$$1.96\frac{\sigma}{\sqrt{n}} \le 0.5 \Rightarrow 1.96\frac{4}{\sqrt{n}} \le 0.5$$

$$\sqrt{n} \ge \frac{1.96(4)}{0.5} = 15.68$$

$$n \ge 245.86 \text{ or } n \ge 246$$

8.7.9 The parameter of interest is $p_1 - p_2$, $n_1 = n_2 = n$, and $B = .03$. Since no prior knowledge is available about p_1 and p_2, we assume the largest possible variation, which occurs if $p_1 = p_2 = .5$. Then

$$z_{.025}\sqrt{\frac{p_1 q_1}{n_1} + \frac{p_2 q_2}{n_2}} \le .03 \Rightarrow 1.96\sqrt{\frac{(.5)(.5)}{n} + \frac{(.5)(.5)}{n}} \le .03$$

$$\sqrt{n} \ge \frac{1.96\sqrt{.5}}{.03} \Rightarrow n \ge 2134.2 \text{ or } n_1 = n_2 = 2135$$

8.7.11 The parameter of interest is $p_1 - p_2$, $n_1 = n_2 = n$, and $B = .03$. Since no prior knowledge is available about p_1 and p_2, we assume the largest possible variation, which occurs if $p_1 = p_2 = .5$. Then

$$z_{.025}\sqrt{\frac{p_1 q_1}{n_1} + \frac{p_2 q_2}{n_2}} \le .03 \Rightarrow 1.96\sqrt{\frac{(.5)(.5)}{n} + \frac{(.5)(.5)}{n}} \le .03$$

$$\sqrt{n} \ge \frac{1.96\sqrt{.5}}{.03} \Rightarrow n \ge 2134.2 \text{ or } n_1 = n_2 = 2135$$

8.7.13 Similar to previous exercises with $B = .1$ and $\sigma \approx .5$. The required sample size is obtained by solving

$$1.96\frac{\sigma}{\sqrt{n}} \le B \Rightarrow 1.96\frac{.5}{\sqrt{n}} \le .1$$

$$\sqrt{n} \ge \frac{1.96(.5)}{.1} = 9.8 \Rightarrow n \ge 96.04 \text{ or } n \ge 97$$

Notice that water specimens should be selected randomly and not necessarily from the same rainfall, in order that all observations are independent.

8.7.15 The parameter of interest is $\mu_1 - \mu_2$, the difference in grade-point averages for the two populations of students. Assume that $n_1 = n_2 = n$, and $\sigma_1^2 \approx \sigma_2^2 \approx (.6)^2 = .36$ and that the desired bound is $.2$. Then

$$1.96\sqrt{\frac{\sigma_1^2}{n_1} + \frac{\sigma_2^2}{n_2}} \le .2 \Rightarrow 1.96\sqrt{\frac{.36}{n} + \frac{.36}{n}} \le .2$$

$$\sqrt{n} \ge \frac{1.96\sqrt{.72}}{.2} \Rightarrow n \ge 69.149$$

Therefore, $n_1 = n_2 = 70$ students should be included in each group.

Reviewing What You've Learned

8.R.1 See Section 7.3 of the text.

8.R.3 **a** The 90% confidence interval for $\mu_1 - \mu_2$ is approximately

$$(\bar{x}_1 - \bar{x}_2) \pm 1.645 \sqrt{\frac{s_1^2}{n_1} + \frac{s_2^2}{n_2}}$$

$$(100.4 - 96.2) \pm 1.645 \sqrt{\frac{.8^2}{50} + \frac{1.3^2}{60}}$$

$$4.2 \pm .333 \quad \text{or} \quad 3.867 < (\mu_1 - \mu_2) < 4.533$$

b We are given $B = .2$, $1 - \alpha = .95$, $n_1 = n_2 = n$, $s_1 = .8$ and $s_2 = 1.3$. Using these values to estimate σ_1 and σ_2, the following inequality must be solved:

$$1.96 \sqrt{\frac{\sigma_1^2}{n_1} + \frac{\sigma_2^2}{n_2}} \leq .2 \quad \Rightarrow \quad 1.96 \sqrt{\frac{(.8)^2}{n} + \frac{(1.3)^2}{n}} \leq .2$$

$$\sqrt{n} \geq 14.959 \quad \Rightarrow \quad n \geq 223.77 \quad \text{or} \quad n_1 = n_2 = 224$$

8.R.5 **a** Calculate $\hat{p}_1 = \frac{17}{40} = .425$, $\hat{p}_2 = \frac{23}{80} = .2875$. The approximate 99% confidence interval for $p_1 - p_2$ is

$$(\hat{p}_1 - \hat{p}_2) \pm 2.58 \sqrt{\frac{\hat{p}_1 \hat{q}_1}{n_1} + \frac{\hat{p}_2 \hat{q}_2}{n_2}}$$

$$(.425 - .2875) \pm 2.58 \sqrt{\frac{.425(.575)}{40} + \frac{.2875(.7125)}{80}}$$

$$.1375 \pm .240 \quad \text{or} \quad -.1025 < (p_1 - p_2) < .3775$$

Intervals constructed in this manner will enclose the true value of $p_1 - p_2$ 99% of the time in repeated sampling. Hence, we are fairly certain that this particular interval encloses $p_1 - p_2$.

b Assuming maximum variation $(p_1 = p_2 = .5)$ and $n_1 = n_2 = n$, the inequality to be solved is

$$z_{.005} \sqrt{\frac{p_1 q_1}{n_1} + \frac{p_2 q_2}{n_2}} \leq .06 \quad \Rightarrow \quad 2.58 \sqrt{\frac{(.5)(.5)}{n} + \frac{(.5)(.5)}{n}} \leq .06$$

$$\sqrt{n} \geq 30.406 \quad \Rightarrow \quad n \geq 924.5 \quad \text{or} \quad n_1 = n_2 = 925$$

8.R.7 **a** The "error margin" is equivalent to the margin of error, the upper limit to the difference between the estimate and the parameter to be estimated.

b If $n = 30$ and $s = .017$, the margin of error is $1.96 \frac{s}{\sqrt{n}} = 1.96 \frac{.017}{\sqrt{30}} = .00608$ and the chemist is correct.

8.R.9 **a** There are problems with non-response and households without "landlines".

b If you use $p = .5$ as an estimate for p, the margin of error is approximately

$$\pm 1.96 \sqrt{\frac{.5(.5)}{2250}} = \pm .021$$

c To reduce the margin of error in part **b** to $\pm .01$, solve for n in the equation

$$1.96\sqrt{\frac{.5(.5)}{n}} = .01 \;\Rightarrow\; \sqrt{n} = \frac{1.96(.5)}{.01} = 98 \;\Rightarrow\; n = 9604$$

8.R.11 It is assumed that $p = .2$ and that the desired bound is .01. Hence,

$$1.96\sqrt{\frac{pq}{n}} \leq .01 \;\Rightarrow\; \sqrt{n} \geq \frac{1.96\sqrt{.05(.95)}}{.01} = 42.72$$

$$n \geq 1824.76 \text{ or } n \geq 1825$$

8.R.13 Ten samples of $n = 400$ printed circuit boards were tested and a $100(1-\alpha)\%$ confidence interval for p was constructed for each of the ten samples. For this exercise, $100(1-\alpha)\% = 90\%$, or $\alpha = .10$. The object is to find the probability that exactly one of the intervals will not enclose the true value of p. Hence, the situation descries a binomial experiment with $n = 10$ and

$p^* = P[$an interval will not contain the true value of $p]$

$x = $ number of intervals which do not enclose p

By definition of a 90% confidence interval, it can be said that 90% of the intervals generated in repeated sampling will contain the true value of p; 10% of the intervals will not contain p. Thus, $p^* = .1$ and the desired probabilities are calculated using the methods of Chapter 4.

1 $P[\text{exactly one of the intervals fails to contain } p] = P[x = 1] = C_1^{10}(.1)^1(.9)^9 = .3874$

2 $P[\text{at least one}] = 1 - P[x = 0] = 1 - C_0^{10}(.1)^0(.9)^{10} = 1 - .349 = .651$

8.R.15 a The approximate 95% confidence interval for μ is

$$\bar{x} \pm 1.96 \frac{s}{\sqrt{n}} = 2.962 \pm 1.96 \frac{.529}{\sqrt{69}} = 2.962 \pm .125$$

or $2.837 < \mu < 3.087$.

b In order to cut the interval in half, the sample size must be multiplied by 4. If this is done, the new half-width of the confidence interval is

$$1.96 \frac{\sigma}{\sqrt{4n}} = \frac{1}{2}\left(1.96 \frac{\sigma}{\sqrt{n}}\right).$$

Hence, in this case, the new sample size is $4(69) = 276$.

8.R.17 The point estimate of μ is $\bar{x} = 21.6$ and the margin of error in estimation is

$$1.96 \, SE = 1.96 \frac{\sigma}{\sqrt{n}} \approx 1.96 \frac{s}{\sqrt{n}} = 1.96 \frac{2.1}{\sqrt{70}} = .49$$

8.R.19 Answers will vary from student to student. Comparisons should be made within the two types of plants for the two treatments (free-standing versus supported). For example, 95% confidence intervals for the two comparisons are calculated as:

Sunflower: $(\bar{x}_1 - \bar{x}_2) \pm 1.96\sqrt{\frac{s_1^2}{n_1} + \frac{s_2^2}{n_2}} \;\Rightarrow\; (\bar{x}_1 - \bar{x}_2) \pm 1.96\sqrt{SE_1^2 + SE_2^2}$

$(35.3 - 32.1) \pm 1.96\sqrt{.72^2 + .72^2} \;\Rightarrow\; 3.2 \pm 2.00 \text{ or } 1.2 < (\mu_1 - \mu_2) > 5.2$

Maize: $(\bar{x}_1 - \bar{x}_2) \pm 1.96\sqrt{\frac{s_1^2}{n_1} + \frac{s_2^2}{n_2}} \Rightarrow (\bar{x}_1 - \bar{x}_2) \pm 1.96\sqrt{SE_1^2 + SE_2^2}$

$(16.2 - 14.6) \pm 1.96\sqrt{.41^2 + .40^2} \Rightarrow 1.6 \pm 1.12$ or $.48 < (\mu_1 - \mu_2) > 2.72$

There is a difference in the mean basal diameters for both types of plant.

9: Large-Sample Tests of Hypotheses

Section 9.1

9.1.1 See Section 9.1.

9.1.3 We want to prove the alternative hypothesis that μ is, in fact, greater than 3. Hence, the alternative hypothesis is $H_a : \mu > 3$ and the null hypothesis is $H_0 : \mu = 3$.

9.1.5 We want to prove the alternative hypothesis that μ is, in fact, different from 100. Hence, the alternative hypothesis is $H_a : \mu \neq 100$ and the null hypothesis is $H_0 : \mu = 100$.

9.1.7 We want to prove the alternative hypothesis that the mean time to recovery μ is less than the current time of 5 days. Hence, the alternative hypothesis is $H_a : \mu < 5$ and the null hypothesis is $H_0 : \mu = 5$.

9.1.9 Define p to be the proportion of small businesses that fail in their first year. If the new tax structure is helping, the proportion of small businesses that fail should be lower than the current proportion, presently 20% (or .20). Hence, the alternative hypothesis is $H_a : p < .20$ and the null hypothesis is $H_0 : p = .20$.

9.1.11 From Exercise 8, we want to test $H_0 : \mu = 70$ versus $H_a : \mu > 70$ and it is given that $\bar{x} = 77, n = 40$ and $s = 12.6$. Assume that H_0 is true, so that $\mu = 70$. The test statistic is $z = \dfrac{\bar{x} - \mu}{\sigma/\sqrt{n}}$ which represents the distance (measured in units of standard deviations) from $\bar{x}$ to the hypothesized mean $\mu = 70$. Using the sample standard deviation s to approximate the population standard deviation σ, the test statistic is calculated as

$$z = \frac{\bar{x} - \mu}{\sigma/\sqrt{n}} \approx \frac{\bar{x} - \mu}{s/\sqrt{n}} = \frac{77 - 70}{12.6/\sqrt{40}} = 3.51$$

Since this z-value is more than three standard deviations from the mean, it is a highly unlikely event, assuming H_0 is true.

9.1.13 In order to dispute the claim of an average of 155 Facebook friends, we would need to show that μ is not equal to 155. Hence, the alternative hypothesis is $H_a : \mu \neq 155$ and the null hypothesis is $H_0 : \mu = 155$.

9.1.15 The probability of observing a value as unlikely or more unlikely than $z = -1.43$ is

$$P(|z| > 1.43) = P(z > 1.43) + P(z < -1.43) = (1 - .9236) + .0764 = .1528$$

This is not an unlikely event, assuming that the mean is actually 155. We would not want to dispute the 155 number.

Section 9.2

9.2.1 The probability of a Type I error is $\alpha = P(\text{reject } H_0 \text{ when } H_0 \text{ is true})$ and the probability of a Type II error is $\beta = P(\text{accept } H_0 \text{ when } H_0 \text{ is false})$.

9.2.3 When the value of α is fixed and the sample size is increased, β decreases.

9.2.5 The power of the test is $1 - \beta = P(\text{reject } H_0 \text{ when } H_0 \text{ is false})$. As μ gets farther from μ_0, the power of the test increases.

9.2.7 For a two-tailed test with $\alpha = .05$, the critical value for the rejection region cuts off $\alpha/2 = .025$ in the two tails of the z distribution in the following figure, so that $z_{.025} = 1.96$. The null hypothesis H_0 will be rejected if $z > 1.96$ or $z < -1.96$ (which you can also write as $|z| > 1.96$).

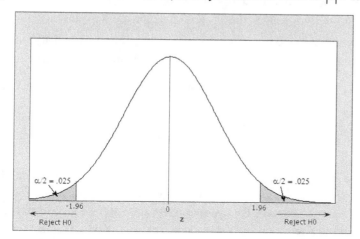

The observed value of the test statistic, $z = 2.16$, falls in the rejection region. Hence, the null hypothesis is rejected with $\alpha = .05$.

9.2.9 The critical value that separates the rejection and nonrejection regions for a lower-tailed test based on a z-statistic will be a value of z that cuts off $\alpha = .05$ in the left tail of the distribution. That is, $z_{.05} = -1.645$. This rejection region is in the lower tail of the z distribution. The null hypothesis H_0 will be rejected if $z < -1.645$. If the observed value of the test statistic is $z = -2.41$, the null hypothesis H_0 will be rejected at the 5% level.

9.2.11 The p-value for a right-tailed test is the area to the right of the observed test statistic $z = 1.15$ or

$$p\text{-value} = P(z > 1.15) = 1 - .8749 = .1251$$

This is the shaded area in the following figure.

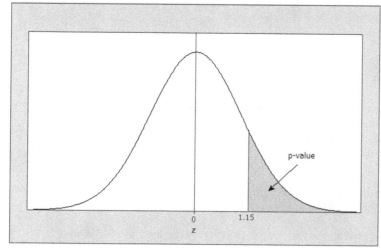

To determine the significance of the test, use the guidelines for statistical significance in Section 9.2. The smaller the p-value, the more evidence there is in favor of rejecting H_0. For this exercise, p-value = .1251 is not statistically significant; H_0 is not rejected.

9.2.13 The p-value for a left-tailed test is the area to the left of the observed test statistic $z = -1.81$ or

$$p\text{-value} = P(z < -1.81) = .0351$$

To determine the significance of the test, use the guidelines for statistical significance in Section 9.2. The smaller the p-value, the more evidence there is in favor of rejecting H_0. For this exercise, p-value $= .0351$ is between .01 and .05. The results are significant at the 5% level, but not at the 1% level (P < .05).

9.2.15-19 In these exercises, the parameter of interest is μ, the population mean. It is given that $\bar{x} = 2.4, n = 35$ and $s = .29$ and the objective of the experiment is to show that the mean exceeds 2.3.

15. To test the hypothesis $H_0 : \mu = 2.3$ versus $H_a : \mu > 2.3$, the best estimator for μ is the sample average $\bar{x}$, and the test statistic is

$$z = \frac{\bar{x} - \mu_0}{\sigma/\sqrt{n}}$$

which represents the distance (measured in units of standard deviations) from $\bar{x}$ to the hypothesized mean μ. Hence, if this value is a large positive number, one of two conclusions may be drawn. Either a very unlikely event has occurred, or the mean exceeds the hypothesized value. Refer to Exercise 14. If $\alpha = .05$, the critical value of z that separates the rejection and non-rejection regions will be a value (denoted by z_0) such that $P(z > z_0) = \alpha = .05$. That is, $z_0 = 1.645$ (see the next figure). Hence, H_0 will be rejected if $z > 1.645$.

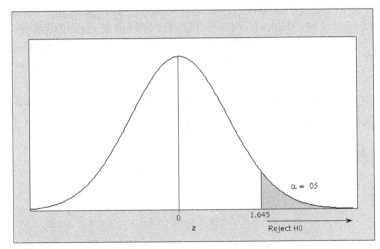

17. The value of the test statistic is $z = \dfrac{\bar{x} - \mu_0}{\sigma/\sqrt{n}} \approx \dfrac{\bar{x} - \mu_0}{s/\sqrt{n}} = \dfrac{2.4 - 2.3}{.049} = 2.04$. Since this is a right-tailed test, the p-value is the area under the standard normal distribution to the right of $z = 2.04$:

$$p\text{-value} = P(z > 2.04) = 1 - .9793 = .0207$$

The p-value, .0207, is less than $\alpha = .05$, and the null hypothesis is rejected at the 5% level of significance. There is sufficient evidence to indicate that $\mu > 2.3$.

19. Refer to Exercise 15, in which the rejection region was given as $z > 1.645$ where

$$z = \frac{\bar{x} - \mu_0}{s/\sqrt{n}} = \frac{\bar{x} - 2.3}{.29/\sqrt{35}}$$

Solving for $\bar{x}$ we obtain the critical value of $\bar{x}$ necessary for rejection of H_0.

$$\frac{\bar{x} - 2.3}{.29/\sqrt{35}} > 1.645 \implies \bar{x} > 1.645 \frac{.29}{\sqrt{35}} + 2.3 = 2.38$$

The probability of a Type II error is defined as $\beta = P(\text{accept } H_0 \text{ when } H_0 \text{ is false})$. Since the acceptance region is $\bar{x} \leq 2.38$, β can be rewritten as

$$\beta = P(\bar{x} \leq 2.38 \text{ when } H_0 \text{ is false}) = P(\bar{x} \leq 2.38 \text{ when } \mu > 2.3)$$

For $\mu = 2.3$,

$$\beta = P(\bar{x} \leq 2.38 \text{ when } \mu = 2.3) = P\left(z \leq \frac{2.38 - 2.3}{.29/\sqrt{35}}\right) = P(z \leq 1.63) = .9484$$

(Notice that the value of β = P(accept H_0 when μ = 2.3) is actually 1 − P(reject H_0 when μ = 2.3) = 1 − α = .95. The above calculation is slightly different due to rounding error in the critical value of $\bar{x}$.)

For $\mu = 2.5$,

$$\beta = P(\bar{x} \leq 2.38 \text{ when } \mu = 2.5) = P\left(z \leq \frac{2.38 - 2.5}{.29/\sqrt{35}}\right) = P(z \leq -2.45) = .0071$$

For $\mu = 2.6$,

$$\beta = P(\bar{x} \leq 2.38 \text{ when } \mu = 2.6) = P\left(z \leq \frac{2.38 - 2.6}{.29/\sqrt{35}}\right) = P(z \leq -4.49) \approx 0$$

9.2.21 The hypotheses to be tested are

$$H_0: \mu = 28 \quad \text{versus} \quad H_a: \mu \neq 28$$

and the test statistic is

$$z = \frac{\bar{x} - \mu_0}{\sigma/\sqrt{n}} \approx \frac{\bar{x} - \mu_0}{s/\sqrt{n}} = \frac{26.8 - 28}{6.5/\sqrt{100}} = -1.85$$

with p-value $= P(|z| > 1.85) = 2(.0322) = .0644$. To draw a conclusion from the p-value, use the guidelines for statistical significance in Section 9.2. Since the p-value is greater than .05, the null hypothesis should not be rejected. There is insufficient evidence to indicate that the mean is different from 28. (Some researchers might report these results as *tending towards significance*.)

9.2.23 **a** In order to make sure that the average weight was one pound, you would test

$$H_0: \mu = 1 \quad \text{versus} \quad H_a: \mu \neq 1$$

b-c The test statistic is

$$z = \frac{\bar{x} - \mu_0}{\sigma/\sqrt{n}} \approx \frac{\bar{x} - \mu_0}{s/\sqrt{n}} = \frac{1.01 - 1}{.18/\sqrt{35}} = .33$$

with p-value $= P(|z| > .33) = 2(.3707) = .7414$. Since the p-value is greater than .05, the null hypothesis should not be rejected. The manager should report that there is insufficient evidence to indicate that the mean is different from 1.

9.2.25 **a-b** We want to test the null hypothesis that μ is, in fact, 80% against the alternative that it is not:

$$H_0: \mu = 80 \quad \text{versus} \quad H_a: \mu \neq 80$$

Since the exercise does not specify $\mu < 80$ or $\mu > 80$, we are interested in a two directional alternative, $\mu \neq 80$.

c The test statistic is $z = \dfrac{\bar{x} - \mu_0}{\sigma/\sqrt{n}} \approx \dfrac{\bar{x} - \mu_0}{s/\sqrt{n}} = \dfrac{79.7 - 80}{.8/\sqrt{100}} = -3.75$. The rejection region with $\alpha = .05$ is determined by a critical value of z such that

$$P(z < -z_0) + P(z > z_0) = \dfrac{\alpha}{2} + \dfrac{\alpha}{2} = .05$$

This value is $z_0 = 1.96$ (see the figure in Exercise 7). Hence, H₀ will be rejected if $z > 1.96$ or $z < -1.96$. The observed value, $z = -3.75$, falls in the rejection region and H₀ is rejected. There is sufficient evidence to refute the manufacturer's claim. The probability that we have made an incorrect decision is $\alpha = .05$.

9.2.27 The hypothesis to be tested is $H_0 : \mu = 5.7$ versus $H_a : \mu < 5.7$

and the test statistic is

$$z = \dfrac{\bar{x} - \mu}{\sigma/\sqrt{n}} \approx \dfrac{\bar{x} - \mu}{s/\sqrt{n}} = \dfrac{3.7 - 5.7}{.5/\sqrt{40}} = -25.298$$

The rejection region with $\alpha = .05$ is $z < -1.645$. The observed value, $z = -25.298$, falls in the rejection region and H₀ is rejected. We conclude that the average pH for rains is more acidic than for pure water.

9.2.29 **a** The hypothesis to be tested is $H_0 : \mu = 98.6$ versus $H_a : \mu \neq 98.6$

and the test statistic is

$$z = \dfrac{\bar{x} - \mu_0}{\sigma/\sqrt{n}} \approx \dfrac{\bar{x} - \mu_0}{s/\sqrt{n}} = \dfrac{98.25 - 98.6}{.73/\sqrt{130}} = -5.47$$

with p-value $= P(z < -5.47) + P(z > 5.47) \approx 2(0) = 0$. With $\alpha = .05$, the p-value is less than α and H₀ is rejected. There is sufficient evidence to indicate that the average body temperature for healthy humans is different from 98.6.

b-c Using the critical value approach, we set the null and alternative hypotheses and calculate the test statistic as in part **a**. The rejection region with $\alpha = .05$ is $|z| > 1.96$. The observed value, $z = -5.47$, does fall in the rejection region and H₀ is rejected. The conclusion is the same is in part **a**.

d How did the doctor record 1 million temperatures in 1868? The technology available at that time makes this a difficult if not impossible task. It may also have been that the instruments used for this research were not entirely accurate.

9.2.31 The hypothesis to be tested is $H_0 : \mu = 238$ versus $H_a : \mu < 238$ and the test statistic is

$$z = \dfrac{\bar{x} - \mu}{\sigma/\sqrt{n}} \approx \dfrac{\bar{x} - \mu}{s/\sqrt{n}} = \dfrac{232 - 238}{20.2/\sqrt{35}} = -1.757$$

The rejection region with $\alpha = .01$ is $z < -2.33$. The observed value, $z = -1.757$, does not fall in the rejection region and H₀ is not rejected. We cannot conclude that the EPA rating is less than claimed.

Section 9.3

9.3.1 The hypothesis of interest is two-tailed: $H_0 : \mu_1 - \mu_2 = 0$ versus $H_a : \mu_1 - \mu_2 \neq 0$ and the test statistic, calculated under the assumption that $\mu_1 - \mu_2 = 0$, is

$$z = \frac{(\bar{x}_1 - \bar{x}_2) - (\mu_1 - \mu_2)}{\sqrt{\dfrac{\sigma_1^2}{n_1} + \dfrac{\sigma_2^2}{n_2}}}$$

with σ_1^2 and σ_2^2 known, or estimated by s_1^2 and s_2^2, respectively. For this exercise,

$$z \approx \frac{(\bar{x}_1 - \bar{x}_2) - 0}{\sqrt{\dfrac{s_1^2}{n_1} + \dfrac{s_2^2}{n_2}}} = \frac{9.7 - 7.4}{\sqrt{\dfrac{10.78}{35} + \dfrac{16.44}{49}}} = 2.87$$

The rejection region, with $\alpha = .05$, is $|z| > 1.96$ (see Exercise 7, Section 9.2). Since the observed value of z falls in the rejection region, H_0 is rejected. There is sufficient evidence to indicate that $\mu_1 - \mu_2 \neq 0$, or $\mu_1 \neq \mu_2$.

9.3.3 Refer to Exercise 1. The p-value for this two-tailed test is

$$p\text{-value} = P(|z| > 2.87) = 2(.0021) = .0042$$

Since the p-value is less than $\alpha = .05$, the null hypothesis can be rejected at the 5% level. There is sufficient evidence to conclude that $\mu_1 - \mu_2 \neq 0$.

9.3.5 The hypothesis of interest is one-tailed: $H_0 : \mu_1 - \mu_2 = 0$ versus $H_a : \mu_1 - \mu_2 > 0$ and the test statistic, calculated under the assumption that $\mu_1 - \mu_2 = 0$, is

$$z \approx \frac{(\bar{x}_1 - \bar{x}_2) - 0}{\sqrt{\dfrac{s_1^2}{n_1} + \dfrac{s_2^2}{n_2}}} = \frac{125.2 - 123.7}{\sqrt{\dfrac{5.6^2}{50} + \dfrac{6.8^2}{50}}} = 1.20$$

with the unknown σ_1^2 and σ_2^2 estimated by s_1^2 and s_2^2, respectively. The rejection region with $\alpha = .01$ is $z > 2.33$. Since the observed value of z does not fall in the rejection region, H_0 is not rejected. There is insufficient evidence to indicate that the mean for population 1 is larger than the mean for population 2.

9.3.7 Refer to Exercise 5. The p-value for this one-tailed test is

$$p\text{-value} = P(z > 1.20) = 1 - .8849 = .1151$$

Since the p-value is greater than $\alpha = .01$, the null hypothesis cannot be rejected at the 1% level. There is insufficient evidence to conclude that $\mu_1 - \mu_2 > 0$.

9.3.9 The hypothesis to be tested is $H_0 : \mu_1 - \mu_2 = 0$ versus $H_a : \mu_1 - \mu_2 \neq 0$ and the test statistic is

$$z \approx \frac{(\bar{x}_1 - \bar{x}_2) - 0}{\sqrt{\dfrac{s_1^2}{n_1} + \dfrac{s_2^2}{n_2}}} = \frac{1925 - 1905}{\sqrt{\dfrac{40^2}{100} + \dfrac{30^2}{100}}} = 4$$

The rejection region, with $\alpha = .05$, is two-tailed or $|z| > 1.96$ and the null hypothesis is rejected. There is difference in mean breaking strengths for the two cables.

9.3.11 a The hypothesis to be tested is $H_0 : \mu_1 - \mu_2 = 0$ versus $H_a : \mu_1 - \mu_2 > 0$ and the test statistic is

$$z \approx \frac{(\bar{x}_1 - \bar{x}_2) - 0}{\sqrt{\dfrac{s_1^2}{n_1} + \dfrac{s_2^2}{n_2}}} = \frac{240 - 227}{\sqrt{\dfrac{980}{200} + \dfrac{820}{200}}} = 4.33$$

The rejection region, with $\alpha = .05$, is one-tailed or $z > 1.645$ and the null hypothesis is rejected. There is a difference in mean yield for the two types of spray.

b An approximate 95% confidence interval for $\mu_1 - \mu_2$ is

$$(\bar{x}_1 - \bar{x}_2) \pm 1.96 \sqrt{\frac{s_1^2}{n_1} + \frac{s_2^2}{n_2}}$$

$$(240 - 227) \pm 1.96 \sqrt{\frac{980}{200} + \frac{820}{200}}$$

$$13 \pm 5.88 \quad \text{or} \quad 7.12 < (\mu_1 - \mu_2) < 18.88$$

9.3.13 a The hypothesis of interest is one-tailed: $H_0 : \mu_1 - \mu_2 = 0$ versus $H_a : \mu_1 - \mu_2 > 0$

b The test statistic, calculated under the assumption that $\mu_1 - \mu_2 = 0$, is

$$z \approx \frac{(\bar{x}_1 - \bar{x}_2) - 0}{\sqrt{\frac{s_1^2}{n_1} + \frac{s_2^2}{n_2}}} = \frac{6.9 - 5.8}{\sqrt{\frac{(2.9)^2}{35} + \frac{(1.2)^2}{35}}} = 2.074$$

The rejection region with $\alpha = .05$, is $z > 1.645$ and H_0 is rejected. There is evidence to indicate that $\mu_1 - \mu_2 > 0$, or $\mu_1 > \mu_2$. That is, there is reason to believe that Vitamin C reduces the mean time to recover.

9.3.15 a The hypothesis of interest is two-tailed: $H_0 : \mu_1 - \mu_2 = 0$ versus $H_a : \mu_1 - \mu_2 \neq 0$ and the test statistic, calculated under the assumption that $\mu_1 - \mu_2 = 0$, is

$$z \approx \frac{(\bar{x}_1 - \bar{x}_2) - 0}{\sqrt{\frac{s_1^2}{n_1} + \frac{s_2^2}{n_2}}} = \frac{34.1 - 36}{\sqrt{\frac{(5.9)^2}{100} + \frac{(6.0)^2}{100}}} = -2.26$$

with *p*-value $= P(|z| > 2.26) = 2(.0119) = .0238$. Since the *p*-value is less than .05, the null hypothesis is rejected. There is evidence to indicate a difference in the mean lead levels for the two sections of the city.

b From Section 8.4, the 95% confidence interval for $\mu_1 - \mu_2$ is approximately

$$(\bar{x}_1 - \bar{x}_2) \pm 1.96 \sqrt{\frac{s_1^2}{n_1} + \frac{s_2^2}{n_2}}$$

$$(34.1 - 36) \pm 1.96 \sqrt{\frac{5.9^2}{100} + \frac{6.0^2}{100}}$$

$$-1.9 \pm 1.65 \quad \text{or} \quad -3.55 < (\mu_1 - \mu_2) < -.25$$

c Since the value $\mu_1 - \mu_2 = 5$ or $\mu_1 - \mu_2 = -5$ is not in the confidence interval in part **b**, it is not likely that the difference will be more than 5 ppm, and hence the statistical significance of the difference is not of practical importance to the engineers.

9.3.17 a To test for a difference in the two population means, a two-tailed hypothesis would be appropriate:

$$H_0 : \mu_1 - \mu_2 = 0 \quad \text{versus} \quad H_a : \mu_1 - \mu_2 \neq 0$$

The test statistic is

$$z \approx \frac{(\bar{x}_1 - \bar{x}_2) - 0}{\sqrt{\frac{s_1^2}{n_1} + \frac{s_2^2}{n_2}}} = \frac{1122 - 1048}{\sqrt{\frac{194^2}{100} + \frac{165^2}{100}}} = 2.91$$

The rejection region with $\alpha = .05$ is $|z| > 1.96$ and H_0 is rejected. There is evidence to indicate that there is a difference in the average SAT scores for Massachusetts and California.

b The *p*-value for this two-tailed test is

$$p\text{-value} = P(z > 2.91) + P(z < -2.91) = 2(.0018) = .0036$$

Since the *p*-value is less than $\alpha = .05$, the null hypothesis can be rejected at the 5% level. There is sufficient evidence to conclude that $\mu_1 - \mu_2 \neq 0$.

9.3.19 a The hypothesis of interest is two-tailed: $H_0: \mu_1 - \mu_2 = 0$ versus $H_a: \mu_1 - \mu_2 \neq 0$ and the test statistic is

$$z \approx \frac{(\bar{x}_1 - \bar{x}_2) - 0}{\sqrt{\frac{s_1^2}{n_1} + \frac{s_2^2}{n_2}}} = \frac{341.5 - 267.76}{\sqrt{\frac{200^2}{100} + \frac{200^2}{100}}} = 2.61$$

with *p*-value $= P(|z| > 2.61) \approx 2(.0045) = .0090$. Since the *p*-value is less than .05, the null hypothesis is rejected. There is evidence to indicate a difference in the mean ticket prices for these two airports.

b The traveler might be more concerned about convenience rather than the difference in cost.

9.3.21 a The hypothesis of interest is two-tailed: $H_0: \mu_1 - \mu_2 = 0$ versus $H_a: \mu_1 - \mu_2 \neq 0$ and the test statistic is

$$z \approx \frac{(\bar{x}_1 - \bar{x}_2) - 0}{\sqrt{\frac{s_1^2}{n_1} + \frac{s_2^2}{n_2}}} = \frac{98.11 - 98.39}{\sqrt{\frac{.7^2}{65} + \frac{.74^2}{65}}} = -2.22$$

with *p*-value $= P(|z| > 2.22) = 2(1 - .9868) = .0264$. Since the *p*-value is between .01 and .05, the null hypothesis is rejected, and the results are significant. There is evidence to indicate a difference in the mean temperatures for men versus women.

b Since the *p*-value $= .0264$, we can reject H_0 at the 5% level (*p*-value $< .05$), but not at the 1% level (*p*-value $> .01$). Using the guidelines for significance given in Section 9.3 of the text, we declare the results statistically *significant*, but not *highly significant*.

Section 9.4

9.4.1 The hypothesis of interest concerns the binomial parameter *p* and is one-tailed:

$$H_0: p = .3 \quad \text{versus} \quad H_a: p < .3$$

It is given that $x = 279$ and $n = 1000$, so that $\hat{p} = \frac{x}{n} = \frac{279}{1000} = .279$. The test statistic is then

$$z = \frac{\hat{p} - p_0}{\sqrt{\frac{p_0 q_0}{n}}} = \frac{.279 - .3}{\sqrt{\frac{.3(.7)}{1000}}} = -1.45$$

The rejection region is one-tailed, with $\alpha = .05$, or $z < -1.645$. Since the observed value does not fall in the rejection region, H_0 is not rejected. We cannot conclude that $p < .3$.

9.4.3 The hypothesis of interest is one-tailed: $H_0: p = .5$ versus $H_a: p > .5$. With $x = 72$ and $n = 120$, so that $\hat{p} = \dfrac{x}{n} = \dfrac{72}{120} = .6$, the test statistic is

$$z = \frac{\hat{p} - p_0}{\sqrt{\dfrac{p_0 q_0}{n}}} = \frac{.6 - .5}{\sqrt{\dfrac{.5(.5)}{120}}} = 2.19$$

The rejection region is one-tailed, with $z > 1.645$ with $\alpha = .05$. Hence, H_0 is rejected at the 5% level. At the 5% significance level, we conclude that $p > .5$.

9.4.5 The test statistic is

$$z = \frac{\hat{p} - p_0}{\sqrt{\dfrac{p_0 q_0}{n}}} = \frac{.378 - .4}{\sqrt{\dfrac{.4(.6)}{1400}}} = -1.68$$

with p-value $= P(|z| > 1.68) = 2(.0465) = .093$. Since the p-value is not less than $\alpha = .05$, the null hypothesis is not rejected. There is insufficient evidence to indicate that p differs from .4.

9.4.7 a The hypothesis to be tested involves the binomial parameter p:

$$H_0: p = .60 \quad \text{versus} \quad H_a: p > .60$$

where p is the proportion of women who consider "improving her health" to be a reason for considering dieting.

b For this test, $\hat{p} = \dfrac{x}{n} = \dfrac{x}{500} = .68$, the test statistic is $z = \dfrac{\hat{p} - p_0}{\sqrt{\dfrac{p_0 q_0}{n}}} = \dfrac{.68 - .6}{\sqrt{\dfrac{.6(.4)}{500}}} = 3.65$ with

p-value $= P(z > 3.65) \approx 0$. The rejection region is one-tailed, with $z > 2.33$ and $\alpha = .01$.

c Since the test statistic falls in the rejection region, the null hypothesis is rejected. There is sufficient evidence to indicate that the proportion of women who consider "improving her health" to be a reason for considering dieting is more than 60%.

9.4.9 a Define p to be the proportion of small businesses that fail in their first year. If the new tax structure is helping, the proportion of small businesses that fail should be lower than the current proportion, presently 20% (or .20). Hence, the hypothesis of interest is one-tailed:

$$H_0: p = .20 \quad \text{versus} \quad H_a: p < .20$$

b With $x = 30$ and $n = 200$, so that $\hat{p} = \dfrac{x}{n} = \dfrac{30}{200} = .15$, the test statistic is

$$z = \frac{\hat{p} - p_0}{\sqrt{\dfrac{p_0 q_0}{n}}} = \frac{.15 - .2}{\sqrt{\dfrac{.2(.8)}{200}}} = -1.77$$

c The rejection region is one-tailed, with $\alpha = .05$ or $z < -1.645$ and H_0 is rejected. There is sufficient evidence to indicate that the new tax structure helps small businesses survive.

d Calculate p-value $= P(z < -1.77) = .0384$. Since the p-value is less than .05, H_0 is rejected. The conclusion is the same as the conclusion in part c.

9.4.11 a-b Since the survival rate without screening is $p = 2/3$, the survival rate with an effective program may be greater than 2/3. Hence, the hypothesis to be tested is

$$H_0: p = 2/3 \quad \text{versus} \quad H_a: p > 2/3$$

c With $\hat{p} = \dfrac{x}{n} = \dfrac{164}{200} = .82$, the test statistic is

$$z = \dfrac{\hat{p} - p_0}{\sqrt{\dfrac{p_0 q_0}{n}}} = \dfrac{.82 - 2/3}{\sqrt{\dfrac{(2/3)(1/3)}{200}}} = 4.6$$

The rejection region is one-tailed, with $\alpha = .05$ or $z > 1.645$ and H_0 is rejected. The screening program seems to increase the survival rate.

d For the one-tailed test, p-value $= P(z > 4.6) < 1 - .9998 = .0002$. That is, H_0 can be rejected for any value of $\alpha \geq .0002$. The results are *highly significant*.

9.4.13 a The hypothesis of interest is

$$H_0: p = .5 \quad \text{versus} \quad H_a: p > .5$$

where p is the probability that the national brand is judged to be better than the store brand.

b With $\hat{p} = \dfrac{x}{n} = \dfrac{8}{35} = .229$, the test statistic is

$$z = \dfrac{\hat{p} - p_0}{\sqrt{\dfrac{p_0 q_0}{n}}} = \dfrac{.229 - .5}{\sqrt{\dfrac{.5(.5)}{35}}} = -3.21$$

The rejection region is one-tailed with $\alpha = .01$, or $z > 2.33$ and H_0 is not rejected. The observed value of the test statistic actually ends up in the wrong tail of the distribution! There is insufficient evidence to indicate that the national brand is preferred to the store brand.

9.4.15 The hypothesis of interest is $H_0: p = .80$ versus $H_a: p < .80$ with $\hat{p} = \dfrac{x}{n} = \dfrac{37}{50} = .74$. The test statistic is

$$z = \dfrac{\hat{p} - p_0}{\sqrt{\dfrac{p_0 q_0}{n}}} = \dfrac{.74 - .80}{\sqrt{\dfrac{.80(.20)}{50}}} = -1.06$$

The rejection region with $\alpha = .05$ is $z < -1.645$ and the null hypothesis is not rejected. (Alternatively, we could calculate p-value $= P(z < -1.06) = .1446$. Since this p-value is greater than .05, the null hypothesis is not rejected.) There is insufficient evidence to refute the experimenter's claim.

9.4.17 The hypothesis of interest is $H_0: p = .35$ versus $H_a: p \neq .35$ with $\hat{p} = \dfrac{x}{n} = \dfrac{114}{300} = .38$, the test statistic is

$$z = \dfrac{\hat{p} - p_0}{\sqrt{\dfrac{p_0 q_0}{n}}} = \dfrac{.38 - .35}{\sqrt{\dfrac{.35(.65)}{300}}} = 1.09$$

with p-value given in the printout as .276. Since this p-value is greater than .05, the null hypothesis is not rejected. There is insufficient evidence to indicate that the proportion of households with large dogs is different from that reported by the Humane Society.

Section 9.5

9.5.1 Calculate $\hat{p}_1 = \dfrac{565}{1250} = .452$, $\hat{p}_2 = \dfrac{621}{1100} = .565$, and $\hat{p} = \dfrac{x_1 + x_2}{n_1 + n_2} = \dfrac{565 + 621}{1250 + 1100} = .505$.

9.5.3 Since it is necessary to detect either $p_1 > p_2$ or $p_1 < p_2$, a two-tailed test is necessary:

$$H_0 : p_1 - p_2 = 0 \quad \text{versus} \quad H_a : p_1 - p_2 \neq 0$$

The test statistic based on the sample data will be (using the rounded values from Exercise 1)

$$z = \dfrac{\hat{p}_1 - \hat{p}_2 - (p_1 - p_2)}{\sqrt{\dfrac{p_1 q_1}{n_1} + \dfrac{p_2 q_2}{n_2}}} \approx \dfrac{\hat{p}_1 - \hat{p}_2}{\sqrt{\hat{p}\hat{q}\left(\dfrac{1}{n_1} + \dfrac{1}{n_2}\right)}} = \dfrac{.452 - .565}{\sqrt{.505(.495)\left(\dfrac{1}{1250} + \dfrac{1}{1100}\right)}} = -5.47$$

The rejection region, with $\alpha = .05$, is $|z| > 1.96$ and H_0 is rejected. There is sufficient evidence to indicate a difference in the population proportions.

9.5.5 Since it is necessary to detect either $p_1 > p_2$ or $p_1 < p_2$, a two-tailed test is necessary:

$$H_0 : p_1 - p_2 = 0 \quad \text{versus} \quad H_a : p_1 - p_2 \neq 0$$

Calculate $\hat{p}_1 = \dfrac{120}{500} = .24$ and $\hat{p}_2 = \dfrac{147}{500} = .294$. and $\hat{p} = \dfrac{x_1 + x_2}{n_1 + n_2} = \dfrac{120 + 147}{500 + 500} = .267$

The test statistic, based on the sample data will be

$$z = \dfrac{\hat{p}_1 - \hat{p}_2 - (p_1 - p_2)}{\sqrt{\dfrac{p_1 q_1}{n_1} + \dfrac{p_2 q_2}{n_2}}} \approx \dfrac{\hat{p}_1 - \hat{p}_2}{\sqrt{\hat{p}\hat{q}\left(\dfrac{1}{n_1} + \dfrac{1}{n_2}\right)}} = \dfrac{.24 - .294}{\sqrt{.267(.733)\left(\dfrac{1}{500} + \dfrac{1}{500}\right)}} = -1.93$$

Calculate the two tailed p-value $= P(|z| > 1.93) = 2(.0268) = .0536$. Since this p-value is greater than .01, H_0 is not rejected. There is insufficient evidence of a difference in the two population proportions.

9.5.7 Since it is necessary to detect either $p_1 > p_2$ or $p_1 < p_2$, a two-tailed test is necessary:

$$H_0 : p_1 - p_2 = 0 \quad \text{versus} \quad H_a : p_1 - p_2 \neq 0$$

Calculate $\hat{p}_1 = \dfrac{74}{140} = .529$ and $\hat{p}_2 = \dfrac{81}{140} = .579$. and $\hat{p} = \dfrac{x_1 + x_2}{n_1 + n_2} = \dfrac{74 + 81}{140 + 140} = .554$

The test statistic based on the sample data will be

$$z = \dfrac{\hat{p}_1 - \hat{p}_2 - (p_1 - p_2)}{\sqrt{\dfrac{p_1 q_1}{n_1} + \dfrac{p_2 q_2}{n_2}}} \approx \dfrac{\hat{p}_1 - \hat{p}_2}{\sqrt{\hat{p}\hat{q}\left(\dfrac{1}{n_1} + \dfrac{1}{n_2}\right)}} = \dfrac{.529 - .579}{.0594} = -.84$$

The rejection region with $\alpha = .10$, or $|z| > 1.645$ and H_0 is not rejected. There is no evidence of a difference in the two population proportions.

9.5.9 The hypothesis of interest is one-tailed:

$$H_0 : p_1 - p_2 = 0 \quad \text{versus} \quad H_a : p_1 - p_2 < 0$$

Calculate $\hat{p}_1 = \dfrac{132}{280} = .471$, $\hat{p}_2 = \dfrac{178}{350} = .509$ and $\hat{p} = \dfrac{x_1 + x_2}{n_1 + n_2} = \dfrac{132 + 178}{280 + 350} = .492$.

The test statistic is then

$$z = \frac{\hat{p}_1 - \hat{p}_2}{\sqrt{\hat{p}\hat{q}\left(\frac{1}{n_1} + \frac{1}{n_2}\right)}} = \frac{.471 - .509}{\sqrt{.492(.508)(1/280 + 1/350)}} = -.95$$

Since no value of α is specified in advance, we calculate p-value = $P(z < -.95) = .1711$. Since this p-value is greater than .10, the null hypothesis is not rejected. There is insufficient evidence to indicate that $p_1 < p_2$.

9.5.11 Let p_1 be the proportion of defectives produced by machine A and p_2 be the proportion of defectives produced by machine B. The hypothesis to be tested is

$$H_0 : p_1 - p_2 = 0 \quad \text{versus} \quad H_a : p_1 - p_2 \neq 0$$

Calculate $\hat{p}_1 = \frac{16}{200} = .08$, $\hat{p}_2 = \frac{8}{200} = .04$, and $\hat{p} = \frac{x_1 + x_2}{n_1 + n_2} = \frac{16 + 8}{200 + 200} = .06$. The test statistic is then

$$z = \frac{\hat{p}_1 - \hat{p}_2}{\sqrt{\hat{p}\hat{q}\left(\frac{1}{n_1} + \frac{1}{n_2}\right)}} = \frac{.08 - .04}{\sqrt{.06(.94)(1/200 + 1/200)}} = 1.684$$

The rejection region, with $\alpha = .05$, is $|z| > 1.96$ and H_0 is not rejected. There is insufficient evidence to indicate that the machines are performing differently in terms of the percentage of defectives being produced.

9.5.13 a The hypothesis of interest is

$$H_0 : p_1 - p_2 = 0 \quad \text{versus} \quad H_a : p_1 - p_2 \neq 0$$

Calculate $\hat{p}_1 = \frac{12}{56} = .214$, $\hat{p}_2 = \frac{8}{32} = .25$, and $\hat{p} = \frac{x_1 + x_2}{n_1 + n_2} = \frac{12 + 8}{56 + 32} = .227$.

The test statistic is then

$$z = \frac{\hat{p}_1 - \hat{p}_2}{\sqrt{\hat{p}\hat{q}\left(\frac{1}{n_1} + \frac{1}{n_2}\right)}} = \frac{.214 - .25}{\sqrt{.227(.773)(1/56 + 1/32)}} = -.39$$

The rejection region, with $\alpha = .05$, is $|z| > 1.96$ and H_0 is not rejected. There is insufficient evidence to indicate a difference in the proportion of red M&Ms for the plain and peanut varieties.

b From Section 8.5, the approximate 95% confidence interval is

$$(\hat{p}_1 - \hat{p}_2) \pm 1.96 \sqrt{\frac{\hat{p}_1\hat{q}_1}{n_1} + \frac{\hat{p}_2\hat{q}_2}{n_2}}$$

$$(.214 - .25) \pm 1.96 \sqrt{\frac{.214(.786)}{56} + \frac{.25(.75)}{32}}$$

$$-.036 \pm .185 \quad \text{or} \quad -.221 < (p_1 - p_2) < .149$$

Since the value $p_1 - p_2 = 0$ falls in the interval, there is insufficient evidence to indicate a difference in the proportions. The results agree with part a.

9.5.15 Refer to Exercise 14. The 99% lower one-sided confidence bound for $p_1 - p_2$ is

$$(\hat{p}_1 - \hat{p}_2) - 2.33\sqrt{\frac{\hat{p}_1\hat{q}_1}{n_1} + \frac{\hat{p}_2\hat{q}_2}{n_2}}$$

$$(.018 - .009) - 2.33\sqrt{\frac{.018(.982)}{2266} + \frac{.009(.991)}{2266}}$$

$$.009 - .008 = .001 \quad \text{or} \quad (p_1 - p_2) > .001$$

The difference in risk between the two groups is at least 1 in 1000. This difference may not be of *practical significance* if the benefits of the hormone replacement therapy to the patient outweigh the risk.

9.5.17 The hypothesis of interest is

$$H_0: p_1 - p_2 = 0 \quad \text{versus} \quad H_a: p_1 - p_2 > 0$$

Calculate $\hat{p}_1 = \frac{93}{121} = .769$, $\hat{p}_2 = \frac{119}{199} = .598$, and $\hat{p} = \frac{x_1 + x_2}{n_1 + n_2} = \frac{93 + 119}{121 + 199} = .6625$. The test statistic is then

$$z = \frac{\hat{p}_1 - \hat{p}_2}{\sqrt{\hat{p}\hat{q}\left(\frac{1}{n_1} + \frac{1}{n_2}\right)}} = \frac{.769 - .598}{\sqrt{.6625(.3375)(1/121 + 1/199)}} = 3.14$$

with *p*-value $= P(z > 3.14) = 1 - .9992 = .0008$. Since the *p*-value is less than .01, the results are reported as highly significant at the 1% level of significance. There is evidence to confirm the researcher's conclusion.

Reviewing What You've Learned

9.R.1 **a-b** Since it is necessary to prove that the average pH level is less than 7.5, the hypothesis to be tested is one-tailed:

$$H_0: \mu = 7.5 \quad \text{versus} \quad H_a: \mu < 7.5$$

d The test statistic is

$$z = \frac{\bar{x} - \mu}{\sigma/\sqrt{n}} \approx \frac{\bar{x} - \mu}{s/\sqrt{n}} = \frac{-.2}{.2/\sqrt{30}} = -5.477$$

and the rejection region with $\alpha = .05$ is $z < -1.645$. The observed value, $z = -5.477$, falls in the rejection region and H_0 is rejected. We conclude that the average pH level is less than 7.5.

9.R.3 **a-b** Since there is no prior knowledge as to which mean should be larger, the hypothesis of interest is two-tailed

$$H_0: \mu_1 - \mu_2 = 0 \quad \text{versus} \quad H_a: \mu_1 - \mu_2 \neq 0$$

c The test statistic is approximately

$$z \approx \frac{(\bar{x}_1 - \bar{x}_2) - 0}{\sqrt{\frac{s_1^2}{n_1} + \frac{s_2^2}{n_2}}} = \frac{2980 - 3205}{\sqrt{\frac{1140^2}{40} + \frac{963^2}{40}}} = -.954$$

The rejection region, with $\alpha = .05$, is two-tailed or $|z| > 1.96$. The null hypothesis is not rejected. There is insufficient evidence to indicate a difference in the two means.

9.R.5 Refer to the following figure, which represents the two probability distributions, one assuming that $p_1 - p_2 = 0$ and one assuming that $p_1 - p_2 = .1$.

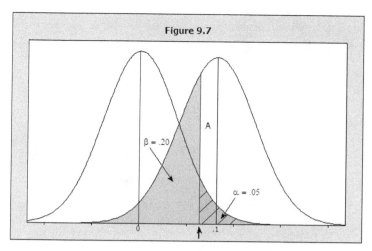

Figure 9.7

The right curve is the true distribution of the random variable $\hat{p}_1 - \hat{p}_2$ and consequently any probabilities that we wish to calculate concerning the random variable must be calculated as areas under the curve to the right. The objective of this exercise is to find a common sample size so that
$\alpha = P[\text{reject } H_0 \text{ when } H_0 \text{ is true}] = .05$ and $\beta = P[\text{accept } H_0 \text{ when } H_0 \text{ is false}] \le .20$. For $\alpha = .05$ consider the critical value of $\hat{p}_1 - \hat{p}_2$ that separates the rejection and acceptance regions. This value will be denoted by $(\hat{p}_1 - \hat{p}_2)_C$. Recall that the random variable $z = (x-\mu)/\sigma$ measures the distance from a particular value x to the mean (in units of standard deviation). Since the z-value corresponding to $(\hat{p}_1 - \hat{p}_2)_C$ is $z = 1.645$, we have

$$1.645 = \frac{(\hat{p}_1 - \hat{p}_2)_C - 0}{\sqrt{\frac{p_1 q_1}{n_1} + \frac{p_2 q_2}{n_2}}} \quad \text{or}$$

$$(\hat{p}_1 - \hat{p}_2)_C = 1.645 \sqrt{\frac{p_1 q_1}{n_1} + \frac{p_2 q_2}{n_2}}$$

Now $\beta = P[\text{accept } H_0 \text{ when } p_1 - p_2 = .1]$ which is the area *under the right-hand curve* to the left of $(\hat{p}_1 - \hat{p}_2)_C$. Since it is required that $\beta = .20$ we must find the z-value corresponding to $A = .2$, which is $z = -.84$ (see Table 3). Then

$$-.84 = \frac{(\hat{p}_1 - \hat{p}_2)_C - .1}{\sqrt{\frac{p_1 q_1}{n_1} + \frac{p_2 q_2}{n_2}}} \quad \text{where } (\hat{p}_1 - \hat{p}_2)_C = 1.645 \sqrt{\frac{p_1 q_1}{n_1} + \frac{p_2 q_2}{n_2}}. \text{ Substituting for } (\hat{p}_1 - \hat{p}_2)_C,$$

$$-.84 = \frac{1.645 \sqrt{\frac{p_1 q_1}{n_1} + \frac{p_2 q_2}{n_2}} - .1}{\sqrt{\frac{p_1 q_1}{n_1} + \frac{p_2 q_2}{n_2}}}$$

$$2.485 = \frac{.1}{\sqrt{\frac{p_1 q_1}{n_1} + \frac{p_2 q_2}{n_2}}}$$

The following two assumptions will allow us to calculate the appropriate sample size:

1. $n_1 = n_2 = n$.

2 The maximum value of $p(1-p)$ will occur when $p=1-p=.5$. Since values of p_1 and p_2 are unknown, the use of $p=.5$ will provide a valid sample size, although it may be slightly larger than necessary.

Then, solving for n, we obtain

$$2.485 = \frac{.1}{\sqrt{.5(.5)\left(\frac{1}{n}+\frac{1}{n}\right)}} \Rightarrow 2.485 = \frac{.1}{.5\sqrt{2/n}}$$

$\sqrt{n} = 17.57$ or $n = 308.76$. Hence, a common sample size for the researcher's test will be $n = 309$.

9.R.7 a The hypothesis to be tested is $H_0 : \mu = 533$ versus $H_a : \mu \neq 533$.

The test statistic is

$$z \approx \frac{\bar{x}-\mu}{s/\sqrt{n}} = \frac{531-533}{98/\sqrt{100}} = -.20$$

and the *p*-value is

$$p\text{-value} = P(z > .20) + P(z < -.20) = 2(.4207) = .8414$$

Since the *p*-value, .8414, is greater than $\alpha = .05$, and H_0 cannot be rejected and we cannot conclude that the average critical reading score for California students in 2017 is different from the national average.

b The hypothesis to be tested is

$$H_0 : \mu = 527 \text{ versus } H_a : \mu \neq 527.$$

The test statistic is

$$z \approx \frac{\bar{x}-\mu}{s/\sqrt{n}} = \frac{524-527}{96/\sqrt{100}} = -.31$$

and the *p*-value is $p\text{-value} = P(z > .31) + P(z < -.31) = 2(.3783) = .7566$. Since the *p*-value, .7566, is greater than $\alpha = .05$, and H_0 cannot be rejected and we cannot conclude that the average math score for California students in 2017 is different from the national average.

9.R.9 The hypothesis to be tested is $H_0 : \mu = 5$ versus $H_a : \mu > 5$ and the test statistic is

$$z = \frac{\bar{x}-\mu_0}{\sigma/\sqrt{n}} \approx \frac{\bar{x}-\mu_0}{s/\sqrt{n}} = \frac{7.2-5}{6.2/\sqrt{38}} = 2.19$$

The rejection region with $\alpha = .01$ is $z > 2.33$. Since the observed value, $z = 2.19$, does not fall in the rejection region and H_0 is not rejected. The data do not provide sufficient evidence to indicate that the mean ppm of PCBs in the population of game birds exceeds the FDA's recommended limit of 5 ppm.

9.R.11 The hypothesis of interest is $H_0 : p = .43$ versus $H_a : p \neq .43$.

With $\hat{p} = \frac{x}{n} = \frac{46}{100} = .46$, the test statistic is

$$z = \frac{\hat{p}-p_0}{\sqrt{\frac{p_0 q_0}{n}}} = \frac{.46-.43}{\sqrt{\frac{.43(.57)}{100}}} = 0.61$$

The rejection region, with $\alpha = .05$, is two-tailed or $|z| > 1.96$ and the null hypothesis is not rejected. There is insufficient evidence to indicate that the percentage reported by *USA Today* is incorrect.

9.R.13 The hypothesis to be tested is

$$H_0 : p_1 - p_2 = 0 \quad \text{versus} \quad H_a : p_1 - p_2 \neq 0$$

Calculate $\hat{p}_1 = \dfrac{x_1}{375} = .55$, $\hat{p}_2 = \dfrac{x_2}{439} = .66$, and $\hat{p} = \dfrac{x_1 + x_2}{n_1 + n_2} = \dfrac{375(.55) + 439(.66)}{375 + 439} = .609$. The test statistic is then

$$z = \frac{\hat{p}_1 - \hat{p}_2}{\sqrt{\hat{p}\hat{q}\left(\dfrac{1}{n_1} + \dfrac{1}{n_2}\right)}} = \frac{.55 - .66}{\sqrt{.609(.391)(1/375 + 1/439)}} = -3.21$$

The rejection region for $\alpha = .01$ is $|z| > 2.58$ and the null hypothesis is rejected. There is sufficient evidence to indicate that the proportion of students who are Advanced or Early-Advanced in English proficiency differs for these two districts.

9.R.15 a An approximate 99% confidence interval for $\mu_1 - \mu_2$ is

$$(\bar{x}_1 - \bar{x}_2) \pm 2.58\sqrt{\frac{s_1^2}{n_1} + \frac{s_2^2}{n_2}}$$

$$(9017 - 5853) \pm 2.58\sqrt{\frac{7162^2}{130} + \frac{1961^2}{80}}$$

$$3164 \pm 1716.51 \quad \text{or} \quad 1447.49 < (\mu_1 - \mu_2) < 4880.51$$

b Pure breaststroke swimmer swim between 1447 and 4881 more meters per week than to individual medley swimmers. This would be reasonable, since swimmers in the individual medley have three other strokes to practice—freestyle, backstroke and butterfly.

9.R.17 No. The agronomist would have to show experimentally that the increase was 3 or more bushels per quadrant in order to achieve practical importance.

10: Inference from Small Samples

Section 10.1

10.1.1 1) The sample must be randomly selected and 2) the population from which you are sampling must be normally distributed.

10.1.3-5 Refer to Table 4, Appendix I, indexing df along the left or right margin and t_a across the top.

    3. $t_{.01} = 2.552$ with 18 df **5.** $t_{.005} = 2.787$ with 25 df

10.1.7-11 Refer to Table 4, Appendix I, indexing df along the left or right margin and t_a across the top.

    7. $-t_{.01} = -2.821$ with 9 df **9.** $-t_{.025} = -2.131$ with 15 df

    11. $-t_{.005} = -2.807$ with 23 df

10.1.13-15 Recall from Section 7.4 that samples from normal distributions should results in histograms and stem and leaf plots exhibiting a relatively mound shape. Boxplots should be relatively symmetric with no outliers and normal probability plots should have points reasonably close to a straight-line pattern with no systematic pattern that is not a straight line.

    13. Approximately normal. The points are relatively close to a straight-line pattern.

    15. Approximately normal. The boxplot is symmetric with no outliers.

Section 10.2

10.2.1 A Student's t test can be used to test the hypothesis about a single population mean when the sample has been randomly selected from a normal population. It will work quite satisfactorily for populations which possess mound-shaped frequency distributions resembling the normal distribution.

10.2.3 For a right-tailed test, the critical value that separates the rejection and nonrejection regions for a right tailed test based on a t-statistic will be a value of t (called t_α) such that $P(t > t_\alpha) = \alpha = .05$ and $df = 16$. That is, $t_{.05} = 1.746$. The null hypothesis H_0 will be rejected if $t > 1.746$.

10.2.5 For a left-tailed test with $\alpha = .01$ and $df = 7$, the critical value that separates the rejection and nonrejection regions will be a value of t (called t_α) such that $P(t < -t_\alpha) = \alpha = .01$, so that $-t_{.01} = -2.998$. H_0 will be rejected if $t < -2.998$.

10.2.7 For a two-tailed test with $\alpha = .05$ and $df = 7$, the critical value for the rejection region cuts off $\alpha/2 = .025$ in the two tails of the t distribution, so that $t_{.025} = 2.365$. Reject H_0 if $|t| > t_{.025} = 2.365$.

10.2.9 The p-value for a two-tailed test is defined as p-value $= P(|t| > 2.43) = 2P(t > 2.43)$ so that

$$P(t > 2.43) = \frac{1}{2} p\text{-value}$$

Refer to Table 4, Appendix I, with $df = 12$. The exact probability, $P(t > 2.43)$ is unavailable; however, it is evident that $t = 2.43$ falls between $t_{.025} = 2.179$ and $t_{.01} = 2.681$. Therefore, the area to the right of $t = 2.43$ must be between .01 and .025. Since

$$.01 < \frac{1}{2} p\text{-value} < .025$$

the p-value can be approximated as $.02 < p\text{-value} < .05$.

10.2.11 For a two-tailed test, $p\text{-value} = P(|t| > 1.19) = 2P(t > 1.19)$, so that $P(t > 1.19) = \frac{1}{2}p\text{-value}$. From Table 4 with $df = 25$, $t = 1.19$ is smaller than $t_{.10} = 1.316$ so that

$$\frac{1}{2}p\text{-value} > .10 \quad \text{and} \quad p\text{-value} > .20$$

10.2.13 Use Table 4 with $df = 5$. Since the value $t = 1.2$ is smaller than $t_{.10} = 1.476$, the area to its right must be greater than .10 and you can bound the p-value as

$$p\text{-value} > .10$$

10.2.15 Use Table 4 with $df = 8$ and remember that $P(t < -3.3) = P(t > 3.3)$. Since the value $t = 3.3$ falls between $t_{.01} = 2.896$ and $t_{.005} = 3.355$, you can bound the p-value as

$$.005 < p\text{-value} < .01$$

10.2.17 The hypothesis to be tested is

$$H_0: \mu = 48 \quad \text{versus} \quad H_a: \mu \neq 48$$

and the test statistic is

$$t = \frac{\bar{x} - \mu}{s/\sqrt{n}} = \frac{47.1 - 48}{\sqrt{\frac{4.7}{12}}} = -1.438$$

The rejection region with $\alpha = .05$ and $n - 1 = 11$ degrees of freedom is located in both tails of the t-distribution and is found from Table 4 as $|t| > t_{.025} = 2.201$. Since the observed value of the test statistic does not fall in the rejection region, H_0 is not rejected and we cannot conclude that μ is different from 48.

10.2.19 a Using the formulas given in Chapter 2, calculate $\sum x_i = 70.5$ and $\sum x_i^2 = 499.27$. Then

$$\bar{x} = \frac{\sum x_i}{n} = \frac{70.5}{10} = 7.05$$

$$s^2 = \frac{\sum x_i^2 - \frac{(\sum x_i)^2}{n}}{n-1} = \frac{499.27 - \frac{(70.5)^2}{10}}{9} = .249444 \quad \text{and} \quad s = .4994$$

b With $df = n - 1 = 9$, the appropriate value of t is $t_{.01} = 2.821$ (from Table 4) and the 99% upper one-sided confidence bound is

$$\bar{x} + t_{.01}\frac{s}{\sqrt{n}} \Rightarrow 7.05 + 2.821\sqrt{\frac{.249444}{10}} \Rightarrow 7.05 + .446$$

or $\mu < 7.496$. Intervals constructed using this procedure will enclose μ 99% of the time in repeated sampling. Hence, we are fairly certain that this particular interval encloses μ.

c The hypothesis to be tested is $H_0: \mu = 7.5$ versus $H_a: \mu < 7.5$ and the test statistic is

$$t = \frac{\bar{x} - \mu}{s/\sqrt{n}} = \frac{7.05 - 7.5}{\sqrt{\frac{.249444}{10}}} = -2.849$$

The rejection region with $\alpha = .01$ and $n - 1 = 9$ degrees of freedom is located in the lower tail of the t-distribution and is found from Table 4 as $t < -t_{.01} = -2.821$. Since the observed value of the test statistic falls in the rejection region, H₀ is rejected and we conclude that μ is less than 7.5.

d Notice that the 99% upper one-sided confidence bound for μ does not include the value $\mu = 7.5$. This would confirm the results of the hypothesis test in part **c**, in which we concluded that μ is less than 7.5.

10.2.21 The 90% confidence interval is

$$\bar{x} \pm t_{.05} \frac{s}{\sqrt{n}} \Rightarrow 11.3 \pm 1.746 \frac{3.4}{\sqrt{17}} \Rightarrow 11.3 \pm 1.440$$

or $9.860 < \mu < 12.740$.

10.2.23 a Using the formulas given in Chapter 2, calculate $\sum x_i = 12.55$ and $\sum x_i^2 = 13.3253$. Then

$$\bar{x} = \frac{\sum x_i}{n} = \frac{12.55}{14} = .8964$$

$$s = \sqrt{\frac{\sum x_i^2 - \frac{(\sum x_i)^2}{n}}{n-1}} = \sqrt{\frac{13.3253 - \frac{(12.55)^2}{14}}{13}} = .3995$$

With $df = n - 1 = 13$, the appropriate value of t is $t_{.025} = 2.16$ (from Table 4) and the 95% confidence interval is

$$\bar{x} \pm t_{.025} \frac{s}{\sqrt{n}} \Rightarrow .8964 \pm 2.160 \frac{.3995}{\sqrt{14}} \Rightarrow .8964 \pm .2306$$

or $.666 < \mu < 1.127$. Intervals constructed using this procedure will enclose μ 95% of the time in repeated sampling. Hence, we are fairly certain that this particular interval encloses μ.

b Calculate $\bar{x} = \frac{\sum x_i}{n} = \frac{4.9}{4} = 1.225$ and $s = \sqrt{\frac{\sum x_i^2 - \frac{(\sum x_i)^2}{n}}{n-1}} = \sqrt{\frac{6.0058 - \frac{(4.9)^2}{4}}{3}} = .0332$

and the 95% confidence interval is

$$\bar{x} \pm t_{.025} \frac{s}{\sqrt{n}} \Rightarrow 1.225 \pm 3.182 \frac{.0332}{\sqrt{4}} \Rightarrow 1.225 \pm .0528 \quad \text{or} \quad 1.172 < \mu < 1.278.$$

The interval is narrower than the interval in part **a**, even though the sample size is smaller, because the data is so much less variable.

c For white tuna in water,

$$\bar{x} = \frac{\sum x_i}{n} = \frac{10.24}{8} = 1.28, \quad s = \sqrt{\frac{\sum x_i^2 - \frac{(\sum x_i)^2}{n}}{n-1}} = \sqrt{\frac{13.235 - \frac{(10.24)^2}{8}}{7}} = .1351$$

and the 95% confidence interval is

$$\bar{x} \pm t_{.025} \frac{s}{\sqrt{n}} \Rightarrow 1.28 \pm 2.365 \frac{.1351}{\sqrt{8}} \Rightarrow 1.28 \pm .113 \quad \text{or} \quad 1.167 < \mu < 1.393.$$

For light tuna in oil,

$$\bar{x} = \frac{\Sigma x_i}{n} = \frac{12.62}{11} = 1.147 \text{ and } s = \sqrt{\frac{\Sigma x_i^2 - \frac{(\Sigma x_i)^2}{n}}{n-1}} = \sqrt{\frac{19.0828 - \frac{(12.62)^2}{11}}{10}} = .6785$$

and the 95% confidence interval is

$$\bar{x} \pm t_{.025} \frac{s}{\sqrt{n}} \Rightarrow 1.147 \pm 2.228 \frac{.6785}{\sqrt{11}} \Rightarrow 1.147 \pm .456 \quad \text{or} \quad .691 < \mu < 1.603.$$

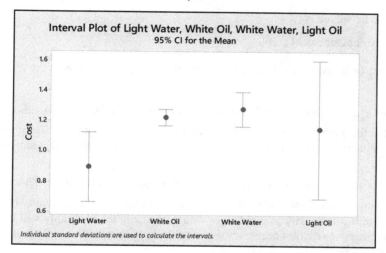

The plot of the four treatment means shows substantial differences in variability. The cost of light tuna in water appears to be the lowest, and quite different from either of the white tuna varieties.

10.2.25 Calculate $\bar{x} = \frac{\Sigma x_i}{n} = \frac{608}{10} = 60.8$

$$s^2 = \frac{\Sigma x_i^2 - \frac{(\Sigma x_i)^2}{n}}{n-1} = \frac{37538 - \frac{(608)^2}{10}}{9} = 63.5111 \text{ and } s = 7.9694$$

The 95% confidence interval based on $df = 9$ is

$$\bar{x} \pm t_{.025} \frac{s}{\sqrt{n}} \Rightarrow 60.8 \pm 2.262 \frac{7.9694}{\sqrt{10}} \Rightarrow 60.8 \pm 5.701 \text{ or } 55.099 < \mu < 66.501.$$

10.2.27 a Notice that the stem and leaf plot has one peaks and two gaps. For a small sample of size $n = 15$, we could imagine a few more points in the "gaps", making the stem and leaf plot look relatively mound-shaped. Given the small number of measurements, there is no reason to conclude that there is a problem with the normality assumption.

b Calculate $\bar{x} = \frac{\Sigma x_i}{n} = \frac{360}{15} = 24$

$$s^2 = \frac{\Sigma x_i^2 - \frac{(\Sigma x_i)^2}{n}}{n-1} = \frac{9522 - \frac{(360)^2}{15}}{14} = 63.00 \text{ and } s = 7.937$$

c The 95% confidence interval with $df = 14$ is

$$\bar{x} \pm t_{.025} \frac{s}{\sqrt{n}} \Rightarrow 24 \pm 2.145 \frac{7.937}{\sqrt{15}} \Rightarrow 24 \pm 4.396$$

or $19.604 < \mu < 28.396$.

10.2.29 The problem of selecting a proper sample size to achieve a given bound is now complicated by the fact that the t value, which is used in calculating the correct sample size, changes as the value of n changes. In Chapter 8, the procedure was to choose n so that the half-width of the $100(1-\alpha)\%$ confidence interval was less than some given bound, B. That is,

$$z_{\alpha/2} \times (\text{std error of estimator}) \leq B$$

Now, the inequality to be solved is

$$t_{\alpha/2} \times (\text{std error of estimator}) \leq B$$

and the $t_{\alpha/2}$ value must be based on $n-1$ degrees of freedom. Since n is unknown, the procedure is as follows:

1 Calculate n using $z_{\alpha/2}$ instead of $t_{\alpha/2}$. If the value for n is large (that is, $n \geq 30$), this sample size will be valid. If the value for n is between 30 and 40, you may choose to adjust the value of n slightly by replacing the z-value with the appropriate t-value and picking larger values of n until the inequality will hold exactly.

2 If the value for n is smaller than 30, we are not justified in using the value $z_{\alpha/2}$. This value must be replaced by the appropriate t-value with $n-1$ degrees of freedom. If the inequality holds, the sample size is valid; if not, it is necessary to pick larger values of n until the inequality will hold. This repetitive procedure is usually not necessary, because the $z_{\alpha/2}$ value will usually yield a satisfactory approximation to the required sample size.

In this exercise, we want to estimate μ to within .06 with probability .95. Hence, the following inequality must hold:

$$t_{.025} \frac{s}{\sqrt{n}} \leq .06$$

Consider the sample size obtained by replacing $t_{.025}$ with $z_{.025}$.

$$1.96 \frac{.1812}{\sqrt{n}} \leq .06 \Rightarrow \sqrt{n} \geq 5.9192 \Rightarrow n \geq 35.04 \text{ or } n = 36$$

Since this sample size is greater than 30, the sample size $n = 36$ is valid. Replacing $z = 1.96$ with $t_{.025} = 2.030$ with $df = n - 1 = 35$, you can calculate

$$2.030 \frac{.1812}{\sqrt{36}} = .0613$$

which is slightly larger than the necessary bound. To actually achieve the necessary bound, you need $t_{.025} = 2.026$ with $df = n - 1 = 37$, so that $n = 38$ and

$$2.026 \frac{.1812}{\sqrt{38}} = .0596$$

The difference in these two sample sizes is minimal.

10.2.31 Calculate $\bar{x} = \dfrac{\sum x_i}{n} = \dfrac{78.6}{15} = 5.24$

$$s^2 = \dfrac{\sum x_i^2 - \dfrac{(\sum x_i)^2}{n}}{n-1} = \dfrac{412.22 - \dfrac{(78.6)^2}{15}}{14} = .02543 \quad \text{and} \quad s = .15946$$

The 95% confidence interval based on $df = 14$ is

$$\bar{x} \pm t_{.025} \dfrac{s}{\sqrt{n}} \Rightarrow 5.24 \pm 2.145 \dfrac{.15946}{\sqrt{15}} \Rightarrow 5.24 \pm .088 \text{ or } 5.152 < \mu < 5.328.$$

10.2.33 a Answers will vary. A typical histogram generated by **Minitab** shows that the data are approximately mound-shaped.

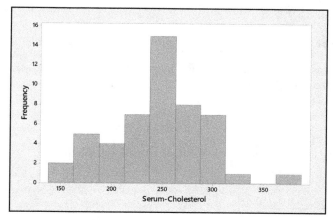

b Calculate $\bar{x} = \dfrac{\sum x_i}{n} = \dfrac{12348}{50} = 246.96$

$$s^2 = \dfrac{\sum x_i^2 - \dfrac{(\sum x_i)^2}{n}}{n-1} = \dfrac{3{,}156{,}896 - \dfrac{(12348)^2}{50}}{49} = 2192.52898 \quad \text{and} \quad s = 46.8244$$

Table 4 does not give a value of t with area .025 to its right. If we are conservative, and use the value of t with $df = 45$, the value of t will be $t_{.025} = 2.014$, and the approximate 95% confidence interval is

$$\bar{x} \pm t_{.025} \dfrac{s}{\sqrt{n}} \Rightarrow 246.96 \pm 2.014 \dfrac{46.8244}{\sqrt{50}} \Rightarrow 246.96 \pm 13.337$$

or $233.623 < \mu < 260.297$.

10.2.35 Since it is necessary to determine whether the injected rats drink more water than noninjected rates, the hypothesis to be tested is $H_0: \mu = 22.0 \quad H_a: \mu > 22.0$ and the test statistic is

$$t = \dfrac{\bar{x} - \mu_0}{s/\sqrt{n}} = \dfrac{31.0 - 22.0}{\dfrac{6.2}{\sqrt{17}}} = 5.985.$$

Using the *critical value approach*, the rejection region with $\alpha = .05$ and $n - 1 = 16$ degrees of freedom is located in the upper tail of the t-distribution and is found from Table 4 as $t > t_{.05} = 1.746$. Since the observed value of the test statistic falls in the rejection region, H_0 is rejected and we conclude that the injected rats do drink more water than the noninjected rats. The 90% confidence interval is

$$\bar{x} \pm t_{.05} \frac{s}{\sqrt{n}} \Rightarrow 31.0 \pm 1.746 \frac{6.2}{\sqrt{17}} \Rightarrow 31.0 \pm 2.625 \quad \text{or} \quad 28.375 < \mu < 33.625.$$

Section 10.3

10.3.1 As in the case of the single population mean, random samples must be independently drawn from two populations which possess normal distributions with a common variance, σ^2. Consequently, it is logical that information in the two sample variances, s_1^2 and s_2^2, should be pooled in order to give the best estimate of the common variance, σ^2. In this way, all of the sample information is being utilized to its best advantage.

10.3.3 The degrees of freedom for s^2, the pooled estimator of σ^2 are $n_1 + n_2 - 2 = 10 + 12 - 2 = 20$.

10.3.5 $s^2 = \dfrac{(n_1-1)s_1^2 + (n_2-1)s_2^2}{n_1 + n_2 - 2} = \dfrac{9(3.4) + 3(4.9)}{10 + 4 - 2} = 3.775$ with 12 degrees of freedom.

10.3.7 As in Chapter 9, the hypothesis to be tested is $H_0: \mu_1 - \mu_2 = 0$ versus $H_a: \mu_1 - \mu_2 \neq 0$

When the actual data are given and s^2 must be calculated, the calculation is done by using your scientific calculator to first obtain s_1 and s_2 and substituting into the formula for s^2. Notice that

$$(n-1)s^2 = \Sigma(x_i - \bar{x})^2 = \Sigma x_i^2 - \frac{(\Sigma x_i)^2}{n}$$

Hence, for the pooled estimator, we can eliminate some rounding error by calculating

$$s^2 = \frac{\Sigma x_1^2 - \dfrac{(\Sigma x_1)^2}{n_1} + \Sigma x_2^2 - \dfrac{(\Sigma x_2)^2}{n_2}}{n_1 + n_2 - 2}$$

The preliminary calculations are shown next:

Sample 1	Sample 2
$\Sigma x = 28$	$\Sigma x = 43$
$\Sigma x^2 = 242$	$\Sigma x^2 = 411$
$n_1 = 4$	$n_2 = 5$

Then

$$s^2 = \frac{242 - \dfrac{(28)^2}{4} + 411 - \dfrac{(43)^2}{5}}{4 + 5 - 2} = \frac{46 + 41.2}{7} = 12.45714286$$

The test statistic, under the assumption that $\sigma_1^2 = \sigma_2^2 = \sigma^2$ is calculated using the pooled value of s^2 in the t-statistic shown next:

$$t = \frac{(\bar{x}_1 - \bar{x}_2) - 0}{\sqrt{s^2\left(\dfrac{1}{n_1} + \dfrac{1}{n_2}\right)}} = \frac{7 - 8.6}{\sqrt{12.4571\left(\dfrac{1}{4} + \dfrac{1}{5}\right)}} = -.676$$

The rejection region is two-tailed, based on $df = n_1 + n_2 - 2 = 7$ degrees of freedom. With $\alpha = .05$ from Table 4, the rejection region is $|t| > t_{.025} = 2.365$.

For a two-tailed test with $df = 7$, $p\text{-value} = P(t > .676) + P(t < -.676) = 2P(t > .676)$, so that $P(t > .676) = \frac{1}{2}p\text{-value}$. From Table 4 with $df = 7$, $t = .676$ is less than the smallest value, $t_{.10} = 1.415$, so that $\frac{1}{2}p\text{-value} > .10$ and $p\text{-value} > .20$. Since the observed value, $t = -.676$ does not fall in the rejection region, and since the p-value is greater than .05, H_0 is not rejected. We do not have sufficient evidence to indicate that $\mu_1 - \mu_2 \neq 0$.

10.3.9 If you check the ratio of the two variances using the rule of thumb given in this section you will find:

$$\frac{\text{larger } s^2}{\text{smaller } s^2} = \frac{(7.1)^2}{(6.08)^2} = 1.36$$

which is less than three. Therefore, it is reasonable to assume that the two population variances are equal. From the **Minitab** printout, the test statistic is $t = -2.937$ with $df = 16$ and $p\text{-value} = .0097$. Since the $p\text{-value} = .0097$ is less than .05, the results are significant. There is sufficient evidence to indicate a difference in the two population means.

10.3.11 a The hypothesis to be tested is: $H_0 : \mu_1 - \mu_2 = 0$ versus $H_a : \mu_1 - \mu_2 \neq 0$.

b The preliminary calculations are shown next:

Method 1	Method 2
$\Sigma x = 685$	$\Sigma x = 736$
$\Sigma x^2 = 94,259$	$\Sigma x^2 = 108,556$
$n_1 = 5$	$n_2 = 5$

Then

$$s^2 = \frac{94259 - \frac{(685)^2}{5} + 108556 - \frac{(736)^2}{5}}{5+5-2} = \frac{414 + 216.8}{8} = 78.85$$

and the test statistic is

$$t = \frac{(\bar{x}_1 - \bar{x}_2) - 0}{\sqrt{s^2\left(\frac{1}{n_1} + \frac{1}{n_2}\right)}} = \frac{(685/5) - (736/5)}{\sqrt{78.85\left(\frac{1}{5} + \frac{1}{5}\right)}} = -1.816$$

The rejection region is two-tailed, based on $df = n_1 + n_2 - 2 = 8$ degrees of freedom. With $\alpha = .01$, from Table 4, the rejection region is $|t| > t_{.005} = 3.355$ and H_0 is not rejected. There is insufficient evidence to indicate a difference in the two population means.

c The p-value is $p\text{-value} = P(|t| > 1.816)$, so that $P(t > 1.816) = \frac{1}{2}p\text{-value}$. From Table 4 with $df = 8$, $t = 1.816$ is between $t_{.10}$ and $t_{.05}$. Therefore, the area to the right of $t = 1.816$ must be between .05 and .10 so that $.05 < \frac{1}{2}p\text{-value} < .10$ and $.10 < p\text{-value} < .20$. H_0 is not rejected and the results in part b are confirmed.

10.3.13 a Calculate

$$\bar{x}_1 = \frac{41.6}{7} = 5.943 \text{ and } \bar{x}_2 = \frac{112.9}{7} = 16.129$$

$$s_1^2 = \frac{255.96 - \frac{(41.6)^2}{7}}{6} = 1.45619 \quad \text{and} \quad s_2^2 = \frac{1832.11 - \frac{(112.9)^2}{7}}{6} = 1.86571$$

Since the ratio of the variances is less than 3, you can use the pooled estimator of σ^2 calculated as

$$s^2 = \frac{(n_1-1)s_1^2 + (n_2-1)s_2^2}{n_1+n_2-2} = \frac{6(1.45619) + 6(1.86571)}{12} = 1.66095$$

A 90% confidence interval for $(\mu_1 - \mu_2)$ is given as

$$(\bar{x}_1 - \bar{x}_2) \pm t_{.05}\sqrt{s^2\left(\frac{1}{n_1} + \frac{1}{n_2}\right)}$$

$$(5.943 - 16.129) \pm 1.782\sqrt{1.66095\left(\frac{1}{7} + \frac{1}{7}\right)}$$

$$-10.186 \pm 1.228 \quad \text{or} \quad -11.414 < (\mu_1 - \mu_2) < -8.958$$

b Since the value $\mu_1 - \mu_2 = 0$ is not in the interval, it is unlikely that $\mu_1 = \mu_2$. Therefore, we conclude that there is evidence of a difference in the average amount of oil required to produce these two crops.

10.3.15 a The hypothesis to be tested is $H_0 : \mu_1 - \mu_2 = 0$ versus $H_a : \mu_1 - \mu_2 \neq 0$. From the **Minitab** printout, the following information is available:

$\bar{x}_1 = .896 \qquad s_1^2 = (.400)^2 \qquad n_1 = 14$

$\bar{x}_2 = 1.147 \qquad s_2^2 = (.679)^2 \qquad n_2 = 11$

and the test statistic is

$$t = \frac{(\bar{x}_1 - \bar{x}_2) - 0}{\sqrt{s^2\left(\frac{1}{n_1} + \frac{1}{n_2}\right)}} = -1.16$$

The rejection region is two-tailed, based on $n_1 + n_2 - 2 = 23$ degrees of freedom. With $\alpha = .05$, from Table 4, the rejection region is $|t| > t_{.025} = 2.069$ and H_0 is not rejected. There is not enough evidence to indicate a difference in the population means.

b It is not necessary to bound the *p*-value using Table 4, since the exact *p*-value is given on the printout as P-Value = .260.

c If you check the ratio of the two variances using the rule of thumb given in this section you will find:

$$\frac{\text{larger } s^2}{\text{smaller } s^2} = \frac{(.679)^2}{(.400)^2} = 2.88$$

which is less than three. Therefore, it is reasonable to assume that the two population variances are equal.

10.3.17 a Check the ratio of the two variances using the rule of thumb given in this section:

$$\frac{\text{larger } s^2}{\text{smaller } s^2} = \frac{2.78095}{.17143} = 16.22$$

which is greater than three. Therefore, it is not reasonable to assume that the two population variances are equal.

b You should use the unpooled t test with Satterthwaite's approximation to the degrees of freedom for testing $H_0: \mu_1 - \mu_2 = 0$ versus $H_a: \mu_1 - \mu_2 \neq 0$. The test statistic is

$$t = \frac{(\bar{x}_1 - \bar{x}_2) - 0}{\sqrt{\frac{s_1^2}{n_1} + \frac{s_2^2}{n_2}}} = \frac{3.73 - 4.8}{\sqrt{\frac{2.78095}{15} + \frac{.17143}{15}}} = -2.412$$

with

$$df = \frac{\left(\frac{s_1^2}{n_1} + \frac{s_2^2}{n_2}\right)^2}{\frac{\left(\frac{s_1^2}{n_1}\right)^2}{n_1 - 1} + \frac{\left(\frac{s_2^2}{n_2}\right)^2}{n_2 - 1}} = \frac{(.185397 + .0114287)^2}{.002455137 + .00000933} = 15.7$$

With $df \approx 15$, the p-value for this test is bounded between .02 and .05 so that H_0 can be rejected at the 5% level of significance. There is evidence of a difference in the mean number of uncontaminated eggplants for the two disinfectants.

10.3.19 a The **Minitab** stem and leaf plots are shown next. Notice the mounded shapes which justify the assumption of normality.

```
                                        Stem-and-leaf of Sunmaid   N = 14
                                        1   22   0
                                        1   23
                                        5   24   0000
Stem-and-leaf of Generic   N = 14       7   25   00
                                        7   26
1   24   0                              7   27   0
4   25   000                            6   28   0000
(5) 26   00000                          2   29   0
5   27   00                             1   30   0
3   28   000

Leaf Unit = 0.1                         Leaf Unit = 0.1
```

b Use your scientific calculator or the computing formulas to find:

$\bar{x}_1 = 26.214 \quad s_1^2 = 1.565934 \quad s_1 = 1.251$

$\bar{x}_2 = 26.143 \quad s_2^2 = 5.824176 \quad s_2 = 2.413$

Since the ratio of the variances is greater than 3, you must use the unpooled t test with Satterthwaite's approximate df.

$$df = \frac{\left(\frac{s_1^2}{n_1} + \frac{s_2^2}{n_2}\right)^2}{\frac{\left(\frac{s_1^2}{n_1}\right)^2}{n_1 - 1} + \frac{\left(\frac{s_2^2}{n_2}\right)^2}{n_2 - 1}} \approx 19$$

c For testing $H_0: \mu_1 - \mu_2 = 0$ versus $H_a: \mu_1 - \mu_2 \neq 0$, the test statistic is

$$t = \frac{(\bar{x}_1 - \bar{x}_2) - 0}{\sqrt{\frac{s_1^2}{n_1} + \frac{s_2^2}{n_2}}} = \frac{26.214 - 26.143}{\sqrt{\frac{1.565934}{14} + \frac{5.824176}{14}}} = .10$$

For a two-tailed test with $df = 19$, the p-value can be bounded using Table 4 so that

$$\frac{1}{2}\text{p-value} > .10 \quad \text{or} \quad \text{p-value} > .20$$

Since the p-value is greater than .10, $H_0 : \mu_1 - \mu_2 = 0$ is not rejected. There is insufficient evidence to indicate that there is a difference in the mean number of raisins per box.

10.3.21 a If swimmer 2 is faster, his(her) average time should be less than the average time for swimmer 1. Therefore, the hypothesis of interest is $H_0 : \mu_1 - \mu_2 = 0$ versus $H_a : \mu_1 - \mu_2 > 0$ and the preliminary calculations are as follows:

Swimmer 1	Swimmer 2
$\Sigma x = 596.46$	$\Sigma x = 596.27$
$\Sigma x^2 = 35576.6976$	$\Sigma x^2 = 35554.1093$
$n_1 = 10$	$n_2 = 10$

Then

$$s^2 = \frac{(n_1-1)s_1^2 + (n_2-1)s_2^2}{n_1+n_2-2}$$

$$= \frac{35576.6976 - \frac{(596.46)^2}{10} + 35554.1093 - \frac{(596.27)^2}{10}}{5+5-2} = .03124722$$

Also, $\bar{x}_1 = \frac{596.46}{10} = 59.646$ and $\bar{x}_2 = \frac{596.27}{10} = 59.627$

The test statistic is

$$t = \frac{(\bar{x}_1 - \bar{x}_2) - 0}{\sqrt{s^2\left(\frac{1}{n_1} + \frac{1}{n_2}\right)}} = \frac{59.646 - 59.627}{\sqrt{.03124722\left(\frac{1}{10} + \frac{1}{10}\right)}} = 0.24$$

For a one-tailed test with $df = n_1 + n_2 - 2 = 18$, the p-value can be bounded using Table 4 so that

p-value $> .10$, and H_0 is not rejected. There is insufficient evidence to indicate that swimmer 2's average time is still faster than the average time for swimmer 1.

10.3.23 a If you check the ratio of the two variances using the rule of thumb given in this section you will find:

$$\frac{\text{larger } s^2}{\text{smaller } s^2} = \frac{7.589^2}{4.464^2} = 2.89$$

which is less than three. Therefore, it is reasonable to assume that the two population variances are equal.

b The hypothesis of interest is $H_0 : \mu_1 - \mu_2 = 0$ versus $H_a : \mu_1 - \mu_2 \neq 0$. From the **TI 84 Plus** printout, the test statistic is $t = 0.325$. The rejection region based on 29 df and $\alpha = .05$ is $|t| > t_{.025} = 2.045$ and the null hypothesis is not rejected; there is insufficient evidence to indicate that the average number of completed passes is different for Alex Smith and Joe Flacco.

c From the **TI 84 Plus** printout, the two-tailed p-value $= .747$. Since the p-value is greater than $\alpha = .05$, the null hypothesis is not rejected as in part **b**.

d The value of $t_{.025}$ with $df = 29$ is 2.045 and the 95% confidence interval for $(\mu_1 - \mu_2)$ is given as

$$(\bar{x}_1 - \bar{x}_2) \pm t_{.025}\sqrt{s^2\left(\frac{1}{n_1} + \frac{1}{n_2}\right)}$$

$$(22.733 - 22) \pm 2.045\sqrt{39.4115\left(\frac{1}{15} + \frac{1}{16}\right)}$$

$$.733 \pm 4.614 \text{ or } -3.881 < (\mu_1 - \mu_2) < 5.347$$

Since the value $\mu_1 - \mu_2 = 0$ is in the interval, it is possible that the two means are equal. We have insufficient evidence to indicate that there is a difference in the means, confirming the results of part **b**.

10.3.25 a The hypothesis to be tested is

$$H_0: \mu_1 - \mu_2 = 0 \qquad H_a: \mu_1 - \mu_2 \neq 0.$$

The following information is available:

$$\bar{x}_1 = 27.2 \qquad s_1^2 = 16.36 \quad \text{and} \quad n_1 = 10$$

$$\bar{x}_2 = 33.5 \qquad s_2^2 = 18.92 \quad \text{and} \quad n_2 = 10$$

Since the ratio of the variances is less than 3, you can use the pooled *t* test. The pooled estimator of σ^2 is calculated as

$$s^2 = \frac{(n_1 - 1)s_1^2 + (n_2 - 1)s_2^2}{n_1 + n_2 - 2} = \frac{9(16.36) + 9(18.92)}{10 + 10 - 2} = 17.64$$

and the test statistic is

$$t = \frac{(\bar{x}_1 - \bar{x}_2) - 0}{\sqrt{s^2\left(\frac{1}{n_1} + \frac{1}{n_2}\right)}} = \frac{27.2 - 33.5}{\sqrt{17.64\left(\frac{1}{10} + \frac{1}{10}\right)}} = -3.354$$

Critical value approach: The rejection region is two-tailed, based on $df = 18$ degrees of freedom. With $\alpha = .05$, from Table 4, the rejection region is $|t| > t_{.025} = 2.101$ and H_0 is rejected. There is sufficient evidence to indicate a difference in the mean absorption rates between the two drugs.

b *p*-value approach: The *p*-value is $2P(t > 3.354)$ for a two-tailed test with 18 degrees of freedom. Since $t = 3.354$ exceeds the largest tabled value, $t_{.005} = 2.878$, we have

$$p\text{-value} < 2(.005) = .01$$

Since the *p*-value is less than $\alpha = .05$, H_0 can be rejected at the 5% level of significance, confirming the results of part **a**.

c A 95% confidence interval for $(\mu_1 - \mu_2)$ is given as

$$(\bar{x}_1 - \bar{x}_2) \pm t_{.025}\sqrt{s^2\left(\frac{1}{n_1} + \frac{1}{n_2}\right)}$$

$$(27.2 - 33.5) \pm 2.101\sqrt{17.64\left(\frac{1}{10} + \frac{1}{10}\right)}$$

$$-6.3 \pm 3.946 \text{ or } -10.246 < (\mu_1 - \mu_2) < -2.354$$

Since the confidence interval does not contain the value $\mu_1 - \mu_2 = 0$, you can conclude that there is a difference in two population means. This confirms the conclusion in part **a**.

10.3.27 a The analysis is identical to that used in previous exercises. To test $H_0: \mu_1 - \mu_2 = 0$ versus $H_a: \mu_1 - \mu_2 \neq 0$, the test statistic is

$$t = \frac{(\bar{x}_1 - \bar{x}_2) - 0}{\sqrt{s^2 \left(\frac{1}{n_1} + \frac{1}{n_2}\right)}} = 9.5641$$

with p-value = .0000. Since this p-value is very small, H_0 is rejected. There is evidence of a difference in the means.

b Since there is a difference in the mean strengths for the two kinds of material, the strongest material (A) should be used (all other factors being equal).

Section 10.4

10.4.1 Paired observations are used to estimate the difference between two population means in preference to an estimation based on independent random samples selected from the two populations because of the increased information caused by blocking the observations. We expect blocking to create a large reduction in the standard deviation, if differences do exist among the blocks.

Paired observations are not always preferable. The degrees of freedom that are available for estimating σ^2 are less for paired than for unpaired observations. If there were no difference between the blocks, the paired experiment would then be less beneficial.

10.4.3 The number of degrees of freedom in a paired difference test is $n - 1 = 12 - 1 = 11$, where n is the number of pairs.

10.4.5 The table of differences, along with the calculation of $\bar{d}$, s_d^2 and s_d is presented next.

d_i	2	−1	−2	5	$\Sigma d_i = 4$
d_i^2	4	1	4	25	$\Sigma d_i^2 = 34$

$$\bar{d} = \frac{\Sigma d_i}{n} = \frac{4}{4} = 1, \quad s_d^2 = \frac{\Sigma d_i^2 - \frac{(\Sigma d_i)^2}{n}}{n-1} = \frac{34 - \frac{(4)^2}{4}}{3} = 10 \text{ and } s_d = \sqrt{10} = 3.162$$

10.4.7 The test statistic is

$$t = \frac{\bar{d} - \mu_d}{s_d / \sqrt{n}} = \frac{.3 - 0}{\sqrt{\frac{.16}{10}}} = 2.372 \text{ with } n-1 = 9 \text{ degrees of freedom.}$$

The p-value is then $P(|t| > 2.372) = 2P(t > 2.372)$ so that $P(t > 2.372) = \frac{1}{2}$ p-value. Since the value $t = 2.372$ falls between two tabled entries for $df = 9$ ($t_{.025} = 2.262$ and $t_{.01} = 2.821$), you can conclude that

$$.01 < \frac{1}{2} p\text{-value} < .025$$
$$.02 < p\text{-value} < .05$$

Since the p-value is less than $\alpha = .05$, the null hypothesis is rejected, and we conclude that there is a difference in the two population means.

10.4.9 It is necessary to use a paired-difference test, since the two samples are not random and independent. The hypothesis of interest is

$$H_0: \mu_1 - \mu_2 = 0 \quad \text{or} \quad H_0: \mu_d = 0$$
$$H_a: \mu_1 - \mu_2 \neq 0 \quad \text{or} \quad H_a: \mu_d \neq 0$$

The table of differences, along with the calculation of $\bar{d}$ and s_d^2, is presented next.

d_i	.1	.1	0	.2	−.1	$\Sigma d_i = .3$
d_i^2	.01	.01	.00	.04	.01	$\Sigma d_i^2 = .07$

$$\bar{d} = \frac{\Sigma d_i}{n} = \frac{.3}{5} = .06 \quad \text{and} \quad s_d^2 = \frac{\Sigma d_i^2 - \frac{(\Sigma d_i)^2}{n}}{n-1} = \frac{.07 - \frac{(.3)^2}{5}}{4} = .013$$

The test statistic is

$$t = \frac{\bar{d} - \mu_d}{s_d/\sqrt{n}} = \frac{.06 - 0}{\sqrt{\frac{.013}{5}}} = 1.177$$

with $n - 1 = 4$ degrees of freedom. The rejection region with $\alpha = .05$ is $|t| > t_{.025} = 2.776$, and H_0 is not rejected. We cannot conclude that the means are different.

A 95% confidence interval for $\mu_1 - \mu_2 = \mu_d$ is

$$\bar{d} \pm t_{.025} \frac{s_d}{\sqrt{n}} \Rightarrow .06 \pm 2.776\sqrt{\frac{.013}{5}} \Rightarrow .06 \pm .142$$

or $-.082 < (\mu_1 - \mu_2) < .202$.

since the value $\mu_1 - \mu_2 = 0$ lies in the interval, it is possible that the two means are equal. We cannot conclude that the means are different, confirming the results of the hypothesis test.

10.4.11 A paired-difference test is used, since the two samples are not random and independent. The hypothesis of interest is $H_0: \mu_1 - \mu_2 = 0 \quad H_a: \mu_1 - \mu_2 > 0$ and the table of differences, along with the calculation of $\bar{d}$ and s_d^2, is presented next.

Pair	1	2	3	4	Totals
d_i	−1	5	11	7	22

$$\bar{d} = \frac{\Sigma d_i}{n} = \frac{22}{4} = 5.5 \quad s_d^2 = \frac{\Sigma d_i^2 - \frac{(\Sigma d_i)^2}{n}}{n-1} = \frac{196 - \frac{(22)^2}{4}}{3} = 25 \quad \text{and} \quad s_d = 5$$

and the test statistic is

$$t = \frac{\bar{d} - \mu_d}{s_d/\sqrt{n}} = \frac{5.5 - 0}{\frac{5}{\sqrt{4}}} = 2.2$$

The one-tailed p-value with $df = 3$ can be bounded between .05 and .10. Since this value is greater than .05, H_0 is not rejected. The results are not significant; there is insufficient evidence to indicate that lack of preschool experience has a depressing effect on IQ scores.

10.4.13 a A paired-difference test is used, since the two samples are not independent (for any given city, Allstate and 21st Century premiums will be related).

b The hypothesis of interest is

$$H_0: \mu_1 - \mu_2 = 0 \quad \text{or} \quad H_0: \mu_d = 0$$
$$H_a: \mu_1 - \mu_2 \neq 0 \quad \text{or} \quad H_a: \mu_d \neq 0$$

where μ_1 is the average for Allstate insurance and μ_2 is the average cost for 21st Century insurance. The table of differences, along with the calculation of $\bar{d}$ and s_d, is presented next.

City	1	2	3	4	Totals
d_i	291	464	283	192	1230
d_i^2	84,681	215,296	80,089	36,864	416,930

$$\bar{d} = \frac{\sum d_i}{n} = \frac{1230}{4} = 307.5 \quad \text{and}$$

$$s_d = \sqrt{\frac{\sum d_i^2 - \frac{(\sum d_i)^2}{n}}{n-1}} = \sqrt{\frac{416{,}930 - \frac{(1230)^2}{4}}{3}} = \sqrt{12901.66667} = 113.5855$$

The test statistic is

$$t = \frac{\bar{d} - \mu_d}{s_d/\sqrt{n}} = \frac{307.5 - 0}{\frac{113.5855}{\sqrt{4}}} = 5.414$$

with $n - 1 = 3$ degrees of freedom. The rejection region with $\alpha = .01$ is $|t| > t_{.005} = 5.841$, and H$_0$ is not rejected. There is insufficient evidence to indicate a difference in the average premiums for Allstate and 21st Century.

c p-value $= P(|t| > 5.414) = 2P(t > 5.414)$. Since $t = 5.414$ is between $t_{.005} = 5.841$ and $t_{.01} = 4.541$,

$$2(.005) < p\text{-value} < 2(.01) \Rightarrow .01 < p\text{-value} < .02.$$

d   A 99% confidence interval for $\mu_1 - \mu_2 = \mu_d$ is

$$\bar{d} \pm t_{.005} \frac{s_d}{\sqrt{n}} \Rightarrow 307.5 \pm 5.841 \frac{113.5855}{\sqrt{4}} \Rightarrow 307.5 \pm 331.726$$

or $-24.226 < (\mu_1 - \mu_2) < 639.226$.

e   The four cities in the study were not necessarily a random sample of cities from throughout the United States. Therefore, you cannot make valid comparisons between Allstate and 21st Century for the United States in general.

10.4.15 a   Each subject was presented with both signs in random order. If his reaction time in general is high, both responses will be high; if his reaction time in general is low, both responses will be low. The large variability from subject to subject will mask the variability due to the difference in sign types. The paired-difference design will eliminate the subject to subject variability.

b   The hypothesis of interest is

$$H_0: \mu_1 - \mu_2 = 0 \quad \text{or} \quad H_0: \mu_d = 0$$
$$H_a: \mu_1 - \mu_2 \neq 0 \quad \text{or} \quad H_a: \mu_d \neq 0$$

Using the *Excel* printout, we locate the test statistic:

$$t = \frac{\bar{d} - \mu_d}{s_d/\sqrt{n}} = 9.150$$

Since this is a two tailed test, the *p*-value is given in the printout as P(T<=t) two-tail = 7.45782E-06 or *p*-value = .00000746. These results are highly significant, and H_0 is rejected. We can conclude that the means are different.

10.4.17 a A paired-difference test is used, since the two samples are not random and independent (at any location, the ground and air temperatures are related). The hypothesis of interest is

$$H_0 : \mu_1 - \mu_2 = 0 \quad H_a : \mu_1 - \mu_2 \neq 0$$

The table of differences, along with the calculation of $\bar{d}$ and s_d^2, is presented next.

Location	1	2	3	4	5	Total
d_i	−.4	−2.7	−1.6	−1.7	−1.5	−7.9

$$\bar{d} = \frac{\sum d_i}{n} = \frac{-7.9}{5} = -1.58$$

$$s_d^2 = \frac{\sum d_i^2 - \frac{(\sum d_i)^2}{n}}{n-1} = \frac{15.15 - \frac{(-7.9)^2}{5}}{4} = .667 \quad \text{and} \quad s_d = .8167$$

and the test statistic is

$$t = \frac{\bar{d} - \mu_d}{s_d/\sqrt{n}} = \frac{-1.58 - 0}{\frac{.8167}{\sqrt{5}}} = -4.326$$

A rejection region with $\alpha = .05$ and $df = n - 1 = 4$ is $|t| > t_{.025} = 2.776$, and H_0 is rejected at the 5% level of significance. We conclude that the air-based temperature readings are biased.

b The 95% confidence interval for $\mu_1 - \mu_2 = \mu_d$ is

$$\bar{d} \pm t_{.025} \frac{s_d}{\sqrt{n}} \Rightarrow -1.58 \pm 2.776 \frac{.8167}{\sqrt{5}} \Rightarrow -1.58 \pm 1.014$$

or $-2.594 < (\mu_1 - \mu_2) < -.566$.

c The inequality to be solved is $t_{\alpha/2} SE \leq B$.

We need to estimate the difference in mean temperatures between ground-based and air-based sensors to within .2 degrees centigrade with 95% confidence. Since this is a paired experiment, the inequality becomes

$$t_{.025} \frac{s_d}{\sqrt{n}} \leq .2$$

With $s_d = .8167$ and *n* represents the number of pairs of observations, consider the sample size obtained by replacing $t_{.025}$ by $z_{.025} = 1.96$.

$$1.96 \frac{.8167}{\sqrt{n}} \leq .2$$

$$\sqrt{n} \geq 8.0019 \Rightarrow n = 64.03 \text{ or } n = 65$$

Since the value of *n* is greater than 30, the use of $z_{\alpha/2}$ for $t_{\alpha/2}$ is justified.

10.4.19 a Use the *Minitab* printout given in the text. The hypothesis of interest is

$$H_0: \mu_1 - \mu_2 = 0 \qquad H_a: \mu_1 - \mu_2 > 0$$

and the test statistic is

$$t = \frac{\bar{d} - \mu_d}{s_d/\sqrt{n}} = \frac{1.4875 - 0}{\frac{1.491}{\sqrt{8}}} = 2.82$$

The *p*-value shown in the printout is *p*-value = .013. Since the *p*-value is less than .05, H_0 is rejected at the 5% level of significance. We conclude that assessor 1 gives higher assessments than assessor 2.

b A 95% lower one-sided confidence bound for $\mu_1 - \mu_2 = \mu_d$ is

$$\bar{d} - t_{.05} \frac{s_d}{\sqrt{n}} \Rightarrow 1.4875 - 1.895 \frac{1.491}{\sqrt{8}} \Rightarrow 1.4875 - .9989 \qquad \text{or} (\mu_1 - \mu_2) > .489.$$

c In order to apply the paired-difference test, the 8 properties must be randomly and independently selected and the assessments must be normally distributed.

10.4.21 A paired-difference analysis must be used. The hypothesis of interest is

$$H_0: \mu_1 - \mu_2 = 0 \quad \text{or} \quad H_0: \mu_d = 0$$
$$H_a: \mu_1 - \mu_2 > 0 \quad \text{or} \quad H_a: \mu_d > 0$$

The table of differences is presented next. Use your scientific calculator to find $\bar{d}$ and s_d,

| d_i | 3 | 3 | −2 | 1 | −1 | 3 | −1 |

Calculate $\bar{d} = .857$, $s_d = 2.193$, and the test statistic is

$$t = \frac{\bar{d} - \mu_d}{s_d/\sqrt{n}} = \frac{.857 - 0}{\frac{2.193}{\sqrt{7}}} = 1.03$$

Since $t = 1.03$ with $df = n - 1 = 6$ is smaller than the smallest tabled value $t_{.10}$, *p*-value > .10 for this one-tailed test and H_0 is not rejected. We cannot conclude that the average time outside the office is less when music is piped in.

10.4.23 A paired-difference test is used. To test $H_0: \mu_1 - \mu_2 = 0$ versus $H_a: \mu_1 - \mu_2 > 0$, where μ_1 is the mean before the safety program and μ_2 is the mean after the program, calculate the differences:

7, 6, −1, 5, 6, 1

Then $\bar{d} = \frac{\Sigma d_i}{n} = \frac{24}{6} = 4 \qquad s_d^2 = \frac{148 - \frac{(24)^2}{6}}{5} = 10.4 \quad \text{and} \quad s_d = 3.2249$

and the test statistic is

$$t = \frac{\bar{d} - \mu_d}{s_d/\sqrt{n}} = \frac{4 - 0}{\frac{3.2249}{\sqrt{6}}} = 3.038$$

For a one-tailed test with $df = 5$, the rejection region with $\alpha = .01$ is $t > t_{.01} = 3.365$, and H_0 is not rejected. There is insufficient evidence to indicate that the safety program was effective in reducing lost-time accidents.

Section 10.5

10.5.1-5 Refer to Table 5, Appendix I, indexing df along the left or right margin and χ_a^2 across the top.

1. $\chi_{.05}^2 = 16.9190$ with 9 df
3. $\chi_{.025}^2 = 59.3417$ with 40 df
5. $\chi_{.90}^2 = 10.0852$ with 17 df

10.5.7 It is necessary to test $H_0 : \sigma^2 = 15$ versus $H_a : \sigma^2 > 15$. This will be done using s^2, the sample variance, which is a good estimator for σ^2. Refer to Section 10.5 of the text and notice that the quantity

$$\chi^2 = \frac{(n-1)s^2}{\sigma^2}$$

has a chi-square distribution in repeated sampling. This distribution is shown next.

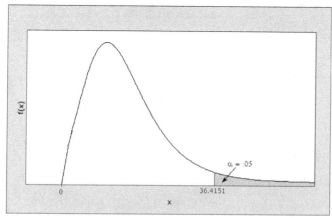

The distribution is nonsymmetrical and the random variable

$$\frac{(n-1)s^2}{\sigma^2}$$

takes on values starting at zero (since s^2, σ^2 and $(n-1)$ are never negative). The test statistic is

$$\chi^2 = \frac{(n-1)s^2}{\sigma_0^2} = \frac{24(21.4)}{15} = 34.24$$

A one-tailed test of hypothesis is required. Hence, a critical value of χ^2 (denoted by χ_C^2) must be found such that $P(\chi^2 > \chi_C^2) = .05$. Indexing $\chi_{.05}^2$ with $n-1 = 24$ degrees of freedom (see Table 5), the critical value is found to be $\chi_{.05}^2 = 36.4151$ (see the figure above). The value of the test statistic does not fall in the rejection region. Hence, H_0 is not rejected. We cannot conclude that the variance exceeds 15.

A 90% confidence interval for σ^2 will be

$$\frac{(n-1)s^2}{\chi_{\alpha/2}^2} < \sigma^2 < \frac{(n-1)s^2}{\chi_{(1-\alpha/2)}^2}$$

where $\chi_{\alpha/2}^2$ represents the value of χ^2 such that 2.5% of the area under the curve lies to its right. Similarly, $\chi_{(1-\alpha/2)}^2$ will be the χ^2 value such that an area .975 lies to its right. Hence, we have located one-half of α in each tail of the distribution. Indexing $\chi_{.025}^2$ and $\chi_{.975}^2$ with $n-1 = 24$ degrees of freedom in Table 5 yields

$$\chi^2_{.025} = 39.3641 \quad \text{and} \quad \chi^2_{.975} = 12.4011$$

and the 95% confidence interval is

$$\frac{24(21.4)}{39.3641} < \sigma^2 < \frac{24(21.4)}{12.4011} \quad \text{or} \quad 13.047 < \sigma^2 < 41.416$$

10.5.9 Calculate $\sum x_i = 17.7$, $\sum x_i^2 = 48.95$ and $n = 7$. Then

$$s^2 = \frac{\sum x_i^2 - \frac{(\sum x_i)^2}{n}}{n-1} = \frac{48.95 - \frac{(17.7)^2}{7}}{6} = .6990476$$

Indexing $\chi^2_{.025}$ and $\chi^2_{.975}$ with $n-1 = 6$ degrees of freedom in Table 5 yields $\chi^2_{.025} = 14.4494$ and $\chi^2_{.975} = 1.237347$ and the 95% confidence interval is

$$\frac{6(.6990476)}{14.4494} < \sigma^2 < \frac{6(.6990476)}{1.237347} \quad \text{or} \quad .290 < \sigma^2 < 3.390$$

It is necessary to test $H_0 : \sigma^2 = .8$ versus $H_a : \sigma^2 \neq .8$ and the test statistic is

$$\chi^2 = \frac{(n-1)s^2}{\sigma_0^2} = \frac{6(.6990476)}{.8} = 5.24$$

The two-tailed rejection region with $\alpha = .05$ and $n-1 = 6$ degrees of freedom is

$$\chi^2 > \chi^2_{.025} = 14.4494 \quad \text{or} \quad \chi^2 < \chi^2_{.975} = 1.237347$$

and H_0 is not rejected. There is insufficient evidence to indicate that σ^2 is different from .8.

10.5.11 a It is necessary to test $H_0 : \sigma^2 = 100$ versus $H_a : \sigma^2 > 100$ and the test statistic is

$$\chi^2 = \frac{(n-1)s^2}{\sigma_0^2} = \frac{24(16)^2}{100} = 61.44$$

The one-tailed rejection region with $\alpha = .05$ and $n-1 = 24$ degrees of freedom is

$$\chi^2 > \chi^2_{.05} = 36.4151$$

and H_0 is rejected. There is sufficient evidence to indicate that σ^2 is greater than 100.

b The p-value is $P(\chi^2 > 61.44)$ with 24 df. Since the value $\chi^2 = 61.44$ is greater than the largest value in Table 5 ($\chi^2_{.005} = 45.5585$) the p-value is less than .005 and the null hypothesis is rejected as in part a.

c Indexing $\chi^2_{.025}$ and $\chi^2_{.975}$ with $n-1 = 24$ degrees of freedom in Table 5 yields $\chi^2_{.025} = 39.3641$ and $\chi^2_{.975} = 12.4011$ and the 95% confidence interval is

$$\frac{24(16^2)}{39.3641} < \sigma^2 < \frac{24(16^2)}{12.4011} \quad \text{or} \quad 156.081 < \sigma^2 < 495.440$$

The value $\sigma^2 = 100$ is not in the interval, indicating that it is unlikely that $\sigma^2 = 100$. This validates the results of parts a and b.

10.5.13 a The hypothesis of interest is $H_0 : \sigma = .7$ versus $H_a : \sigma > .7$ or equivalently $H_0 : \sigma^2 = .49$ versus $H_a : \sigma^2 > .49$. Calculate

$$s^2 = \frac{\Sigma x_i^2 - \frac{(\Sigma x_i)^2}{n}}{n-1} = \frac{36 - \frac{(10)^2}{4}}{3} = 3.6667$$

The test statistic is

$$\chi^2 = \frac{(n-1)s^2}{\sigma_0^2} = \frac{3(3.6667)}{.49} = 22.449$$

The one-tailed rejection region with $\alpha = .05$ and $n-1 = 3$ degrees of freedom is $\chi^2 > \chi^2_{.05} = 7.81$ and H_0 is rejected. There is sufficient evidence to indicate that σ^2 is greater than .49.

b Indexing $\chi^2_{.05}$ and $\chi^2_{.95}$ with $n-1 = 3$ degrees of freedom in Table 5 yields $\chi^2_{.05} = 7.81473$ and $\chi^2_{.95} = .351846$ and the 90% confidence interval is

$$\frac{(n-1)s^2}{\chi^2_{.05}} < \sigma^2 < \frac{(n-1)s^2}{\chi^2_{.95}}$$

$$\frac{3(3.6667)}{7.81473} < \sigma^2 < \frac{3(3.6667)}{.351846} \qquad \text{or} \qquad 1.408 < \sigma^2 < 31.264$$

10.5.15 a The hypothesis to be tested is $H_0 : \mu = 5 \quad H_a : \mu \neq 5$.

Calculate $\bar{x} = \frac{\Sigma x_i}{n} = \frac{19.96}{4} = 4.99$, $s^2 = \frac{\Sigma x_i^2 - \frac{(\Sigma x_i)^2}{n}}{n-1} = \frac{99.6226 - \frac{(19.96)^2}{4}}{3} = .0074$

and the test statistic is

$$t = \frac{\bar{x} - \mu_0}{s/\sqrt{n}} = \frac{4.99 - 5}{\sqrt{\frac{.0074}{4}}} = -.232$$

The rejection region with $\alpha = .05$ and $n-1 = 3$ degrees of freedom is found from Table 4 as $|t| > t_{.025} = 3.182$. Since the observed value of the test statistic does not fall in the rejection region, H_0 is not rejected. There is insufficient evidence to show that the mean differs from 5 mg/cc.

b The manufacturer claims that the range of the potency measurements will equal .2. Since this range is given to equal 6σ, we know that $\sigma \approx .0333$. Then

$$H_0 : \sigma^2 = (.0333)^2 = .0011 \qquad H_a : \sigma^2 > .0011$$

The test statistic is

$$\chi^2 = \frac{(n-1)s^2}{\sigma_0^2} = \frac{3(.0074)}{.0011} = 20.18$$

and the one-tailed rejection region with $\alpha = .05$ and $n-1 = 3$ degrees of freedom is $\chi^2 > \chi^2_{.05} = 7.81$. H_0 is rejected; there is sufficient evidence to indicate that the range of the potency will exceed the manufacturer's claim.

10.5.17 a The force transmitted to a wearer, x, is known to be normally distributed with $\mu = 800$ and $\sigma = 40$. Hence,

$$P(x > 1000) = P\left(z > \frac{1000 - 8000}{40}\right) = P(z > 5) \approx 0$$

It is highly improbable that any particular helmet will transmit a force in excess of 1000 pounds.

b Since $n = 40$, a large sample test will be used to test

$$H_0 : \mu = 800 \qquad H_a : \mu > 800$$

The test statistic is

$$t = \frac{\bar{x} - \mu_0}{s/\sqrt{n}} = \frac{825 - 800}{\sqrt{\frac{2350}{40}}} = 3.262$$

and the rejection region with $\alpha = .05$ is $z > 1.645$. H_0 is rejected and we conclude that $\mu > 800$.

10.5.19 The hypothesis of interest is $H_0 : \sigma = 150$ versus $H_a : \sigma < 150$. Calculate

$$(n-1)s^2 = \Sigma x_i^2 - \frac{(\Sigma x_i)^2}{n} = 92,305,600 - \frac{(42,812)^2}{20} = 662,232.8$$

and the test statistic is $\chi^2 = \frac{(n-1)s^2}{\sigma_0^2} = \frac{662,232.8}{150^2} = 29.433$. The one-tailed rejection region with $\alpha = .01$

and $n - 1 = 19$ degrees of freedom is $\chi^2 < \chi^2_{.99} = 7.63273$, and H_0 is not rejected. There is insufficient evidence to indicate that he is meeting his goal.

Section 10.6

10.6.1 In order to use the F statistic to test the hypothesis concerning the equality of two population variances, we must assume that independent random samples have been drawn from two normal populations.

10.6.3-7 Refer to Table 6, Appendix I, indexing df_1 across the top and df_2 along the left or right margin. The value of a is in the column next to df_2 and the tabled value of F_a is in the interior of the the table.

3. $F_{.01} = 15.21$ with $df_1 = 6$ and $df_2 = 4$ **5.** $F_{.005} = 3.68$ with $df_1 = 15$ and $df_2 = 18$

7. $F_{.025} = 4.03$ with $df_1 = 9$ and $df_2 = 9$

10.6.9-11 The student will need to find critical values of F for various levels of α in order to find the approximate p-value.

9. The critical values with $df_1 = 4$ and $df_2 = 13$ from Table 6 follow.

a	.10	.05	.025	.01	.005
F_a	2.43	3.18	4.00	5.21	6.23

Then p-value $= P(F > 6.16)$ and $.005 < p\text{-value} < .01$.

11. The critical values with $df_1 = 8$ and $df_2 = 16$ from Table 6 follow.

a	.10	.05	.025	.01	.005
F_a	2.09	2.59	3.12	3.89	4.52

Then p-value $= P(F > 2.85)$ and $.025 < p\text{-value} < .05$.

10.6.13 When the assumptions for the F distribution are met, then s_1^2/s_2^2 possesses an F distribution with $df_1 = n_1 - 1$ and $df_2 = n_2 - 1$ degrees of freedom. Note that df_1 and df_2 are the degrees of freedom associated with s_1^2 and s_2^2, respectively. The F distribution is non-symmetrical with the degree of skewness dependent on the above-mentioned degrees of freedom. Table 6 presents the critical values of F (depending on the degrees of freedom) such that $P(F > F_a) = a$ for $a = .10, .05, .025, .01$ and $.005$, respectively. Because right-hand tail areas correspond to an upper-tailed test of an hypothesis, we will always identify the larger sample variance as s_1^2 (that is, we will always place the larger sample variance in the numerator of $F = s_1^2/s_2^2$). Hence, an upper-tailed test is implied and the critical values of F will determine the rejection region.

The hypothesis of interest is $H_0 : \sigma_1^2 = \sigma_2^2$ versus $H_a : \sigma_1^2 > \sigma_2^2$ and the test statistic is

$$F = \frac{s_1^2}{s_2^2} = \frac{18.3}{7.9} = 2.316.$$

The rejection region (one-tailed) with $\alpha = .05$ and $df_1 = df_2 = 12$ degrees of freedom is $F > 2.69$, and H_0 is not rejected. We cannot conclude that the variances are different.

The critical values of F for various values of a are given next.

a	.10	.05	.025	.01	.005
F_a	2.15	2.69	3.28	4.16	4.91

Hence, the p-value $= P(F > 2.316)$ lies between .05 and .10, and the results are not significant.

From Table 6, $F_{df_1, df_2} = 3.28$ and $F_{df_2, df_1} = 3.28$. The 95% confidence interval for σ_1^2/σ_2^2 is

$$\frac{s_1^2}{s_2^2} \frac{1}{F_{df_1, df_2}} < \frac{\sigma_1^2}{\sigma_2^2} < \frac{s_1^2}{s_2^2} F_{df_2, df_1}$$

$$\frac{18.3}{7.9}\left(\frac{1}{3.28}\right) < \frac{\sigma_1^2}{\sigma_2^2} < \frac{18.3}{7.9}(3.28) \quad \text{or} \quad .320 < \frac{\sigma_1^2}{\sigma_2^2} < 7.598$$

10.6.15 a The hypothesis of interest is $H_0 : \sigma_1^2 = \sigma_2^2$ versus $H_a : \sigma_1^2 \neq \sigma_2^2$ and the test statistic is

$$F = \frac{s_1^2}{s_2^2} = \frac{3.1}{1.4} = 2.21.$$

Since we wish to test the hypothesis

$$H_0 : \sigma_1^2 = \sigma_2^2 \quad \text{versus} \quad H_a : \sigma_1^2 \neq \sigma_2^2$$

There will be another portion of the rejection region in the lower tail of the distribution. The area to the right of the critical value will represent only $\alpha/2$, and the probability of a Type I error is $2(\alpha/2) = \alpha$.

The two-sided rejection region with $df_1 = 24$ and $df_2 = 24$ and $\alpha = .05$ is $F > F_{.025} = 2.27$ and the null hypothesis is not rejected. There is insufficient evidence to indicate a difference in the precision of the two machines.

b The 95% confidence interval for the ratio of the two variances is

$$\frac{s_1^2}{s_2^2} \frac{1}{F_{df_1, df_2}} < \frac{\sigma_1^2}{\sigma_2^2} < \frac{s_1^2}{s_2^2} F_{df_2, df_1}$$

$$\frac{3.1}{1.4}\left(\frac{1}{2.27}\right) < \frac{\sigma_1^2}{\sigma_2^2} < \frac{3.1}{1.4}(2.27) \quad \text{or} \quad .975 < \frac{\sigma_1^2}{\sigma_2^2} < 5.03$$

Since the possible values for σ_1^2/σ_2^2 includes the value 1, it is possible that there is no difference in the precision of the two machines. This confirms the results of part **a**.

10.6.17 a The hypothesis of interest is $H_0: \sigma_1^2 = \sigma_2^2$ versus $H_a: \sigma_1^2 > \sigma_2^2$ and the test statistic is

$$F = \frac{s_1^2}{s_2^2} = \frac{.25^2}{.12^2} = 4.34.$$

The rejection region (one-tailed) with $\alpha = .05$ and $df_1 = df_2 = 49$ degrees of freedom by interpolation in Table 6. The value $F_{49,49}$ is roughly halfway between $F_{40,40} = 1.69$ and $F_{60,60} = 1.53$; therefore, we reject H_0 if $F > F_{49,49} \approx 1.61$. The observed value of the test statistic falls in the rejection region and we conclude that the "suspect line" possesses a larger variance.

b The student must obtain various critical levels of F from Table 6. We "roughly" interpolate $F_{49,49}$ as halfway between $F_{40,40}$ and $F_{60,60}$.

a	.05	.025	.01	.005
F_a	1.61	1.775	1.975	2.13

In any event, p-value $= P(F > 4.34) < .005$.

c Noting that $F_{df_1, df_2} = F_{df_2, df_1} \approx 1.61$, the 90% confidence interval for σ_1^2/σ_2^2 is

$$\frac{s_1^2}{s_2^2} \frac{1}{F_{df_1, df_2}} < \frac{\sigma_1^2}{\sigma_2^2} < \frac{s_1^2}{s_2^2} F_{df_2, df_1}$$

$$\frac{.25^2}{.12^2}\left(\frac{1}{1.61}\right) < \frac{\sigma_1^2}{\sigma_2^2} < \frac{.25^2}{.12^2}(1.61) \quad \text{or} \quad 2.696 < \frac{\sigma_1^2}{\sigma_2^2} < 6.988$$

10.6.19 a The assumption of equal variances $\left(\sigma_1^2 = \sigma_2^2\right)$ was made.

b The hypothesis of interest is $H_0: \sigma_1^2 = \sigma_2^2$ versus $H_a: \sigma_1^2 \neq \sigma_2^2$ and the test statistic is

$$F = \frac{s_1^2}{s_2^2} = \frac{.679^2}{.400^2} = 2.88.$$

The upper portion of the rejection region with $\alpha = .05$, $df_1 = 10$ and $df_2 = 13$ is $F > F_{.025} = 3.25$ and H_0 is not rejected. There is no reason to believe that the assumption has been violated.

10.6.21 a The hypothesis of interest is $H_0: \sigma_1^2 = \sigma_2^2$ versus $H_a: \sigma_1^2 \neq \sigma_2^2$ and the test statistic is

$$F = \frac{s_1^2}{s_2^2} = \frac{.273}{.094} = 2.904.$$

The upper portion of the rejection region with $\alpha = 2(.005) = .01$ is $F > F_{.005} = 6.54$ (from Table 6) and H_0 is not rejected. There is insufficient evidence to indicate that the supplier's shipments differ in variability.

b The 99% confidence interval for σ_2^2 is

$$\frac{(n_2-1)s^2}{\chi_{.005}^2} < \sigma_2^2 < \frac{(n_2-1)s^2}{\chi_{.995}^2}$$

$$\frac{9(.094)}{23.5893} < \sigma_2^2 < \frac{9(.094)}{1.734926} \quad \text{or} \quad .036 < \sigma_2^2 < .488$$

Intervals constructed in this manner enclose σ_2^2 99% of the time. Hence, we are fairly certain that σ_2^2 is between .036 and .488.

10.6.23 a The range of the first sample is 47 while the range of the second sample is only 16. There is probably a difference in the variances.

b The hypothesis of interest is $H_0 : \sigma_1^2 = \sigma_2^2$ versus $H_a : \sigma_1^2 \neq \sigma_2^2$.

Calculate $s_1^2 = \dfrac{177{,}294 - \dfrac{(838)^2}{4}}{3} = 577.6667 \qquad s_2^2 = \dfrac{192{,}394 - \dfrac{(1074)^2}{6}}{5} = 29.6$

and the test statistic is

$$F = \dfrac{s_1^2}{s_2^2} = \dfrac{577.6667}{29.6} = 19.516.$$

The critical values with $df_1 = 3$ and $df_2 = 5$ are shown next from Table 6.

a	.10	.05	.025	.01	.005
F_a	3.62	5.41	7.76	12.06	16.53

Hence, $p\text{-value} = 2P(F > 19.516) < 2(.005) = .01$. Since the p-value is smaller than .01, H_0 is rejected at the 1% level of significance. There is a difference in variability.

c Since the Student's t test requires the assumption of equal variance, it would be inappropriate in this instance. You should use the unpooled t test with Satterthwaite's approximation to the degrees of freedom.

Reviewing What You've Learned

10.R.1 a The hypothesis to be tested is $H_0 : \mu = .05 \qquad H_a : \mu > .05$ and the test statistic is

$$t = \dfrac{\bar{x} - \mu_0}{s/\sqrt{n}} = \dfrac{.058 - .05}{\dfrac{.012}{\sqrt{10}}} = 2.108.$$

Critical value approach: The rejection region with $\alpha = .05$ and $df = n - 1 = 9$ degrees of freedom is located in the upper tail of the t-distribution and is found from Table 4 as $t > t_{.05} = 1.833$. Since the observed value falls in the rejection region, H_0 is rejected and we conclude that μ is greater than .05.

p-value approach: The p-value $= P(t > 2.108)$. Since the value $t = 2.108$ falls between $t_{.025}$ and $t_{.05}$, the p-value can be bounded as $.025 < p\text{-value} < .05$. In any event, the p-value is less than .05 and H_0 can be rejected at the 5% level of significance.

10.R.3 Using the formulas given in Chapter 2 or your scientific calculator, calculate

$$\bar{x} = \dfrac{\sum x_i}{n} = \dfrac{42.6}{10} = 4.26$$

$$s^2 = \dfrac{\sum x_i^2 - \dfrac{(\sum x_i)^2}{n}}{n-1} = \dfrac{190.46 - \dfrac{(42.6)^2}{10}}{9} = .998 \quad \text{and} \quad s = .999$$

The 95% confidence interval is then

$$\bar{x} \pm t_{.025} \frac{s}{\sqrt{n}} \Rightarrow 4.26 \pm 2.262 \frac{.999}{\sqrt{10}} \Rightarrow 4.26 \pm .715$$

or $3.545 < \mu < 4.975$.

10.R.5 From Exercise 4, the best estimate for σ is

$$s = \sqrt{\frac{\sum x_i^2 - \frac{(\sum x_i)^2}{n}}{n-1}} = \sqrt{\frac{344,567 - \frac{(1845)^2}{10}}{9}} = \sqrt{462.7222} = 21.511.$$

Then, with $B = 5$, solve for n in the following inequality:

$$t_{.025} \frac{s}{\sqrt{n}} \le 5$$

which is approximately

$$1.96 \frac{21.511}{\sqrt{n}} \le 5 \Rightarrow \sqrt{n} \ge \frac{1.96(21.511)}{5} = 8.432$$

$n \ge 71.10$ or $n \ge 72$

Since n is greater than 30, the sample size, $n = 72$, is valid.

10.R.7 a Use the computing formulas or your scientific calculator to calculate

$$\bar{x} = \frac{\sum x_i}{n} = \frac{27.53}{10} = 2.753 \qquad s^2 = \frac{75.8155 - \frac{(27.53)^2}{10}}{9} = .0028233$$

$s = .053135$ and the 99% confidence interval is

$$\bar{x} \pm t_{.005} \frac{s}{\sqrt{n}} \Rightarrow 2.753 \pm 3.25 \frac{.053135}{\sqrt{10}} \Rightarrow 2.753 \pm .0546 \text{ or } 2.6984 < \mu < 2.8076.$$

b Intervals constructed using this procedure will enclose μ 99% of the time in repeated sampling. Hence, we are fairly certain that this particular interval encloses μ.

c The sample must be randomly selected, or at least behave as a random sample from the population of interest. For the chemist performing the analysis, this means that he or she must be certain that there is no unknown factor which is affecting the measurements, thus causing a biased sample rather than a representative sample.

10.R.9 a A paired-difference analysis is used. The hypothesis of interest is $H_0 : \mu_1 - \mu_2 = 0 \quad H_a : \mu_1 - \mu_2 \ne 0$

and the differences are shown next.

.54, .54, .42, .42, .54, .36, .54, 1.00, .55

Calculate

$$\bar{d} = \frac{\sum d_i}{n} = \frac{4.91}{9} = .545556, \quad s_d = \sqrt{\frac{\sum d_i^2 - \frac{(\sum d_i)^2}{n}}{n-1}} = \sqrt{\frac{2.9513 - \frac{(4.91)^2}{9}}{8}} = \sqrt{.0340778} = .184602$$

and the test statistic is

$$t = \frac{\bar{d} - \mu_d}{s_d / \sqrt{n}} = \frac{.545556 - 0}{\frac{.184602}{\sqrt{9}}} = 8.866$$

The two-tailed p-value with $df = 8$ is less than $2(.005) = .01$. Since this value is less than .05, H_0 is rejected. The results are highly significant; there is sufficient evidence to indicate a difference in the average cost for brand versus generic Buspirone tablets.

b The estimated savings per tablet by purchasing the generic brand of Buspirone is the average difference between the costs of the brand name and generic forms, or $\mu_1 - \mu_2 = \mu_d$. The 95% confidence interval for $\mu_1 - \mu_2 = \mu_d$ is

$$\bar{d} \pm t_{.025} \frac{s_d}{\sqrt{n}} \Rightarrow .5456 \pm 2.306 \frac{.184602}{\sqrt{9}} \Rightarrow .5456 \pm .1419$$

or $.4037 < (\mu_1 - \mu_2) < .6875$. The average savings is between $0.41 and $0.68 per tablet.

10.R.11 The object is to determine whether or not there is a difference between the mean responses for the two different stimuli to which the people have been subjected. The samples are independently and randomly selected, and the assumptions necessary for the t test of Section 10.4 are met. The hypothesis to be tested is

$$H_0 : \mu_1 - \mu_2 = 0 \quad H_a : \mu_1 - \mu_2 \neq 0$$

and the preliminary calculations are as follows:

$$\bar{x}_1 = \frac{15}{8} = 1.875 \text{ and } \bar{x}_2 = \frac{21}{8} = 2.625$$

$$s_1^2 = \frac{33 - \frac{(15)^2}{8}}{7} = .69643 \text{ and } s_2^2 = \frac{61 - \frac{(21)^2}{8}}{7} = .83929$$

Since the ratio of the variances is less than 3, you can use the pooled t test. The pooled estimator of σ^2 is calculated as

$$s^2 = \frac{(n_1 - 1)s_1^2 + (n_2 - 1)s_2^2}{n_1 + n_2 - 2} = \frac{4.875 + 5.875}{14} = .7679$$

and the test statistic is

$$t = \frac{(\bar{x}_1 - \bar{x}_2) - 0}{\sqrt{s^2\left(\frac{1}{n_1} + \frac{1}{n_2}\right)}} = \frac{1.875 - 2.625}{\sqrt{.7679\left(\frac{1}{8} + \frac{1}{8}\right)}} = -1.712$$

The two-tailed rejection region with $\alpha = .05$ and $df = 14$ is $|t| > t_{.025} = 2.145$, and H_0 is not rejected. There is insufficient evidence to indicate that there is a difference in means.

10.R.13 Refer to Exercise 11 and 12. For the unpaired design, the 95% confidence interval for $(\mu_1 - \mu_2)$ is

$$(\bar{x}_1 - \bar{x}_2) \pm t_{.025} \sqrt{s^2\left(\frac{1}{n_1} + \frac{1}{n_2}\right)}$$

$$-.75 \pm 2.145 \sqrt{.7679\left(\frac{1}{8} + \frac{1}{8}\right)}$$

$$-.75 \pm .94 \text{ or } -1.69 < (\mu_1 - \mu_2) < 0.19$$

while for the paired design, the 95% confidence interval is

$$\bar{d} \pm t_{.025} \frac{s_d}{\sqrt{n}} \Rightarrow -.75 \pm 2.365 \frac{.88641}{\sqrt{8}} \Rightarrow -.75 \pm .74$$

or $-1.49 < (\mu_1 - \mu_2) < -.01$. Although the width of the confidence interval has decreased slightly, it does not appear that blocking has increased the amount of information by much.

10.R.15 To test the hypothesis $H_0: \mu = 16$ versus $H_a: \mu < 16$, the test statistic is

$$t = \frac{\bar{x} - \mu_0}{s/\sqrt{n}} = \frac{15.7 - 16}{.5/\sqrt{9}} = -1.8$$

The p-value with 8 degrees of freedom is bounded as $.05 < p\text{-value} < .10$. Hence, the null hypothesis is not rejected. There is insufficient evidence to indicate that the mean weight is less than claimed.

10.R.17 The hypothesis to be tested is $H_0: \mu_1 - \mu_2 = 0$ $H_a: \mu_1 - \mu_2 < 0$ and the preliminary calculations are as follows:

$$\bar{x}_1 = \frac{.6}{6} = .1 \quad \text{and} \quad \bar{x}_2 = \frac{.83}{6} = .1383$$

$$s_1^2 = \frac{.0624 - \frac{(.6)^2}{6}}{5} = .00048 \quad \text{and} \quad s_2^2 = \frac{.1175 - \frac{(.83)^2}{6}}{5} = .00053667$$

Since the ratio of the variances is less than 3, you can use the pooled t test. The pooled estimator of σ^2 is calculated as

$$s^2 = \frac{(n_1 - 1)s_1^2 + (n_2 - 1)s_2^2}{n_1 + n_2 - 2} = \frac{.0024 + .00268}{10} = .0005083$$

and the test statistic is

$$t = \frac{(\bar{x}_1 - \bar{x}_2) - 0}{\sqrt{s^2\left(\frac{1}{n_1} + \frac{1}{n_2}\right)}} = \frac{-.0383}{\sqrt{.0005083\left(\frac{1}{6} + \frac{1}{6}\right)}} = -2.945$$

The p-value for a one-tailed test with 10 degrees of freedom is bounded as $.005 < p\text{-value} < .010$. Hence, the null hypothesis H_0 is rejected. There is sufficient evidence to indicate that $\mu_1 < \mu_2$.

10.R.19 Calculate $\bar{x} = \frac{\Sigma x_i}{n} = \frac{104.9}{25} = 4.196$

$$s^2 = \frac{\Sigma x_i^2 - \frac{(\Sigma x_i)^2}{n}}{n-1} = \frac{454.81 - \frac{(104.9)^2}{25}}{24} = .6104 \quad \text{and} \quad s = .7813$$

The 95% confidence interval based on $df = 24$ is

$$\bar{x} \pm t_{.025} \frac{s}{\sqrt{n}} \Rightarrow 4.196 \pm 2.064 \frac{.7813}{\sqrt{25}} \Rightarrow 4.196 \pm .323$$

or $3.873 < \mu < 4.519$. Intervals constructed in this manner enclose the true value of μ 95% of the time. Hence, we are fairly certain that this interval contains the true value of μ.

10.R.21 a The hypothesis of interest is $H_0: \sigma_1^2 = \sigma_2^2$ versus $H_a: \sigma_1^2 \neq \sigma_2^2$ and the test statistic is

$$F = \frac{s_1^2}{s_2^2} = \frac{22^2}{20^2} = 1.21.$$

The upper portion of the two-tailed rejection region with $\alpha = .05$ is $F > F_{19,19} \approx F_{20,19} = 2.51$ and H$_0$ is not rejected. There is insufficient evidence to indicate that the population variances are different.

b The hypothesis to be tested is $H_0 : \mu_1 - \mu_2 = 0$ $H_a : \mu_1 - \mu_2 \neq 0$. Based on the results of part **a**, you can use the pooled t test. The pooled estimator of σ^2 is calculated as

$$s^2 = \frac{(n_1 - 1)s_1^2 + (n_2 - 1)s_2^2}{n_1 + n_2 - 2} = \frac{19(22^2) + 19(20^2)}{38} = 442$$

and the test statistic is

$$t = \frac{(\bar{x}_1 - \bar{x}_2) - 0}{\sqrt{s^2 \left(\frac{1}{n_1} + \frac{1}{n_2}\right)}} = \frac{78 - 67}{\sqrt{442 \left(\frac{1}{20} + \frac{1}{20}\right)}} = 1.65$$

The rejection region with $\alpha = .05$ and $df = 20 + 20 - 2 = 38$ (approximated with $df = 29$) is $|t| > 2.045$ and H$_0$ is not rejected. There is insufficient evidence to indicate a difference in the two populaton means.

10.R.23 The hypothesis of interest is $H_0 : \sigma_A^2 = \sigma_B^2$ versus $H_a : \sigma_A^2 < \sigma_B^2$ and the test statistic is

$$F = \frac{s_B^2}{s_A^2} = \frac{.065}{.027} = 2.407 .$$

The rejection region (one-tailed) will be determined by a critical value of F based on $df_1 = 9$ and $df_2 = 29$ degrees of freedom, with area .05 to its right. That is, from Table 6, you will reject H$_0$ if $F > 2.22$. The observed value of F falls in the rejection region, and we conclude that $\sigma_A^2 < \sigma_B^2$.

10.R.25 The underlying populations are ratings and can only take on the finite number of values, 1, 2,..., 9, 10. Neither population has a normal distribution, but both are discrete. Further, the samples are not independent, since the same person is asked to rank each car design. Hence, two of the assumptions required for the Student's t test have been violated.

11: The Analysis of Variance

Section 11.1

11.1.1 The assumptions are given in this section.
- The observations within each population are normally distributed with at common variance σ^2.
- Assumptions regarding the sampling procedures specified for each design must be followed.

11.1.3-7—Definitions for these three terms can be found in section 11.1 of the text.

11.1.9 There is only one *factor*, dose of vitamin C, at three levels—200, 500 and 1000 mg. Since there is only one factor, the three levels of the factor also represent the treatments in the experiment.

11.1.11 There is only one *factor*, technology tool, at three levels—*TI 84* calculator, iPad and laptop. Since there is only one factor, the three levels of the factor also represent the treatments in the experiment.

11.1.13 Test scores very often have mound-shaped or approximately normal distributions.

11.1.15 Since the lumber is being cut to the nearest foot, the amount left can take any value between 0 and 12 inches (not including 0 or 12). This data will have a continuous uniform distribution.

Section 11.2

11.2.1 In comparing 6 populations, there are $k-1$ degrees of freedom for treatments and $n = 6(5) = 30$. The ANOVA table is shown next. The given sums of squares are inserted and missing entries found by subtraction. The mean squares are found as $MS = SS/df$.

Source	df	SS	MS	F
Treatments	5	5.2	1.04	1.541
Error	24	16.2	0.675	
Total	29	21.4		

The hypothesis of interest is $H_0: \mu_1 = \mu_2 = \cdots = \mu_6$ versus H_a: at least one of the means is different from the others and the test statistic is

$F = MST/MSE = 1.541$ with $df_1 = 5$ and $df_2 = 24$ degrees of freedom.

With $\alpha = .05$ and 5 and 24 degrees of freedom, H_0 is rejected if $F > F_{.05} = 2.62$. Since $F = 1.541$ does not fall in the rejection region, the null hypothesis is not rejected. There is insufficient evidence to indicate a difference among the means.

The critical values of F with $df_1 = 5$ and $df_2 = 24$ (Table 6) for bounding the *p*-value for this one-tailed test are shown next.

a	.10	.05	.025	.01	.005
F_a	2.10	2.62	3.15	3.90	4.49

Since the observed value $F = 1.541$ is less than $F_{.10}$, *p*-value $> .10$ and H_0 is not rejected as before.

11.2.3 The following preliminary calculations are necessary:

$T_1 = 14 \quad T_2 = 19 \quad T_3 = 5 \quad G = 38$

$$\text{CM} = \frac{\left(\sum x_{ij}\right)^2}{n} = \frac{(38)^2}{14} = 103.142857$$

$$\text{Total SS} = \sum x_{ij}^2 - \text{CM} = 3^2 + 2^2 + \cdots + 2^2 + 1^2 - \text{CM} = 130 - 103.142857 = 26.8571$$

$$\text{SST} = \sum \frac{T_i^2}{n_i} - \text{CM} = \frac{14^2}{5} + \frac{19^2}{5} + \frac{5^2}{4} - \text{CM} = 117.65 - 103.142857 = 14.5071$$

and $\text{MST} = \dfrac{\text{SST}}{k-1} = \dfrac{14.5071}{2} = 7.2536$. By subtraction,

$$\text{SSE} = \text{Total SS} - \text{SST} = 26.8571 - 14.5071 = 12.3500$$

and the degrees of freedom, by subtraction, are $13 - 2 = 11$. Then

$$\text{MSE} = \frac{\text{SSE}}{11} = \frac{12.3500}{11} = 1.1227$$

The information is consolidated in an ANOVA table.

Source	df	SS	MS	F
Treatments	2	14.5071	7.2536	6.46
Error	11	12.3500	1.1227	
Total	13	26.8571		

The hypothesis to be tested is

$H_0 : \mu_1 = \mu_2 = \mu_3$ versus H_a : at least one pair of means are different

The test statistic is $F = \dfrac{\text{MST}}{\text{MSE}} = \dfrac{7.2536}{1.1227} = 6.46$ and the rejection region, based on an F-distribution with 2 and 11 degrees of freedom is $F > F_{.05} = 3.98$ and H_0 is rejected. The critical values of F for bounding the p-value for this one-tailed test are shown next.

a	.10	.05	.025	.01	.005
F_a	2.86	3.98	5.26	7.21	8.91

Since the observed value $F = 6.46$ is between $F_{.01}$ and $F_{.025}$, $.01 < p\text{-value} < .025$ and H_0 is rejected at the 5% level of significance. There is a difference among the means.

11.2.5 Refer to Exercise 1, where $n_1 = n_2 = 5$. The 95% confidence intervals are

a $\bar{x}_1 \pm t_{.025}\sqrt{\dfrac{\text{MSE}}{n_1}} \Rightarrow 3.07 \pm 2.064\sqrt{\dfrac{.675}{5}} \Rightarrow 3.07 \pm .758$

or $2.312 < \mu_1 < 3.828$.

and $(\bar{x}_1 - \bar{x}_2) \pm t_{.025}\sqrt{\text{MSE}\left(\dfrac{1}{n_1} + \dfrac{1}{n_2}\right)}$

$(3.07 - 2.52) \pm 2.064\sqrt{0.675\left(\dfrac{2}{5}\right)}$

$.55 \pm 1.072$ or $-.522 < \mu_1 - \mu_2 < 1.622$

11.2.7 Use the fact that both the sums of squares and the degrees of freedom are additive. Then find the mean squares and the test statistic and fill in the table next.

Source	df	SS	MS	F
Treatments	3	9.75	3.25	4.457
Error	24	17.5	.7292	
Total	27	27.25		

The test statistic is $F = \dfrac{MST}{MSE} = \dfrac{3.25}{.7292} = 4.457$ and the rejection region, based on an F-distribution with 3 and 24 degrees of freedom is $F > F_{.01} = 4.72$ and H_0 is not rejected. The critical values of F for bounding the p-value for this one-tailed test are shown next.

a	.10	.05	.025	.01	.005
F_a	2.33	3.01	3.72	4.72	5.52

Since the observed value $F = 4.457$ is between $F_{.01}$ and $F_{.025}$, $.01 < p$-value $< .025$ and H_0 is not rejected at the 1% level of significance. There is insufficient evidence to indicate that there is a difference among the means.

11.2.9 **a** The 95% confidence interval for μ_A is

$$\bar{x}_A \pm t_{.025}\sqrt{\dfrac{MSE}{n_A}} \Rightarrow 76 \pm 2.306\sqrt{\dfrac{62.333}{5}} \Rightarrow 76 \pm 8.142 \quad \text{or} \quad 67.86 < \mu_A < 84.14.$$

b The 95% confidence interval for μ_B is

$$\bar{x}_B \pm t_{.025}\sqrt{\dfrac{MSE}{n_B}} \Rightarrow 66.33 \pm 2.306\sqrt{\dfrac{62.333}{3}} \Rightarrow 66.33 \pm 10.51 \quad \text{or} \quad 55.82 < \mu_B < 76.84.$$

c The 95% confidence interval for $\mu_A - \mu_B$ is

$$(\bar{x}_A - \bar{x}_B) \pm t_{.025}\sqrt{MSE\left(\dfrac{1}{n_A} + \dfrac{1}{n_B}\right)}$$

$$(76 - 66.33) \pm 2.306\sqrt{62.333\left(\dfrac{1}{5} + \dfrac{1}{3}\right)}$$

$$9.667 \pm 13.296 \quad \text{or} \quad -3.629 < \mu_A - \mu_B < 22.963$$

d Note that these three confidence intervals cannot be jointly valid because all three employ the same value of $s = \sqrt{MSE}$ and are dependent.

11.2.11 **a** We would be reasonably confident that the data satisfied the normality assumption because each measurement represents the average of 10 continuous measurements. The Central Limit Theorem assures us that this mean will be approximately normally distributed.

b We have a completely randomized design with four treatments, each containing 6 measurements. The analysis of variance table is given in the *Minitab* printout. The F test is

$$F = \dfrac{MST}{MSE} = \dfrac{6.580}{.115} = 57.38$$

with p-value $= .000$ (in the column marked "P"). Since the p-value is very small (less than .01), H_0 is rejected. There is a significant difference in the mean leaf length among the four locations with $P < .01$ or even $P < .001$.

c The hypothesis to be tested is $H_0 : \mu_1 = \mu_4$ versus $H_a : \mu_1 \neq \mu_4$ and the test statistic is

$$t = \frac{\bar{x}_1 - \bar{x}_4}{\sqrt{MSE\left(\frac{1}{n_1} + \frac{1}{n_4}\right)}} = \frac{6.0167 - 3.65}{\sqrt{.11467\left(\frac{1}{6} + \frac{1}{6}\right)}} = 12.11$$

The p-value with $df = 20$ is $2P(t > 12.11)$ is bounded (using Table 4) as

p-value $< 2(.005) = .01$

and the null hypothesis is rejected. We conclude that there is a difference between the means.

d The 99% confidence interval for $\mu_1 - \mu_4$ is

$$(\bar{x}_1 - \bar{x}_4) \pm t_{.005}\sqrt{MSE\left(\frac{1}{n_1} + \frac{1}{n_4}\right)}$$

$$(6.0167 - 3.65) \pm 2.845\sqrt{.11467\left(\frac{1}{6} + \frac{1}{6}\right)}$$

$2.367 \pm .556$ or $1.811 < \mu_1 - \mu_4 < 2.923$

11.2.13 The design is completely randomized with 3 treatments and 5 replications per treatment. The *Minitab* printout shows the analysis of variance for this experiment.

One-way ANOVA: Calcium versus Method

Analysis of Variance

Source	DF	Adj SS	Adj MS	F-Value	P-Value
Method	2	0.04116	0.0205800	16.38	0.000
Error	12	0.01508	0.0012567		
Total	14	0.05624			

Model Summary

S	R-sq	R-sq(adj)	R-sq(pred)
0.0354495	73.19%	68.72%	58.10%

The test statistic, $F = 16.38$ with p-value $= .000$ indicates the results are highly significant; there is a difference in the mean calcium contents for the three methods. All assumptions appear to have been satisfied.

11.2.15 **a** The design is a completely randomized design (four independent samples).

b The following preliminary calculations are necessary:

$T_1 = 1311 \quad T_2 = 1174 \quad T_3 = 1258 \quad T_4 = 1343 \quad G = 5086$

$$CM = \frac{\left(\sum x_{ij}\right)^2}{n} = \frac{(5086)^2}{20} = 1,293,369.8 \quad \text{Total SS} = \sum x_{ij}^2 - CM = 1,297,302 - CM = 3932.2$$

$$SST = \sum \frac{T_i^2}{n_i} - CM = \frac{1311^2}{5} + \frac{1174^2}{5} + \frac{1258^2}{5} + \frac{1343^2}{5} - CM = 3272.2$$

Calculate $MS = SS/df$ and consolidate the information in an ANOVA table.

Source	df	SS	MS
Treatments	3	3272.2	1090.7333
Error	16	660	41.25
Total	19	3932.2	

   c The hypothesis to be tested is

$$H_0 : \mu_1 = \mu_2 = \mu_3 = \mu_4 \quad \text{versus} \quad H_a : \text{at least one pair of means are different}$$

and the F test to detect a difference in average prices is

$$F = \frac{MST}{MSE} = 26.44.$$

The rejection region with $\alpha = .05$ and 3 and 16 df is approximately $F > 3.24$ and H_0 is rejected. [Alternatively, we could bound the p-value using Table 6 as p-value $< .005$.] There is enough evidence to indicate a difference in the average prices for the four states.

11.2.17 a The design is a completely randomized design with three samples, each having a different number of measurements.

     b Use the computing formulas in Section 11.5 or the *Minitab* printout that follows.

One-way ANOVA: Iron versus Site

Analysis of Variance

Source	DF	Adj SS	Adj MS	F-Value	P-Value
Site	2	132.28	66.1386	126.85	0.000
Error	21	10.95	0.5214		
Total	23	143.23			

Model Summary

S	R-sq	R-sq(adj)	R-sq(pred)
0.722087	92.36%	91.63%	90.31%

Means

Site	N	Mean	StDev	95% CI
A	5	1.512	0.736	(0.840, 2.184)
I	5	1.712	0.436	(1.040, 2.384)
L	14	6.372	0.786	(5.971, 6.773)

Pooled StDev = 0.722087

The F test for treatments has a test statistic $F = 126.85$ with p-value = .000. The null hypothesis is rejected and we conclude that there is a significant difference in the average percentage of iron oxide at the three sites.

Section 11.3

11.3.1 Sample means must be independent and based upon samples of equal size.

11.3.3-7 Use Tables 11(a) and 11(b).

    3. $q_{.05}(3,9) = 3.95$ **5.** $q_{.01}(6,24) = 5.37$

7. $q_{.01}(3,15) = 4.84$

11.3.9 $\omega = q_{.01}(6,30)\dfrac{s}{\sqrt{6}} = 5.24\sqrt{\dfrac{8}{6}} = 6.0506$

11.3.11 $\omega = q_{.05}(6,18)\sqrt{\dfrac{9.12}{4}} = 4.49\sqrt{\dfrac{9.12}{4}} = 6.780$

11.3.13 With $k = 4$, $df = 20$, $n_t = 6$,

$$\omega = q_{.01}(4,20)\dfrac{\sqrt{MSE}}{\sqrt{n_t}} = 5.02\sqrt{\dfrac{.115}{6}} = .69$$

The ranked means are shown next.

3.65	5.35	5.65	6.0167
$\bar{x}_4$	$\bar{x}_3$	$\bar{x}_2$	$\bar{x}_1$

11.3.15 The design is completely randomized with 3 treatments and 5 replications per treatment. The *Minitab* printout next shows the analysis of variance for this experiment.

One-way ANOVA: mg/dl versus Lab

Analysis of Variance

Source	DF	Adj SS	Adj MS	F-Value	P-Value
Lab	2	42.56	21.28	0.60	0.562
Error	12	422.46	35.21		
Total	14	465.02			

Model Summary

S	R-sq	R-sq(adj)	R-sq(pred)
5.93341	9.15%	0.00%	0.00%

Means

Lab	N	Mean	StDev	95% CI
1	5	108.86	7.47	(103.08, 114.64)
2	5	105.04	6.01	(99.26, 110.82)
3	5	105.60	3.70	(99.82, 111.38)

Pooled StDev = 5.93341

Tukey Pairwise Comparisons

Tukey Simultaneous Tests for Differences of Means

Difference of Levels	Difference of Means	SE of Difference	95% CI	T-Value	Adjusted P-Value
2 - 1	-3.82	3.75	(-13.82, 6.18)	-1.02	0.580
3 - 1	-3.26	3.75	(-13.26, 6.74)	-0.87	0.669
3 - 2	0.56	3.75	(-9.44, 10.56)	0.15	0.988

Individual confidence level = 97.94%

a The analysis of variance F test for $H_0: \mu_1 = \mu_2 = \mu_3$ is $F = .60$ with p-value $= .562$. The results are not significant and H_0 is not rejected. There is insufficient evidence to indicate a difference in the treatment means.

b Since the treatment means are not significantly different, there is no need to use Tukey's test to search for the pairwise differences. Notice that all three intervals generated by *Minitab* contain zero, indicating that the pairs cannot be judged different.

11.3.17 a The following preliminary calculations are necessary:

$$T_1 = 6080 \quad T_2 = 6530 \quad T_3 = 5320 \quad G = 17,930$$

$$CM = \dfrac{\left(\sum x_{ij}\right)^2}{n} = \dfrac{(17,930)^2}{30} = 10,716,163.333333$$

Total SS $= \sum x_{ij}^2 - CM = 11,016,900 - CM = 300,736.666667$

$$\text{SST} = \sum \frac{T_i^2}{n_i} - \text{CM} = \frac{6080^2}{10} + \frac{6530^2}{10} + \frac{5320^2}{10} - \text{CM} = 74{,}806.666667$$

Calculate $\text{MS} = \text{SS}/df$ and consolidate the information in an ANOVA table.

Source	df	SS	MS
Treatments	2	74,806.666667	37,403.333333
Error	27	225,930.000000	8367.777778
Total	29	300,736.666667	

The hypothesis to be tested is

$H_0 : \mu_1 = \mu_2 = \mu_3$ versus H_a : at least one pair of means are different

and the F test to detect a difference in average scores is

$$F = \frac{\text{MST}}{\text{MSE}} = 4.47.$$

The rejection region with $\alpha = .05$ and 2 and 27 df is $F > 3.35$ and H_0 is rejected. There is evidence of a difference in the average scores for the three graduate programs.

b The 95% confidence interval for $\mu_1 - \mu_2$ is

$$(\bar{x}_{LS} - \bar{x}_{PS}) \pm t_{.025}\sqrt{\text{MSE}\left(\frac{1}{n_{LS}} + \frac{1}{n_{PS}}\right)}$$

$$\left(\frac{6080}{10} - \frac{6530}{10}\right) \pm 2.052\sqrt{8367.777778\left(\frac{1}{10} + \frac{1}{10}\right)}$$

-45 ± 83.946 or $-128.946 < \mu_{LS} - \mu_{PS} < 38.946$

c With $k = 3$, $df = 27$, $n_t = 10$,

$$\omega = q_{.05}(3, 27)\frac{\sqrt{\text{MSE}}}{\sqrt{n_t}} \approx 3.53\sqrt{\frac{8367.777778}{10}} = 102.113$$

The ranked means are shown next.

532   608   653

$\bar{x}_{SS}$ $\bar{x}_{LS}$ $\bar{x}_{PS}$

There is no significant difference between Social Sciences and Life Sciences, or between Life Sciences and Physical Sciences; but the average scores for Social Sciences and Physical Sciences are different from each other.

Section 11.4

11.4.1 For a randomized block design, we assume that the effect of the treatment will be the same, regardless of which block you are using. That is, the blocks and treatments do not *interact*.

11.4.3 There are $4 \times 3 = 12$ measurements with 4 treatments within 3 blocks. There are $k - 1 = 3$ treatment degrees of freedom and $b - 1 = 2$ block df. The analysis of variance table is shown next.

Source	df
Treatments	3
Blocks	2
Error	6
Total	11

11.4.5 Use the computing formulas to calculate the sum of squares as follows:

$$CM = \frac{\left(\sum x_{ij}\right)^2}{n} = \frac{(113)^2}{12} = 1064.08333$$

$$\text{Total SS} = \sum x_{ij}^2 - CM = 6^2 + 10^2 + \cdots + 14^2 - CM = 1213 - CM = 148.91667$$

$$SST = \sum \frac{T_j^2}{3} - CM = \frac{22^2 + 34^2 + 27^2 + 30^2}{3} - CM = 25.58333$$

$$SSB = \sum \frac{B_i^2}{4} - CM = \frac{33^2 + 25^2 + 55^2}{4} - CM = 120.66667 \text{ and}$$

$$SSE = \text{Total SS} - SST - SSB = 2.6667$$

Calculate $MS = SS/df$ and consolidate the information in an ANOVA table.

Source	df	SS	MS	F
Treatments	3	25.5833	8.5278	19.19
Blocks	2	120.6667	60.3333	135.75
Error	6	2.6667	0.4444	
Total	11	148.9167		

11.4.7 To test the difference among treatment means, the test statistic is

$$F = \frac{MST}{MSE} = \frac{8.528}{.4444} = 19.19$$

and the rejection region with $\alpha = .01$ and 3 and 6 df is $F > 9.78$. There is a significant difference among the treatment means. From Table 6 with 3 and 6 df, the observed value of the test statistic is greater than $F_{.005} = 12.62$, so that p-value < .005.

To test the difference among block means, the test statistic is

$$F = \frac{MSB}{MSE} = \frac{60.3333}{.4444} = 135.75$$

and the rejection region with $\alpha = .01$ and 2 and 6 df is $F > 10.92$. There is a significant difference among the block means.

With $k = 4$, $df = 6$, $n_t = 3$,

$$\omega = q_{.01}(4,6)\frac{\sqrt{MSE}}{\sqrt{n_t}} = 7.03\sqrt{\frac{.4444}{3}} = 2.706$$

The ranked means are shown next.

$$\begin{array}{cccc} 7.33 & 9.00 & 10.00 & 11.33 \\ \bar{x}_1 & \bar{x}_3 & \bar{x}_4 & \bar{x}_2 \end{array}$$

11.4.9-11 The given sums of squares are inserted and missing entries found by subtraction. The mean squares are found as $MS = SS/df$.

Source	df	SS	MS	F
Treatments	2	11.4	5.70	4.01
Blocks	5	17.1	3.42	2.41
Error	10	14.2	1.42	
Total	17	42.7		

9. To compare the treatment means, the test statistic $F = MST/MSE = 4.01$ and the rejection region with 2 and 10 df is $F > F_{.05} = 4.10$. The null hypothesis is not rejected. There is insufficient evidence to indicate a difference between treatment means.

11. To test for differences among block means, the test statistic is $F = MSB/MSE = 2.41$. The critical values of F from Table 6 with 5 and 10 df are shown next.

a	.10	.05	.025	.01	.005
F_a	2.52	3.33	4.24	5.64	6.87

Since the observed value $F = 2.41$ is less than $F_{.10}$, p-value $> .10$ and the null hypothesis is not rejected. There is insufficient evident to indicate differences among block means. We would conclude that blocking has not been effective.

11.4.13-15 The given sums of squares are inserted and missing entries found by subtraction. The mean squares are found as $MS = SS/df$.

Source	df	SS	MS	F
Treatments	4	14.2	3.55	9.68
Blocks	6	18.9	3.15	8.59
Error	24	8.8	0.3667	
Total	34	41.9		

13. There are always $b = 7$ observations in a treatment total.

15. To test the difference among treatment means, the test statistic is

$$F = \frac{MST}{MSE} = \frac{3.55}{.3667} = 9.68$$

and the rejection region with $\alpha = .05$ and 4 and 24 df is $F > 2.78$. There is a significant difference among the treatment means.

11.4.17 Use *Minitab* or *MS Excel* to obtain an ANOVA printout, or use the following calculations:

$$CM = \frac{\left(\sum x_{ij}\right)^2}{n} = \frac{(325.2)^2}{12} = 8812.92 \quad \text{Total SS} = \sum x_{ij}^2 - CM = 8819.68 - CM = 6.76$$

$$SS(\text{formulations}) = \sum \frac{T_j^2}{b} - CM = \frac{(106.1)^2 + (110.9)^2 + (108.2)^2}{4} - CM = 2.895$$

$$SS(\text{auto}) = \sum \frac{B_i^2}{k} - CM = \frac{(79)^2 + (82.6)^2 + (82)^2 + (81.6)^2}{3} - CM = 2.520 \text{ and}$$

$$SSE = \text{Total SS} - SST - SSB = 1.345$$

Calculate $MS = SS/df$ and consolidate the information in an ANOVA table.

Source	df	SS	MS	F
Treatments	2	2.895	1.4475	6.46
Blocks	3	2.520	0.8400	3.75
Error	6	1.345	0.2242	
Total	11	6.760		

a To test the null hypothesis that there is no difference in mean mileage per gallon for the three formulations, the test statistic is

$$F = \frac{MST}{MSE} = 6.46$$

The critical values of F from Table 6 with 2 and 6 df are shown next.

a	.10	.05	.025	.01	.005
F_a	3.46	5.14	7.26	10.92	14.54

Since the observed value $F = 6.46$ is between $F_{.025}$ and $F_{.05}$, $.025 < p$-value $< .05$ and the null hypothesis is rejected at the 5% level of significance. There is a significant difference among the treatment means.

b To test the null hypothesis that there is no difference in mean mileage for the four automobiles, the test statistic is

$$F = \frac{MSB}{MSE} = 3.75$$

and the p-value with 3 and 6 df is $.05 < p$-value $< .10$. There is no evidence of a significant difference among the automobiles.

c The 90% confidence interval is

$$(\bar{x}_A - \bar{x}_B) \pm t_{.05}\sqrt{MSE\left(\frac{2}{b}\right)}$$

$$(26.525 - 27.725) \pm 1.943\sqrt{.224167\left(\frac{2}{4}\right)}$$

$$-1.2 \pm .650 \quad \text{or} \quad -1.85 < \mu_A - \mu_B < -.55$$

d To determine where the treatment differences lie, use Tukey's test with

$$\omega = q_{.05}(3,6)\frac{\sqrt{MSE}}{\sqrt{n_t}} = 4.34\sqrt{\frac{.224167}{4}} = 1.027$$

The ranked means are shown next.

$$\begin{array}{ccc} 26.525 & 27.05 & 27.725 \\ \bar{x}_A & \bar{x}_C & \bar{x}_B \end{array}$$

Only gasoline formulations A and B are significantly different from each other.

11.4.19 The factor of interest is "soil preparation", and the blocking factor is "locations". A randomized block design is used and the analysis of variance table can be obtained using the computer printout.

a The F statistic to detect a difference due to soil preparations is

$$F = \frac{MST}{MSE} = 10.06$$

with p-value $= .012$. The null hypothesis can be rejected at the 5% level of significance; there is a significant difference among the treatment means.

b The F statistic to detect a difference due to locations is

$$F = \frac{\text{MSB}}{\text{MSE}} = 10.88$$

with p-value $= .008$. The null hypothesis can be rejected at the 1% level of significance; there is a highly significant difference among the block means.

c Tukey's test can be used to determine where the differences lie:

$$\omega = q_{.05}(3,6)\frac{\sqrt{\text{MSE}}}{\sqrt{n_t}} = 4.34\sqrt{\frac{1.8889}{4}} = 2.98$$

The ranked means are shown next.

$$\begin{array}{ccc} 12.0 & 12.5 & 16.0 \\ \bar{x}_3 & \bar{x}_1 & \bar{x}_2 \end{array}$$

Preparations 1 and 3 are the only two treatments that cannot be declared significantly different.

d The 95% confidence interval is

$$(\bar{x}_B - \bar{x}_A) \pm t_{.025}\sqrt{\text{MSE}\left(\frac{2}{b}\right)}$$

$$(16.0 - 12.5) \pm 2.447\sqrt{1.89\left(\frac{2}{4}\right)}$$

$$3.5 \pm 2.38 \quad \text{or} \quad 1.12 < \mu_B - \mu_A < 5.88$$

11.4.21 A randomized block design has been used with "estimators" as treatments and "construction job" as the block factor. The analysis of variance table is found in the following *Minitab* printout.

ANOVA: Cost versus Estimator, Job

Analysis of Variance for Cost

Source	DF	SS	MS	F	P
Estimator	2	10.862	5.4308	7.20	0.025
Job	3	37.607	12.5358	16.61	0.003
Error	6	4.528	0.7547		
Total	11	52.997			

Model Summary

S	R-sq	R-sq(adj)
0.868748	91.46%	84.34%

Means

Estimator	N	Cost
A	4	32.6125
B	4	34.8875
C	4	34.1875

Both treatments and blocks are significant. The treatment means can be further compared using Tukey's test with

$$\omega = q_{.05}(3,6)\frac{\sqrt{MSE}}{\sqrt{n_t}} = 4.34\sqrt{\frac{.7547}{4}} = 1.885$$

The ranked means are shown next.

$$\begin{array}{ccc} 32.6125 & 34.1875 & 34.8875 \\ \overline{x}_A & \overline{x}_C & \overline{x}_B \end{array}$$

Estimators A and B show a significant difference in average costs.

11.4.23 a A randomized block design has been used with "bonding agent" as treatments and "batches" as blocks. The analysis of variance table is found in the *Minitab* printout as follows.

ANOVA: Pressure versus Agent, Batch

Analysis of Variance for Pressure

Source	DF	SS	MS	F	P
Agent	2	332.8360	166.4180	4.55	0.048
Batch	4	50.7107	12.6777	0.35	0.840
Error	8	292.8573	36.6072		
Total	14	676.4040			

Model Summary

S	R-sq	R-sq(adj)
6.05039	56.70%	24.23%

Means

Agent	N	Pressure
1	5	66.44
2	5	77.82
3	5	70.48

b-c There is a significant difference due to agents ($F = 4.55$ with p-value $= .048$) but not due to batches (blocks) ($F = 0.35$ with p-value $= .840$).

d The treatment means can be further compared using Tukey's test with

$$\omega = q_{.05}(3,8)\frac{\sqrt{MSE}}{\sqrt{n_t}} = 4.04\sqrt{\frac{36.6072}{5}} = 10.93$$

The ranked means are shown next.

66.44	70.48	77.82
Agent 1	Agent 3	Agent 2

11.4.25 a The treatment means can be further compared using Tukey's test. From Table 11(a), we use a conservative estimate with $df = 21$ and $q_{.05}(4,21) \approx q_{.05}(4,20) = 3.96$.

b Calculate $\omega = q_{.05}(4,20)\dfrac{\sqrt{MSE}}{\sqrt{n_t}} = 3.96\sqrt{\dfrac{.1658}{8}} = .570$

c The ranked means are shown next.

WinCo	Staters	Vons	Ralphs
2.68125	3.8525	4.1350	4.3400

The average price at WinCo is significantly lower than the other three stores, which are not significantly different from each other.

Section 11.5

11.5.1 **Interaction** is the tendency for one factor to behave differently, depending on the particular level setting of the other factor.

11.5.3 There are $4 \times 2 \times r = 8r$ measurements with factor A at 4 levels, factor B at 2 levels and r replications. The sources of variation and associated degrees of freedom are given next.

Source	df
A	3
B	1
Interaction	3
Error	$8r - 8$
Total	$8r - 1$

11.5.5 The complete ANOVA table is shown next. Since factor A is run at 3 levels, it must have 2 df. Other entries are found by similar reasoning.

Source	df	SS	MS	F
A	2	5.3	2.6500	1.30
B	3	9.1	3.0333	1.49
Interaction	6	4.8	0.8000	0.39
Error	12	24.5	2.0417	
Total	23	43.7		

The test statistic for testing the interaction is $F = MS(AB)/MSE = 0.39$ and the rejection region is $F > 3.00$. The p-value is greater than .10 since $F = 0.39$ is less than $F_{.10} = 2.33$. Hence, H_0 is not rejected. There is insufficient evidence to indicate interaction between A and B. The test statistic for testing factor A is $F = 1.30$ with p-value $> .10$. The test statistic for factor B is $F = 1.49$ with p-value $> .10$. Neither A nor B are significant.

11.5.7 Refer to Exercise 5. The 95% confidence interval is

$$(\bar{x}_1 - \bar{x}_2) \pm t_{.025}\sqrt{MSE\left(\dfrac{2}{r}\right)}$$

$$(8.3 - 6.3) \pm 2.179\sqrt{2.0417\left(\dfrac{2}{2}\right)}$$

2.0 ± 3.11 or $-1.11 < \mu_1 - \mu_2 < 5.11$

11.5.9 The nine treatment (cell) totals needed for calculation are shown in the table.

| | Factor A | | | |
Factor B	1	2	3	Total
1	12	16	10	38
2	15	25	17	57
3	25	17	27	69
Total	52	58	54	164

$$CM = \frac{164^2}{18} = 1492.2222 \qquad \text{Total SS} = 1662 - CM = 167.7778$$

$$SSA = \frac{52^2 + 58^2 + 54^2}{6} - CM = 3.1111 \qquad SSB = \frac{38^2 + 57^2 + 69^2}{6} - CM = 81.4444$$

$$SS(AB) = \frac{12^2 + 16^2 + \cdots + 27^2}{2} - SSA - SSB - CM = 62.2222$$

Source	df	SS	MS	F
A	2	3.1111	1.5556	
B	2	81.4444	40.7222	
Interaction	4	62.2222	15.5556	6.67
Error	9	21.0000	2.3333	
Total	17	167.7778		

The test statistic for interaction is $F = MS(AB)/MSE = 6.67$ and the rejection region is $F > 3.63$. There is evidence of a significant interaction. That is, the effect of factor A depends upon the level of factor B at which A is measured.

11.5.11 Refer to Exercise 9. Since the interaction is significant, the differences in the nine factor-level combinations should be explored individually, using an interaction plot such as the one generated by *Minitab* that follows.

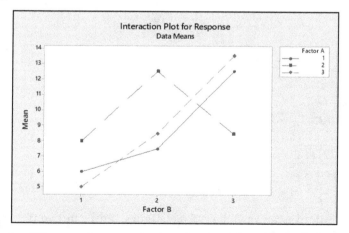

Look at the differences between the three levels of factor A when factor B changes from level 1 to level 2. Levels 2 and 3 behave very similarly while level 1 behaves quite differently. When factor B changes from level 2 to level 3, levels 1 and 3 of factor A behave similarly, and level 2 behaves differently.

11.5.13 Use the computing formulas given in this section or a computer software package to generate the ANOVA table for this 2×3 factorial experiment. The printout that follows was generated using *Minitab*.

ANOVA: Percent Gain versus Markup, Location

Analysis of Variance for Percent Gain

Source	DF	SS	MS	F	P
Markup	2	835.17	417.58	11.87	0.008
Location	1	280.33	280.33	7.97	0.030
Markup*Location	2	85.17	42.58	1.21	0.362
Error	6	211.00	35.17		
Total	11	1411.67			

Model Summary

S	R-sq	R-sq(adj)
5.93015	85.05%	72.60%

a From the printout, $F = 1.21$ with p-value $= .362$. Hence, at the $\alpha = .05$ level, H_0 is not rejected. There is insufficient evidence to indicate interaction.

b Since no interaction is found, the effects of A and B can be tested individually. Both A and B are significant.

c The interaction plot generated by *Minitab* is shown next. Notice that the lines, although not exactly parallel, do not indicate a significant difference in the behavior of the mean responses for the two different locations.

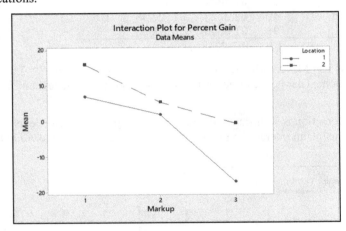

d The 95% confidence interval is

$$(\bar{x}_{31} - \bar{x}_{32}) \pm t_{.025}\sqrt{MSE\left(\frac{2}{r}\right)}$$

$$(-17 + .5) \pm 2.447\sqrt{\frac{211}{6}\left(\frac{2}{2}\right)}$$

$$-16.5 \pm 14.51 \quad \text{or} \quad -31.01 < \mu_{31} - \mu_{32} < -1.99$$

11.5.15 a The design is a 2×4 factorial experiment with $r = 5$ replications. There are two factors, Gender and School, one at two levels and one at four levels.

b The test statistic is $F = MS(School*Gender)/MSE = 1.19$ with p-value $= .329$. Hence, H_0 is not rejected. There is insufficient evidence to indicate interaction between gender and schools.

c You can see in the interaction plot that there is a small difference between the average scores for male and female students at schools 1 and 2, but no difference to speak of at the other two schools. The interaction is not significant. Plots show that schools 2 and 3 have higher scores and school 4 lower scores. The gender difference is slight.

d The test statistic for testing gender is $F = 2.09$ with p-value $= .158$. The test statistic for schools is $F = 27.75$ with p-value $= .000$. There is a significant effect due to schools. Using Tukey's method of paired comparisons with $\alpha = .01$, calculate

$$\omega = q_{.01}(4, 32) \frac{\sqrt{MSE}}{\sqrt{n_t}} = 4.80 \sqrt{\frac{2963.3}{10}} = 82.63$$

The ranked means are shown next.

$$\begin{array}{cccc} 487.1 & 600.1 & 667.4 & 688.6 \\ \overline{x}_4 & \overline{x}_1 & \overline{x}_3 & \overline{x}_2 \end{array}$$

11.5.17 a The analysis of variance table can be found using a computer printout or the following calculations:

Training (A)

Situation (B)	Trained	Not Trained	Total
Standard	334	185	185
Emergency	296	177	473
Total	630	362	992

$CM = \dfrac{992^2}{16} = 61504$ Total SS $= 66640 - CM = 5136$

$SSA = \dfrac{630^2 + 362^2}{8} - CM = 65993 - 61504 = 4489$ $SSB = \dfrac{519^2 + 473^2}{8} - CM = 132.25$

$SS(A \times B) = \dfrac{334^2 + 296^2 + \cdots + 117^2}{4} - SSA - SSB - CM = 56.25$

Source	df	SS	MS	F
A	1	4489.00	4489	117.49
B	1	132.25	132.25	3.46
A×B	1	56.25	56.25	1.47
Error	12	458.50	38.2083	
Total	15	5136.00		

b The test statistic is $F = MS(A \times B)/MSE = 1.47$ and the rejection region is $F > 4.75$ (with $\alpha = .05$). Alternately, you can bound the p-value $> .10$. Hence, H_0 is not rejected. The interaction term is not significant.

c The test statistic is $F = MSB/MSE = 3.46$ and the rejection region is $F > 4.75$ (with $\alpha = .05$). Alternately, you can bound the $.05 < p$-value $< .10$. Hence, H_0 is not rejected. Factor B (Situation) is not significant.

d The test statistic is $F = MSA/MSE = 117.49$ and the rejection region is $F > 4.75$ (with $\alpha = .05$) Alternately, you can bound the p-value $< .005$. Hence, H_0 is rejected. Factor A (Training) is highly significant.

e The interaction plot follows. The response is much higher for the supervisors who have been trained. You can see very little change in the response for the two different situations (standard or emergency). The parallel lines indicate that there is no interaction between the two factors.

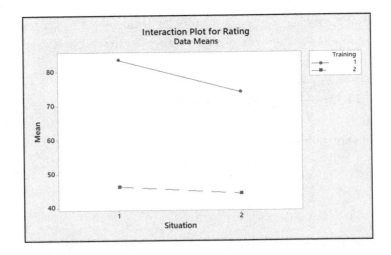

Reviewing What You've Learned

11.R.1 The design is completely randomized with five treatments, containing four, seven, six, five and five measurements, respectively. The analysis of variance table can be found using the computer printout or the following calculations:

$$CM = \frac{\left(\sum x_{ij}\right)^2}{n} = \frac{(20.6)^2}{27} = 15.717$$

$$\text{Total SS} = \sum x_{ij}^2 - CM = 17.500 - CM = 1.783$$

$$SST = \sum \frac{T_i^2}{n_i} - CM = \frac{(2.5)^2}{4} + \frac{(4.7)^2}{7} + \cdots + \frac{(2.4)^2}{5} - CM = 1.212$$

$$SSE = \text{Total SS} - SST = .571$$

a The F test is $F = 11.67$ with p-value $= .000$. The results are highly significant, and H_0 is rejected. There is a difference in mean reaction times due to the five stimuli.

b The hypothesis to be tested is $H_0 : \mu_A = \mu_D$ versus $H_a : \mu_A \neq \mu_D$ and the test statistic is

$$t = \frac{\bar{x}_A - \bar{x}_D}{\sqrt{MSE\left(\frac{1}{n_A} + \frac{1}{n_D}\right)}} = \frac{.625 - .920}{\sqrt{.02596\left(\frac{1}{4} + \frac{1}{5}\right)}} = -2.73$$

The rejection region with $\alpha = .05$ and 22 degrees of freedom is $|t| > t_{.025} = 2.074$ and the null hypothesis is rejected. We conclude that there is a difference between the means.

11.R.3 The residuals in the upper tail of the normal probability plot are smaller than expected, but overall, there is not a problem with normality. The spreads of the residuals when plotted against the fitted values is relatively constant.

11.R.5 Answers will vary from student to student. A completely randomized design has been used. The analysis of variance table is shown in the printout.

One-way ANOVA: 1, 2, 3, 4

Analysis of Variance

Source	DF	Adj SS	Adj MS	F-Value	P-Value
Factor	3	1386	461.93	9.84	0.000
Error	23	1079	46.93		
Total	26	2465			

Model Summary

S	R-sq	R-sq(adj)	R-sq(pred)
6.85061	56.21%	50.50%	39.79%

Means

Factor	N	Mean	StDev	95% CI
1	6	80.33	8.59	(74.55, 86.12)
2	8	91.88	4.91	(86.86, 96.89)
3	5	80.40	4.93	(74.06, 86.74)
4	8	73.50	7.96	(68.49, 78.51)

Pooled StDev = 6.85061

Tukey Pairwise Comparisons

Tukey Simultaneous Tests for Differences of Means

Difference of Levels	Difference of Means	SE of Difference	95% CI	T-Value	Adjusted P-Value
2 - 1	11.54	3.70	(1.31, 21.77)	3.12	0.023
3 - 1	0.07	4.15	(-11.40, 11.54)	0.02	1.000
4 - 1	-6.83	3.70	(-17.06, 3.40)	-1.85	0.278
3 - 2	-11.47	3.91	(-22.27, -0.68)	-2.94	0.035
4 - 2	-18.38	3.43	(-27.85, -8.90)	-5.36	0.000
4 - 3	-6.90	3.91	(-17.70, 3.90)	-1.77	0.314

Individual confidence level = 98.90%

The student should recognize the significant difference in the mean responses for the four training programs, and should further investigate these differences using Tukey's test with ranked means shown next:

4	1	3	2
73.5	80.33	80.4	91.875

11.R.7 The completely randomized design has been used. The analysis of variance table can be obtained using a computer program or the computing formulas.

One-way ANOVA: A, B, C, D

Analysis of Variance

Source	DF	Adj SS	Adj MS	F-Value	P-Value
Factor	3	0.4649	0.15496	5.20	0.011
Error	16	0.4768	0.02980		
Total	19	0.9417			

Model Summary

S	R-sq	R-sq(adj)	R-sq(pred)
0.172627	49.37%	39.87%	20.89%

Means

Factor	N	Mean	StDev	95% CI
A	5	1.5680	0.1366	(1.4043, 1.7317)
B	5	1.7720	0.2160	(1.6083, 1.9357)
C	5	1.5460	0.1592	(1.3823, 1.7097)
D	5	1.9160	0.1689	(1.7523, 2.0797)

Pooled StDev = 0.172627

a To test the difference in treatment means, use $F = \dfrac{\text{MST}}{\text{MSE}} = 5.20$ with p-value $= .011$. H_0 is rejected at the 5% level of significance; there is evidence to suggest a difference in mean discharge for the four plants.

b The hypothesis to be tested is $H_0 : \mu_A = 1.5$ versus $H_a : \mu_A > 1.5$ and the test statistic is

$$t = \frac{\bar{x}_A - \mu_A}{\sqrt{\frac{MSE}{n_A}}} = \frac{1.568 - 1.5}{\sqrt{\frac{.0298}{5}}} = .88$$

The rejection region with $\alpha = .05$ and 16 df is $t > t_{.05} = 1.746$ and the null hypothesis is not rejected. We cannot conclude that the limit is exceeded at plant A.

c The 95% confidence interval for $\mu_A - \mu_D$ is

$$(\bar{x}_A - \bar{x}_D) \pm t_{.025} \sqrt{MSE \left(\frac{1}{n_A} + \frac{1}{n_D} \right)}$$

$$(1.568 - 1.916) \pm 2.12 \sqrt{.0298 \left(\frac{2}{5} \right)}$$

$$-.348 \pm .231 \quad \text{or} \quad -.579 < \mu_A - \mu_D < -.117$$

11.R.9 Answers will vary from student to student. The students should mention the significance of both block and treatment effects. There appear to be no violations of the normality and common variance assumptions. Since the treatment means were significantly different, Tukey's test is used to explore the differences with

$$\omega = q_{.05}(5, 20) \frac{\sqrt{MSE}}{\sqrt{n_t}} = 4.23 \sqrt{\frac{1.9165}{6}} = 2.39$$

The ranked means are shown next.

	E	B	A	C	D
	31.20	32.28	34.35	36.30	36.78

11.R.11 a The experiment is designed as a 2×4 factorial experiment, with cost at $a = 2$ levels, suppliers at $b = 4$ levels and with $r = 3$ replications per factor-level combination.

b Use the computing formulas in Section 11.5 or the *Minitab* printout that follows.

ANOVA: Ratings versus Supplier, Cost

Analysis of Variance for Ratings

Source	DF	SS	MS	F	P
Supplier	3	81.125	27.0417	4.08	0.025
Cost	1	92.042	92.0417	13.89	0.002
Supplier*Cost	3	33.458	11.1528	1.68	0.211
Error	16	106.000	6.6250		
Total	23	312.625			

Model Summary

S	R-sq	R-sq(adj)
2.57391	66.09%	51.26%

c The F test for interaction has a test statistic $F = 1.68$ with p-value $= .211$. The null hypothesis is not rejected and we cannot conclude that there is a significant interaction effect.

d The F test for suppliers has a test statistic $F = 4.08$ with p-value $= .025$. The null hypothesis is rejected and we conclude that there is a significant difference in the ratings for the four suppliers.

e The F test for cost has a test statistic $F = 13.89$ with p-value $= .002$. The null hypothesis is rejected and we conclude that there is a significant difference in the ratings for the two cost levels.

f Answers will vary from student to student.

11.R.13 a The experiment is a 2×3 factorial experiment, with two factors (gender and rank). There are $r = 10$ replications per factor-level combination.

b Use the computing formulas in Section 11.5 or the following *Minitab* printout.

ANOVA: Salary versus Gender, Rank

Analysis of Variance for Salary

Source	DF	SS	MS	F	P
Gender	1	1325.4	1325.4	20.42	0.000
Rank	2	39132.8	19566.4	301.49	0.000
Gender*Rank	2	232.6	116.3	1.79	0.176
Error	54	3504.5	64.9		
Total	59	44195.3			

Model Summary

S	R-sq	R-sq(adj)
8.05598	92.07%	91.34%

c The F test for interaction has a test statistic $F = 1.79$ with p-value $= .176$. The null hypothesis is not rejected and we conclude that there is no significant interaction between rank and gender.

d The F test for rank has a test statistic $F = 301.49$ with p-value $= .000$, and the F test for gender has a test statistic $F = 20.42$ with p-value $= .000$. Both factors are highly significant. We conclude that there is a difference in average salary due to both gender and rank.

e The interaction plot is shown next. Notice the differences in salary due to both rank and gender.

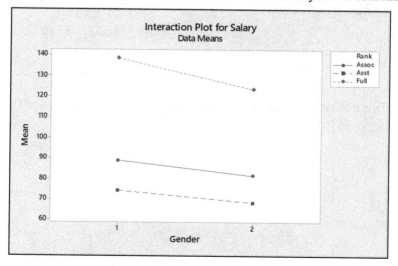

Using Tukey's test is used to explore the differences with

$$\omega = q_{.01}(3, 54) \frac{\sqrt{MSE}}{\sqrt{n_t}} \approx 4.37 \sqrt{\frac{64.9}{20}} = 7.872$$

The ranked means are shown next. All three of the ranks have significantly different average salaries.

Assistant	Associate	Full
71.25	85.18	131.03

On Your Own

11.R.15 The *Minitab* printout for this randomized block experiment follows.

ANOVA: Measurements versus Blocks, Chemicals

Analysis of Variance for Measurements

Source	DF	SS	MS	F	P
Blocks	2	7.1717	3.58583	40.21	0.000
Chemicals	3	5.2000	1.73333	19.44	0.002
Error	6	0.5350	0.08917		
Total	11	12.9067			

Model Summary

S	R-sq	R-sq(adj)
0.298608	95.85%	92.40%

Both the treatment and block means are significantly different. Since the four chemicals represent the treatments in this experiment, Tukey's test can be used to determine where the differences lie:

$$\omega = q_{.05}(4, 6) \frac{\sqrt{MSE}}{\sqrt{n_t}} = 4.90 \sqrt{\frac{.08917}{3}} = .845$$

The ranked means are shown next

11.20	11.40	12.33	12.80
$\bar{x}_3$	$\bar{x}_1$	$\bar{x}_2$	$\bar{x}_4$

The chemical falls into two significantly different groups – A and C versus B and D.

12: Linear Regression and Correlation

Section 12.1

12.1.1 The line corresponding to the equation $y = 2x+1$ can be graphed by locating the y values corresponding to $x = 0, 1,$ and 2.

$$\text{When } x = 0, y = 2(0)+1 = 1$$
$$\text{When } x = 1, y = 2(1)+1 = 3$$
$$\text{When } x = 2, y = 2(2)+1 = 5$$

The graph follows.

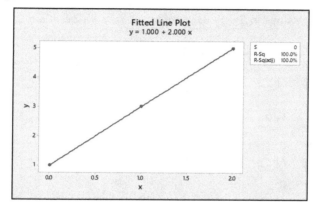

Note that the equation is in the form

$$y = \alpha + \beta x.$$

Thus, the slope of the line is $\beta = 2$ and the y-intercept is $\alpha = 1$.

12.1.3 Similar to Exercise 1. When $x = 0, y = 2(0)+3 = 3$. When $x = 1, y = 2(1)+3 = 5$, and when $x = 2, y = 2(2)+3 = 7$. Since the line is in the form $y = \alpha + \beta x$, the slope of the line is $\beta = 2$ and the y-intercept is $\alpha = 3$. The graph follows.

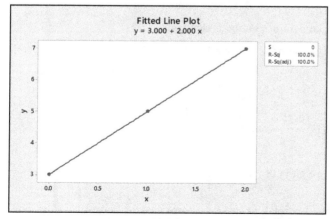

12.1.5 If the y-intercept is $\alpha = -3$ and the slope is $\beta = 1$, the straight line is $y = -3 + x = x - 3$. The graph follows.

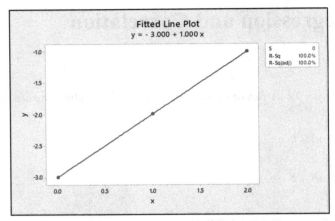

12.1.7 If the y-intercept is $\alpha = -2.5$ and the slope is $\beta = 5$, the straight line is $y = -2.5 + 5x = 5x - 2.5$. The graph follows.

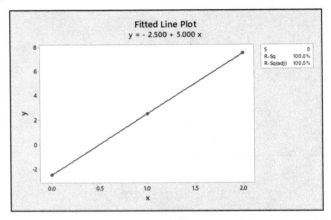

12.1.9 The random errors must 1) be independent in a probabilistic sense 2) have a mean of 0 and a common variance equal to σ^2 and 3) have a normal probability distribution.

12.1.11-13 Use the phrase "y depends on x" or "x is used to predict y" to determine the independent and dependent variables.

    11. y = number of calories burned and x = number of minutes running.

    13. y = number of cones sold and x = temperature on the given day.

12.1.15 The equations for calculating the quantities a and b are found in Section 12.1 of the text and involve the preliminary calculations:

$$\sum x_i = 19;\ \sum y_i = 37;\ \sum x_i^2 = 71;\ \sum y_i^2 = 237;\ \sum x_i y_i = 126$$

$$S_{xy} = \sum x_i y_i - \frac{(\sum x_i)(\sum y_i)}{n} = 126 - \frac{19(37)}{6} = 8.833333$$

$$S_{xx} = \sum x_i^2 - \frac{(\sum x_i)^2}{n} = 71 - \frac{19^2}{6} = 10.833333$$

12.1.17 It is necessary to obtain a prediction equation relating y to x that provides the "best fit" to the data. The "best fitting" line is one which minimizes the sum of squares of the deviations of the observed y-values from the prediction equation. This line, called the "least squares" line, is denoted by

$$\hat{y} = a + bx.$$

The equations for calculating the quantities a and b are found in Section 12.1 of the text and involve the preliminary calculations:

$$S_{xy} = \sum x_i y_i - \frac{(\sum x_i)(\sum y_i)}{n} = 12 - \frac{0(15)}{5} = 12 - 0 = 12$$

$$S_{xx} = \sum x_i^2 - \frac{(\sum x_i)^2}{n} = 10 - \frac{0^2}{5} = 10$$

Calculate $b = \frac{S_{xy}}{S_{xx}} = \frac{12}{10} = 1.2$ and $a = \bar{y} - b\bar{x} = \frac{15}{5} - 1.2(0) = 3$. The least squares line is

$$\hat{y} = a + bx = 3 + 1.2x$$

It is graphed along with the five points. Since the line provides a good fit, the above calculations are probably correct.

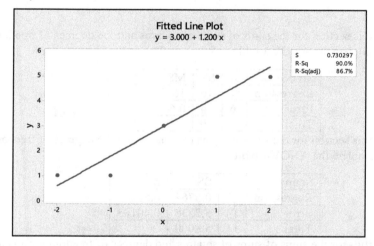

12.1.19 Answers will vary, depending on the type of calculator the student is using. The results should agree with the results of Exercise 17.

12.1.21 a Calculate

$\sum x_i = 104.5$ $\sum y_i = 107.25$ $\sum x_i y_i = 2316.5625$

$\sum x_i^2 = 2319.625$ $\sum y_i^2 = 2322.1875$ $n = 10$

Then

$$S_{xy} = \sum x_i y_i - \frac{(\sum x_i)(\sum y_i)}{n} = 1195.8$$

$$S_{xx} = \sum x_i^2 - \frac{(\sum x_i)^2}{n} = 1227.6$$

$$b = \frac{S_{xy}}{S_{xx}} = \frac{1195.8}{1227.6} = 0.9741 \text{ and } a = \bar{y} - b\bar{x} = 10.725 - (0.9741)(10.45) = 0.546$$

and the least squares line is $\hat{y} = a + bx = 0.546 + 0.974x$.

b The fitted line and the plotted points are shown next, and the fit is very good.

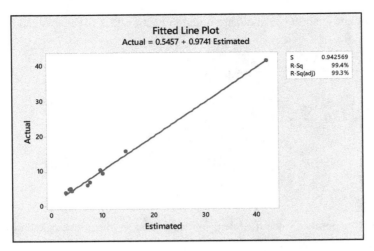

Section 12.2

12.2.1 Use the formulas given in this section for the sums of sums of squares and the degrees of freedom, and the fact that $MS = SS/df$ to complete the ANOVA table.

Source	df	SS	MS
Regression	1	16	16
Error	6	4	0.66667
Total	7	20	

12.2.3 Use the formulas given in this section for the sums of sums of squares and the degrees of freedom, and the fact that $MS = SS/df$ to complete the ANOVA table.

Source	df	SS	MS
Regression	1	0.5762	0.5762
Error	13	5.2238	0.40183
Total	14	5.8000	

12.2.5 Using the additivity properties for the sums of sums of squares and degrees of freedom for an analysis of variance, and the fact that $MS = SS/df$, the completed ANOVA table is shown next.

Source	df	SS	MS
Regression	1	3	3
Error	14	28	2
Total	15	31	

12.2.7 Refer to Exercise 18 (Section 12.1). Calculate

$$S_{xy} = \sum x_i y_i - \frac{(\sum x_i)(\sum y_i)}{n} = 75.3 - \frac{21(24.3)}{6} = 75.3 - 85.05 = -9.75$$

$$S_{xx} = \sum x_i^2 - \frac{(\sum x_i)^2}{n} = 91 - \frac{21^2}{6} = 17.5 \qquad S_{yy} = \sum y_i^2 - \frac{(\sum y_i)^2}{n} = 103.99 - \frac{(24.3)^2}{6} = 5.575$$

Then Total $SS = S_{yy} = \sum y_i^2 - \frac{(\sum y_i)^2}{n} = 5.575$. $\qquad SSR = \frac{(S_{xy})^2}{S_{xx}} = \frac{(-9.75)^2}{17.5} = 5.432143$ and

$$SSE = \text{Total SS} - SSR = S_{yy} - \frac{(S_{xy})^2}{S_{xx}} = 5.575 - 5.432143 = .142857$$

The ANOVA table with 1 df for regression and $n - 2$ df for error follows. Remember that the mean squares are calculated as $MS = SS/df$.

Source	df	SS	MS
Regression	1	5.432143	5.432143
Error	4	0.142857	0.035714
Total	5	5.575	

12.2.9 a The equations for calculating the quantities a and b are found using the preliminary calculations:

$$\sum x_i = 18 \qquad \sum y_i = 2644 \qquad \sum x_i y_i = 8922$$

$$\sum x_i^2 = 64 \qquad \sum y_i^2 = 1,267,528 \qquad n = 6$$

Then

$$S_{xy} = \sum x_i y_i - \frac{(\sum x_i)(\sum y_i)}{n} = 8922 - \frac{18(2644)}{6} = 990$$

$$S_{xx} = \sum x_i^2 - \frac{(\sum x_i)^2}{n} = 64 - \frac{18^2}{6} = 10$$

$$b = \frac{S_{xy}}{S_{xx}} = \frac{990}{10} = 99 \text{ and } a = \bar{y} - b\bar{x} = 440.6667 - 99(3) = 143.6667$$

and the least squares line is $\hat{y} = a + bx = 143.6667 + 99x$.

b The graph of the least squares line and the six data points are shown next.

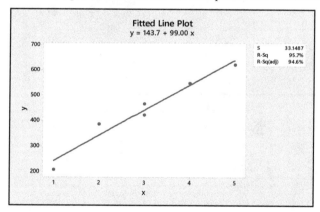

c Calculate Total $SS = S_{yy} = \sum y_i^2 - \frac{(\sum y_i)^2}{n} = 1,267,528 - \frac{(2644)^2}{6} = 102,405.3333$. Then

$$SSR = \frac{(S_{xy})^2}{S_{xx}} = \frac{990^2}{10} = 98,010 \text{ and } \qquad SSE = \text{Total SS} - SSR = S_{yy} - \frac{(S_{xy})^2}{S_{xx}} = 4395.3333$$

The ANOVA table with 1 df for regression and $n - 2$ df for error is shown next. Remember that the mean squares are calculated as $MS = SS/df$.

Source	df	SS	MS
Regression	1	98,010.0000	98,010.0000
Error	4	4395.3333	1098.8333
Total	5	102,405.3333	

12.2.11 a The equations for calculating the quantities a and b are found using the preliminary calculations:

$$\sum x_i = 1490 \qquad \sum y_i = 1978 \qquad \sum x_i y_i = 653,830$$

$$\sum x_i^2 = 540,100 \qquad \sum y_i^2 = 827,504 \qquad n = 5$$

Then

$$S_{xy} = \sum x_i y_i - \frac{(\sum x_i)(\sum y_i)}{n} = 653,830 - \frac{1490(1978)}{5} = 64,386$$

$$S_{xx} = \sum x_i^2 - \frac{(\sum x_i)^2}{n} = 540,100 - \frac{1490^2}{5} = 96,080$$

$$b = \frac{S_{xy}}{S_{xx}} = \frac{64,386}{96,080} = 0.670129 \text{ and } a = \bar{y} - b\bar{x} = 395.6 - 0.670129(298) = 195.902$$

and the least squares line is $\hat{y} = a + bx = 195.90 + 0.67x$.

b The graph of the least squares line and the six data points are shown next.

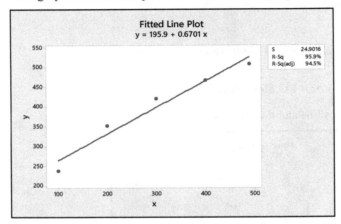

c Calculate Total SS $= S_{yy} = \sum y_i^2 - \frac{(\sum y_i)^2}{n} = 827,504 - \frac{(1978)^2}{5} = 45007.2$. Then

$$SSR = \frac{(S_{xy})^2}{S_{xx}} = \frac{64386^2}{96080} = 43146.9296$$

and $\quad$ SSE = Total SS − SSR $= S_{yy} - \frac{(S_{xy})^2}{S_{xx}} = 1860.2704$

The ANOVA table with 1 df for regression and $n - 2$ df for error is shown next. Remember that the mean squares are calculated as MS = SS/df.

Source	df	SS	MS
Regression	1	43,146.9296	43,146.9296
Error	3	1860.2704	620.0901
Total	4	45,007.2000	

12.2.13 a-b There are $n = 2(5) = 10$ pairs of observations in the experiment, so that the total number of degrees of freedom are $n - 1 = 9$.

c Using the additivity properties for the sums of sums of squares and degrees of freedom for an analysis of variance, and the fact that MS = SS/df, the completed ANOVA table is shown next.

ANOVA	df	SS	MS	F	Significance F
Regression	1	72.2	72.2	14.368	0.0053
Residual	8	40.2	5.025		
Total	9	112.4			

	Coefficients	Standard Error	t Stat	P-value
Intercept	3	2.127	1.411	0.1960
x	0.475	0.125	3.791	0.0053

    d   From the computer printout the least squares line is $\hat{y}=a+bx=3.00+0.475x$.

    e   When $x=10$, the value for y can be predicted using the least squares line as

$$\hat{y}=a+bx=3.00+0.475(10)=7.75.$$

12.2.15 a   The scatterplot generated by *Minitab* follows. The assumption of linearity is reasonable.

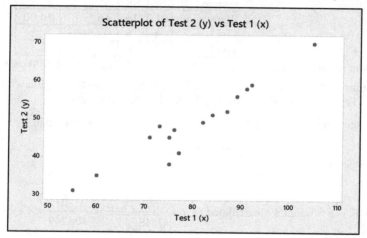

    b   Calculate

$\sum x_i = 1192$ $\sum y_i = 725$ $\sum x_i y_i = 59,324$

$\sum x_i^2 = 96,990$ $\sum y_i^2 = 36,461$ $n = 15$

Then

$$S_{xy} = \sum x_i y_i - \frac{(\sum x_i)(\sum y_i)}{n} = 1710.6667$$

$$S_{xx} = \sum x_i^2 - \frac{(\sum x_i)^2}{n} = 2265.7333 \quad S_{yy} = \sum y_i^2 - \frac{(\sum y_i)^2}{n} = 1419.3333$$

$$b = \frac{S_{xy}}{S_{xx}} = \frac{1710.6667}{2265.7333} = .75502 \text{ and } a = \bar{y} - b\bar{x} = 48.3333 - (0.75502)(79.4667) = -11.665$$

(using full accuracy) and the least squares line is $\hat{y} = a + bx = -11.665 + 0.755x$.

    c   When $x = 85$, the value for y can be predicted using the least squares line as

$$\hat{y} = a + bx = -11.665 + .755(85) = 52.51.$$

d Calculate $SSR = \dfrac{(S_{xy})^2}{S_{xx}} = \dfrac{(1710.666667)^2}{2265.733333} = 1291.58202$ and

$SSE = \text{Total SS} - SSR = S_{yy} - \dfrac{(S_{xy})^2}{S_{xx}} = 1419.333333 - 1291.582024 = 127.751309$

The ANOVA table with 1 *df* for regression and $n - 2$ *df* for error is shown next. Remember that the mean squares are calculated as $MS = SS/df$.

Source	df	SS	MS
Regression	1	1291.582	1291.582
Error	13	127.751	9.827
Total	14	1419.333	

Section 12.3

12.3.1 Refer to the solution for Exercise 1 (Section 12.2) where the ANOVA table was constructed as shown next.

Source	df	SS	MS	F
Regression	1	16	16	24.00
Error	6	4	0.66667	
Total	7	20		

One test statistic for testing $H_0: \beta = 0$ versus $H_a: \beta \neq 0$ is $F = MSR/MSE = 16/0.66667 = 24.0$, shown in the last column of the ANOVA table. With 1 and 6 *df*, the rejection region with $\alpha = .05$ is $F > 5.99$ and H_0 is rejected. There is a significant linear relationship between x and y.

To use a *t* statistic for testing $H_0: \beta = 0$ versus $H_a: \beta \neq 0$, first calculate

$b = \dfrac{S_{xy}}{S_{xx}} = \dfrac{8}{4} = 2$ and $SE = \sqrt{\dfrac{MSE}{S_{xx}}} = \sqrt{\dfrac{0.66667}{4}} = .408249$

The test statistic is a Student's *t*, calculated as $t = \dfrac{b - \beta_0}{\sqrt{MSE/S_{xx}}} = \dfrac{2 - 0}{.408249} = 4.899$

The critical value of *t* is based on $n - 2 = 6$ degrees of freedom and the rejection region for $\alpha = 0.05$ is $|t| > t_{.025} = 2.447$. Since the observed value of *t* falls in the rejection region, we reject H_0 and conclude that $\beta \neq 0$. That is, x is useful in the prediction of y. Notice that $t^2 = (4.899)^2 = 24.000 = F$.

12.3.3 Using the additivity properties for the sums of sums of squares and degrees of freedom for an analysis of variance, and the fact that $F = MSR/MSE$, the completed ANOVA table is shown next.

Source	df	SS	MS	F
Regression	1	4.3	4.3	9.44
Error	18	8.2	0.45556	
Total	19	12.5		

The test statistic for testing $H_0: \beta = 0$ versus $H_a: \beta \neq 0$ is $F = MSR/MSE = 9.44$, shown in the last column of the ANOVA table. With 1 and 18 *df*, the rejection region with $\alpha = .05$ is $F > 4.41$ and H_0 is rejected. There is evidence to suggest a significant linear regression. The proportion of the total variation that is explained by the linear regression of y on x is given by

$r^2 = \dfrac{SSR}{\text{Total SS}} = \dfrac{4.3}{12.5} = .344$

12.3.5 Use the calculations from Exercise 17 (Section 12.1). The hypothesis to be tested is

$$H_0 : \beta = 0 \quad \text{versus} \quad H_a : \beta \neq 0$$

and the test statistic is a Student's t, calculated as

$$t = \frac{b - \beta_0}{\sqrt{MSE/S_{xx}}} = \frac{1.2 - 0}{\sqrt{0.533/10}} = 5.20$$

The critical value of t is based on $n - 2 = 3$ degrees of freedom and the rejection region for $\alpha = 0.01$ is $|t| > t_{.005} = 5.841$. Since the observed value of t does not fall in the rejection region, we do not reject H_0 and do not conclude that $\beta \neq 0$. That is, there is insufficient evidence to conclude that x is useful in the prediction of y.

A $100(1 - \alpha)\%$ confidence interval for β is given as

$$b \pm t_{\alpha/2} \times (\text{std error of } b)$$
$$b \pm t_{\alpha/2} \sqrt{MSE/S_{xx}}$$

For this exercise, the 99% confidence interval is

$$b \pm t_{\alpha/2} \sqrt{MSE/S_{xx}} \Rightarrow 1.2 \pm 5.841 \sqrt{0.5333/10} \Rightarrow 1.2 \pm 1.35$$

or $-0.15 < \beta < 2.55$. Intervals constructed in this manner will enclose β 99% of the time in repeated sampling. Hence, we are fairly confident that this particular interval encloses β.

12.3.7 Use the formula for r^2 given in this section and the calculations in Exercise 12.2.6:

$$r^2 = \frac{SSR}{\text{Total SS}} = \frac{14.4}{16} = 0.90.$$

The coefficient of determination measures the proportion of the total variation in y that is accounted for using the independent variable x. That is, the total variation in y is reduced by 90% by using $\hat{y} = a + bx$ rather than $\bar{y}$ to predict the response y. Note that this fit accounts for a high fraction of the variability, even though we declared the slope to be insignificant in Exercise 12.3.5, where the conclusion was based on a relatively small value of α.

12.3.9 a The scatterplot generated by *Minitab* is shown next. The pattern is fairly linear.

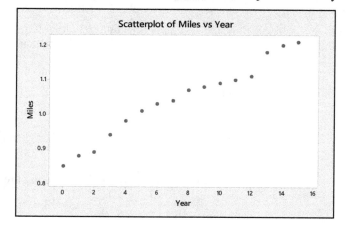

b Calculate

$\sum x_i = 120$ $\qquad \sum y_i = 16.66$ $\qquad \sum x_i y_i = 132.8$

$\sum x_i^2 = 1240$ $\qquad \sum y_i^2 = 17.5336$ $\qquad n = 16$

Then

$$S_{xy} = \sum x_i y_i - \frac{(\sum x_i)(\sum y_i)}{n} = 7.85$$

$$S_{xx} = \sum x_i^2 - \frac{(\sum x_i)^2}{n} = 340 \qquad S_{yy} = \sum y_i^2 - \frac{(\sum y_i)^2}{n} = .186375$$

$$b = \frac{S_{xy}}{S_{xx}} = \frac{7.85}{340} = .023088 \text{ and } a = \bar{y} - b\bar{x} = 1.04125 - (0.023088)(7.5) = .86809$$

and the least squares line is $\hat{y} = a + bx = .86809 + .023088x$. Calculate

$$\text{SSE} = S_{yy} - \frac{(S_{xy})^2}{S_{xx}} = .186375 - \frac{(7.85)^2}{340} = .00513235$$

and $\text{MSE} = \dfrac{\text{SSE}}{n-2} = \dfrac{.00513235}{14} = .000366597$.

The hypothesis to be tested is $H_0 : \beta = 0$ versus $H_a : \beta \neq 0$ and the test statistic is

$$t = \frac{b - \beta_0}{\sqrt{\text{MSE}/S_{xx}}} = \frac{.023088 - 0}{\sqrt{.000366597/340}} = 22.235$$

The critical value of t is based on $n - 2 = 14$ degrees of freedom and the rejection region for $\alpha = 0.05$ is $|t| > t_{.025} = 2.145$. Since the observed value of t falls in the rejection region, we reject H_0 and conclude that $\beta \neq 0$. That is, x is useful in the prediction of y.

c Using the additivity properties for the sums of sums of squares and degrees of freedom for an analysis of variance, and the fact that SSR = $(S_{xy})^2/S_{xx}$, the completed ANOVA table is shown next.

Source	df	SS	MS	F
Regression	1	.181243	.181243	494.389
Error	14	.005132	.000367	
Total	15	.186375		

The test statistic for testing $H_0 : \beta = 0$ versus $H_a : \beta \neq 0$ is $F = \text{MSR}/\text{MSE} = 494.38898$, shown in the last column of the ANOVA table. With 1 and 14 df, the rejection region with $\alpha = .05$ is $F > 4.60$ and H_0 is rejected. There is evidence to suggest a significant linear regression. Notice that

$$\sqrt{F} = \sqrt{494.38898} = 22.235 = t$$

d Calculate $r^2 = \dfrac{\text{SSR}}{\text{Total SS}} = \dfrac{.181243}{.186375} = 0.9725$.

The total variation in y is reduced by 97.25% by using $\hat{y} = a + bx$ rather than $\bar{y}$ to predict the response y. The linear regression is very effective.

12.3.11 a The equations for calculating the quantities a and b involve the following preliminary calculations:

$$\sum x_i = 797 \qquad \sum y_i = 169 \qquad \sum x_i y_i = 13,586$$

$$\sum x_i^2 = 64,063 \qquad \sum y_i^2 = 2887 \qquad n = 10$$

Then

$$S_{xy} = \sum x_i y_i - \frac{(\sum x_i)(\sum y_i)}{n} = 13,586 - \frac{797(169)}{10} = 116.7$$

$$S_{xx} = \sum x_i^2 - \frac{(\sum x_i)^2}{n} = 64,063 - \frac{797^2}{10} = 542.1$$

$$b = \frac{S_{xy}}{S_{xx}} = .215274 \text{ and } a = \bar{y} - b\bar{x} = 16.9 - .215274(79.7) = -.2573$$

and the least squares line is $\hat{y} = a + bx = -.2573 + .2153x$.

b Calculate Total SS $= S_{yy} = \sum y_i^2 - \frac{(\sum y_i)^2}{n} = 2887 - \frac{(169)^2}{10} = 30.9$. Then

$$SSE = S_{yy} - \frac{(S_{xy})^2}{S_{xx}} = 30.9 - \frac{(116.7)^2}{542.1} = 5.777532$$

and $MSE = \frac{SSE}{n-2} = \frac{5.777532}{8} = 0.72219$. The hypothesis to be tested is

$$H_0: \beta = 0 \text{ versus } H_a: \beta \neq 0$$

and the test statistic is $t = \frac{b - \beta_0}{\sqrt{MSE/S_{xx}}} = \frac{.2153 - 0}{\sqrt{0.72219/542.1}} = 5.90$. The critical value of t is based on $n - 2 = 8$ degrees of freedom and the rejection region for $\alpha = 0.05$ is $|t| > t_{.025} = 2.306$, and H_0 is rejected. There is evidence at the 5% level to indicate that x and y are linearly related.

c Calculate $r^2 = \frac{(S_{xy})^2}{S_{xx} S_{yy}} = \frac{(116.7)^2}{(542.1)(30.9)} = 0.813$

Then 81.3% of the total variation in y is accounted for by the independent variable x. That is, the total variation in y is reduced by 81.3% by using $\hat{y} = a + bx$ rather than $\bar{y}$ to predict the response y.

12.3.13 a Refer to Exercise 12.2.11 to find

$$\text{Total SS} = S_{yy} = \sum y_i^2 - \frac{(\sum y_i)^2}{n} = 827,504 - \frac{(1978)^2}{5} = 45,007.2$$

$S_{xx} = 96,080$ and $S_{xy} = 64,386$. Then

$$SSE = S_{yy} - \frac{(S_{xy})^2}{S_{xx}} = 45,007.2 - \frac{(64,386)^2}{96,080} = 1860.2704$$

and $MSE = \frac{SSE}{n-2} = \frac{1860.2704}{3} = 620.09013$. The hypothesis to be tested is

$$H_0: \beta = 0 \text{ versus } H_a: \beta \neq 0$$

and the test statistic is $t = \frac{b - \beta_0}{\sqrt{MSE/S_{xx}}} = \frac{0.670129 - 0}{\sqrt{620.09013/96,080}} = 8.34$

The critical value of t is based on $n - 2 = 3$ degrees of freedom and the observed value $t = 8.34$ is larger than $t_{.005}$ so that p-value < 0.005. Hence, we reject H_0 and conclude that $\beta \neq 0$. That is, x is useful in the prediction of y.

b Calculate $r^2 = \dfrac{S_{xy}^2}{S_{xx}S_{yy}} = \dfrac{64,386^2}{(96,080)(45,007.2)} = 0.959$

The total variation has been reduced by 95.9% by using the linear model.

c Refer to the plot in the solution to Exercise 12.2.11. The points show a curvilinear rather than a linear pattern. Although the fit as measured by r^2 is quite good, it may be that we have fit the wrong type of model to the data.

12.3.15 Refer to Exercise 12.2.16. Calculate

$$S_{xy} = \sum x_i y_i - \dfrac{(\sum x_i)(\sum y_i)}{n} = -14 \qquad S_{xx} = \sum x_i^2 - \dfrac{(\sum x_i)^2}{n} = 16$$

$$S_{yy} = \sum y_i^2 - \dfrac{(\sum y_i)^2}{n} = 12.5 \qquad b = \dfrac{S_{xy}}{S_{xx}} = \dfrac{-14}{16} = -.875 \text{ and } a = \bar{y} - b\bar{x} = 2 - (0.825)(0) = 2$$

$$\text{SSR} = \dfrac{(S_{xy})^2}{S_{xx}} = \dfrac{(-14)^2}{16} = 12.25 \quad \text{and} \quad \text{SSE} = \text{Total SS} - \text{SSR} = S_{yy} - \dfrac{(S_{xy})^2}{S_{xx}} = 12.5 - 12.25 = .25$$

a The best estimate of σ^2 is MSE = .25/3 = .08333.

b The hypothesis to be tested is

$$H_0: \beta = 0 \text{ versus } H_a: \beta \neq 0$$

and the test statistic is a Student's t, calculated as

$$t = \dfrac{b - \beta_0}{\sqrt{\text{MSE}/S_{xx}}} = \dfrac{-.875 - 0}{\sqrt{0.08333/16}} = -12.124$$

The critical value of t is based on $n - 2 = 3$ degrees of freedom and the rejection region for $\alpha = 0.05$ is $|t| > t_{.025} = 3.182$. Since the observed value of t falls in the rejection region, we reject H_0 and conclude that $\beta \neq 0$. That is, texture and storage temperature are linearly related.

c Calculate $r^2 = \dfrac{\text{SSR}}{\text{Total SS}} = \dfrac{12.25}{12.5} = 0.98$

d The total variation has been reduced by 98% by using the linear model.

12.3.17 Refer Exercise 12.1.22. Calculate

$$S_{xy} = \sum x_i y_i - \dfrac{(\sum x_i)(\sum y_i)}{n} = 74.5 \qquad S_{xx} = \sum x_i^2 - \dfrac{(\sum x_i)^2}{n} = 91.375$$

$$S_{yy} = \sum y_i^2 - \dfrac{(\sum y_i)^2}{n} = 67.875 \qquad b = \dfrac{S_{xy}}{S_{xx}} = \dfrac{74.5}{91.375} = .8153215$$

$$\text{SSE} = S_{yy} - \dfrac{(S_{xy})^2}{S_{xx}} = 67.875 - \dfrac{74.5^2}{91.375} = 7.13355 \text{ and MSE} = \dfrac{\text{SSE}}{6} = 1.18892499.$$

a The hypothesis to be tested is $H_0: \beta = 0$ versus $H_a: \beta \neq 0$

and the test statistic is a Student's t, calculated as

$$t = \dfrac{b - \beta_0}{\sqrt{\text{MSE}/S_{xx}}} = \dfrac{.8153}{\sqrt{\dfrac{1.18892499}{91.375}}} = 7.15$$

The rejection region, with $\alpha = .05$, is $|t| > t_{.025,6} = 2.447$ and we reject H_0. That is, there is a linear relationship between arm span and height.

b The 95% confidence interval for the slope β is

$$b \pm t_{\alpha/2}\sqrt{MSE/S_{xx}} \Rightarrow .8153 \pm 2.447\sqrt{\frac{1.18892499}{91.375}} \Rightarrow .8153 \pm .2791$$

or $.5362 < \beta < 1.0944$.

c Since the value $\beta = 1$ is in the confidence interval, da Vinci's supposition in confirmed by the confidence interval in part b.

Section 12.4

12.4.1 Use a normal probability plot of the residuals. The residuals should approximate a straight line, sloping upward.

12.4.3 Use a plot of residuals versus fits. The plot should appear as a random scatter of points, free of any patterns.

12.4.5 The normal probability plot is not too unusual, except for two possible outliers in the upper and lower tails. However, the residuals versus fits plot clearly shows not only the two possible outliers but two distinct clusters. It is possible that the equal variance assumption has been violated.

12.4.7 a If you look carefully, there appears to be a slight curve to the five points.

b The fit of the regression line, measured as $r^2 = 0.959$ indicates that 95.9% of the overall variation can be explained by the straight line model.

c When we look at the residuals there is a strong curvilinear pattern that has not been explained by the straight line model. The relationship between time in months and number of books appears to be curvilinear.

12.4.9 a Since $r^2 = .027$, we know that there is almost *no* linear relationship between price and screen size.

b The normal probability plot does not look unusual, but the residual plot shows an unusual pattern, with five groups of points, depending on the screen size. Within these groups, the variation is similar, but the deviations are not randomly scattered below and above the zero line.

c The scatterplot is shown next. You can see that there is no linear relationship between price and screen size. The unusual pattern in the residual plot probably stems from the fact that you are trying to fit a linear model, when the data does not have a linear pattern.

Section 12.5

12.5.1 The fitted line may not adequately describe the relationship between x and y outside the experimental region.

12.5.3 In order to obtain an estimate for the expected value of y for a given value of x (or for a particular value of y), it would seem reasonable to use the prediction equation, $\hat{y} = a + bx$. Notice that x_0 represents the given value of x for which we are estimating $E(y)$. Calculate

$$S_{xx} = \sum x_i^2 - \frac{(\sum x_i)^2}{n} = 397 - \frac{59^2}{10} = 48.9 \text{ and } MSE = \frac{SSE}{8} = \frac{24}{8} = 3$$

The point estimator for $E(y)$ when $x = 5$ is $\hat{y} = .074 + .46(5) = 2.374$ and the 90% confidence interval is

$$\hat{y} \pm t_{.05}\sqrt{MSE\left(\frac{1}{n} + \frac{(x_0 - \bar{x})^2}{S_{xx}}\right)}$$

$$2.374 \pm 1.860\sqrt{(3)\left(\frac{1}{10} + \frac{(5-5.9)^2}{48.9}\right)}$$

$$2.374 \pm 1.100$$

or $1.274 < E(y) < 3.474$.

12.5.5 Refer to Exercise 3. It is necessary to find a 90% prediction interval for y when $x = 5$. The interval used in predicting a particular value of y is

$$\hat{y} \pm t_{.05}\sqrt{MSE\left(1 + \frac{1}{n} + \frac{(x_0 - \bar{x})^2}{S_{xx}}\right)}$$

$$2.374 \pm 1.860\sqrt{(3)\left(1 + \frac{1}{10} + \frac{(5-5.9)^2}{48.9}\right)}$$

$$2.374 \pm 3.404$$

or $-1.030 < y < 5.778$. Note that the above interval is much wider than the interval calculated for the expected value of y in Exercise 3. The variability of predicting a particular value of y when $x = 5$ is greater than the variability of predicting the average value of y when $x = 5$.

12.5.7 Similar to previous exercises. The point estimator for $E(y)$ when $x = 1$ is $\hat{y} = 3 + 1.2(1) = 4.2$ and the 90% confidence interval is

$$\hat{y} \pm t_{.05}\sqrt{MSE\left(\frac{1}{n} + \frac{(x_0 - \bar{x})^2}{S_{xx}}\right)}$$

$$4.2 \pm 2.353\sqrt{(0.5333)\left(\frac{1}{5} + \frac{(1-0)^2}{10}\right)}$$

$$4.2 \pm 0.941$$

or $3.259 < E(y) < 5.141$.

12.5.9 a From Exercise 12.1.18, the least squares equation is $\hat{y} = a + bx = 6 - 0.557x$. When $x = 2$,

$$\hat{y} = a + bx = 6 - 0.557(2) = 4.886$$

shown in the first line in the section of *Minitab* output labeled "Prediction". The 95% confidence interval for $E(y)$ appears in the third column as $4.60061 < E(y) < 5.17081$.

12.5.11 The value $\hat{y} = 1.54286$ corresponds to $x = \frac{y-a}{b} = \frac{1.54286-6}{-0.557} = 8.00$. This value is far from the average value of x, $\bar{x} = 3.5$, and outside the experimental region $1 \leq x \leq 6$. This *extrapolation* can cause a problem with inaccurate predictions.

12.5.13 a Although very slight, the student might notice a slight curvature to the data points.

b The fit of the linear model is very good, assuming that this is *indeed* the correct model for this data set.

c The normal probability plot follows the correct pattern for the assumption of normality. However, the residuals show the pattern of a quadratic curve, indicating that a quadratic rather than a linear model may have been the correct model for this data.

12.5.15 Use the preliminary calculations from Exercise 12.3.15.

a The point estimator for $E(y)$ when $x = -1$ is $\hat{y} = 2 - .875(-1) = 2.875$ and the 99% confidence interval is

$$\hat{y} \pm t_{.005} \sqrt{\text{MSE}\left(\frac{1}{n} + \frac{(x_0 - \bar{x})^2}{S_{xx}}\right)}$$

$$2.875 \pm 5.841 \sqrt{(.08333)\left(\frac{1}{5} + \frac{(-1-0)^2}{16}\right)}$$

$$2.875 \pm .864$$

or $2.011 < E(y) < 3.739$.

b The point estimator for y when $x = 1$ is $\hat{y} = 2 - .875(1) = 1.125$ and the 99% prediction interval is

$$\hat{y} \pm t_{.005} \sqrt{\text{MSE}\left(1 + \frac{1}{n} + \frac{(x_0 - \bar{x})^2}{S_{xx}}\right)}$$

$$1.125 \pm 5.841 \sqrt{(.08333)\left(1 + \frac{1}{5} + \frac{(1-0)^2}{16}\right)}$$

$$1.125 \pm 1.895$$

or $-.77 < y < 3.02$.

c The width will be the narrowest when $x_0 = \bar{x} = 0$.

12.5.17 Refer to Exercise 12.5.16 and calculate

$$S_{xy} = \sum x_i y_i - \frac{(\sum x_i)(\sum y_i)}{n} = 3573.5 \qquad S_{xx} = \sum x_i^2 - \frac{(\sum x_i)^2}{n} = 332$$

$$S_{yy} = \sum y_i^2 - \frac{(\sum y_i)^2}{n} = 77,824.4375$$

$$b = \frac{S_{xy}}{S_{xx}} = \frac{3573.5}{332} = 10.7635542 \text{ and } a = \bar{y} - b\bar{x} = 282.1875 - 10.7635542(22.5) = 40.00753$$

$$\text{SSE} = S_{yy} - \frac{(S_{xy})^2}{S_{xx}} = 77824.4375 - \frac{3573.5^2}{332} = 39360.87651$$

and $\text{MSE} = \frac{\text{SSE}}{n-2} = \frac{39360.87651}{14} = 2811.491179$.

a The point estimator for $E(y)$ when $x = 20$ is $\hat{y} = 40.00753 + 10.76355(20) = 255.279$ and the 95% confidence interval is

$$\hat{y} \pm t_{.025} \sqrt{MSE\left(\frac{1}{n} + \frac{(x_0 - \bar{x})^2}{S_{xx}}\right)}$$

$$255.279 \pm 2.145 \sqrt{2811.491179 \left(\frac{1}{16} + \frac{(20-22.5)^2}{332}\right)}$$

$$255.279 \pm 32.435$$

or $222.844 < E(y) < 287.714$.

b The point estimator for y when $x = 20$ is still $\hat{y} = 40.00753 + 10.76355(20) = 255.279$ and the 95% prediction interval is

$$\hat{y} \pm t_{.025} \sqrt{MSE\left(1 + \frac{1}{n} + \frac{(x_0 - \bar{x})^2}{S_{xx}}\right)}$$

$$255.279 \pm 2.145 \sqrt{2811.491179 \left(1 + \frac{1}{16} + \frac{(20-22.5)^2}{332}\right)}$$

$$255.279 \pm 118.270$$

or $137.009 < y < 373.549$.

c This would not be advisable, since you are trying to estimate outside the range of experimentation.

Section 12.6

12.6.1 The significance of the algebraic sign and the magnitude of r, the coefficient of correlation, will be discussed in Exercise 12.6.2. Note, however, that r^2 provides a meaningful measure of the strength of the linear relationship between two variables, y and x. It is the ratio of the reduction in the sum of squares of deviations obtained using the model, $y = \alpha + \beta x + \varepsilon$, to the sum of squares of deviations that would be obtained if the variable x were ignored. That is, r^2 measures the amount of variation that can be attributed to the variable x.

12.6.3 When all the points fall exactly on the fitted line, $SSE = S_{yy} - \frac{(S_{xy})^2}{S_{xx}} = 0$ so that $(S_{xy})^2 = S_{xx}S_{yy}$. Then

$$\frac{(S_{xy})^2}{S_{xx}S_{yy}} = r^2 = 1 \text{ and } r = \pm 1.$$

a If the line has a positive slope, $r = +1$

b If the line has a negative slope, $r = -1$

12.6.5 The scatterplot is shown next. The sample correlation coefficient will be negative.

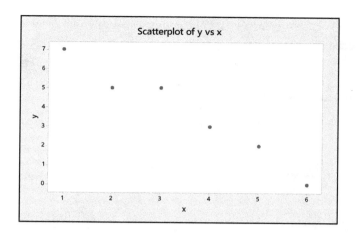

Calculate

$$S_{xy} = \sum x_i y_i - \frac{(\sum x_i)(\sum y_i)}{n} = 54 - \frac{21(22)}{6} = -23$$

$$S_{xx} = \sum x_i^2 - \frac{(\sum x_i)^2}{n} = 91 - \frac{21^2}{6} = 17.5 \qquad S_{yy} = \sum y_i^2 - \frac{(\sum y_i)^2}{n} = 112 - \frac{22^2}{6} = 31.33333$$

Then $r = \dfrac{S_{xy}}{\sqrt{S_{xx}S_{yy}}} = \dfrac{-23}{\sqrt{17.5(31.3333)}} = -0.982$.

The coefficient of determination is $r^2 = (-0.982)^2 = 0.9647$. This value implies that the sum of squares of deviations is reduced by 96.47% using the linear model $\hat{y} = a + bx$ instead of $\bar{y}$ to predict values of y.

12.6.7 The data from Exercise 12.6.5 are reused here, except that the y observations are reordered. The only calculation that has changed from the previous exercise is

$$S_{xy} = \sum x_i y_i - \frac{(\sum x_i)(\sum y_i)}{n} = 100 - \frac{21(22)}{6} = 23$$

Refer to the scatterplot. The sample correlation coefficient will be positive.

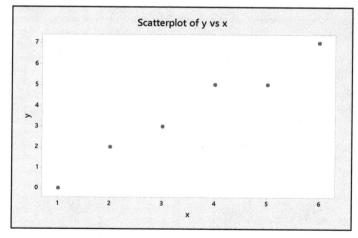

Calculate $r = \dfrac{S_{xy}}{\sqrt{S_{xx}S_{yy}}} = \dfrac{23}{\sqrt{17.5(31.3333)}} = 0.982$ and the coefficient of determination

$r^2 = (0.982)^2 = 0.9647$. This value implies that the sum of squares of deviations is reduced by 96.47%

using the linear model $\hat{y} = a + bx$ instead of $\bar{y}$ to predict values of y. Since the value of r is near 1, a strong positive linear association between the two variables is implied. Note that this value of r is the negative of the value calculated for r in Exercise 12.6.5.

12.6.9 a When x is large, y should be small if the barnacles compete for space on the lobster's surface. Hence, we would expect to find negative correlation.

b-c The test of hypothesis is $H_0 : \rho = 0$ versus $H_a : \rho < 0$

Calculate

$$S_{xy} = \sum x_i y_i - \frac{(\sum x_i)(\sum y_i)}{n} = 42,556 - \frac{2379(652)}{10} = -112,554.8$$

$$S_{xx} = \sum x_i^2 - \frac{(\sum x_i)^2}{n} = 973,255 - \frac{2379^2}{10} = 407,290.9 \quad S_{yy} = \sum y_i^2 - \frac{(\sum y_i)^2}{n} = 114,624 - \frac{652^2}{10} = 102,113.4$$

Then $r = \frac{S_{xy}}{\sqrt{S_{xx} S_{yy}}} = \frac{-112,544.8}{\sqrt{407,290.9(102,113.6)}} = -0.5519$ and the test statistic is

$$t = \frac{r\sqrt{n-2}}{\sqrt{1-r^2}} = \frac{-0.5519\sqrt{8}}{\sqrt{1-(-0.5519)^2}} = -1.872$$

The rejection region for $\alpha = 0.05$ is $t < -t_{.05} = -1.860$ and H_0 is rejected. There is evidence of negative correlation.

12.6.11 a The hypothesis of interest is $H_0 : \rho = 0$ versus $H_a : \rho \neq 0$ and the test statistic is

$$t = \frac{r\sqrt{n-2}}{\sqrt{1-r^2}} = \frac{-0.37\sqrt{67}}{\sqrt{1-(-0.37)^2}} = -3.260$$

The rejection region is $|t| > t_{.025} \approx 1.96$ and H_0 is rejected. There is evidence of correlation between x and y.

b The p-value can be bounded using Table 4 as

$$p\text{-value} = 2P(t > 3.26) < 2(0.005) = 0.01$$

c The negative correlation observed above implies that, if the skater's stride is large, his time to completion will be small.

12.6.13 Using a computer program, your scientific calculator or the computing formulas given in the text to calculate the correlation coefficient r.

$$S_{xy} = \sum x_i y_i - \frac{(\sum x_i)(\sum y_i)}{n} = 1,901,500 - \frac{8050(2100)}{9} = 23,166.667$$

$$S_{xx} = \sum x_i^2 - \frac{(\sum x_i)^2}{n} = 7,802,500 - \frac{8050^2}{9} = 602,222.22 \quad S_{yy} = \sum y_i^2 - \frac{(\sum y_i)^2}{n} = 498,200 - \frac{2100^2}{9} = 8200$$

Then $r = \frac{S_{xy}}{\sqrt{S_{xx} S_{yy}}} = \frac{23,166.667}{\sqrt{602,222.22(8200)}} = 0.3297$. The test of hypothesis is

$H_0 : \rho = 0$ versus $H_a : \rho > 0$

and the test statistic is $t = \frac{r\sqrt{n-2}}{\sqrt{1-r^2}} = \frac{0.3297\sqrt{7}}{\sqrt{1-(0.3297)^2}} = 0.92$ with p-value bounded as

p-value $= P(t > 0.92) > 0.10$

The results are not significant; H$_0$ is not rejected. There is insufficient evidence to indicate a positive correlation between average maximum drill hole depth and average maximum temperature.

12.6.15 a Using a computer program, your scientific calculator or the computing formulas given in the text to calculate the correlation coefficient r.

$$S_{xy} = \sum x_i y_i - \frac{(\sum x_i)(\sum y_i)}{n} = 88,140.6 - \frac{1180.3(896)}{12} = 11.5333$$

$$S_{xx} = \sum x_i^2 - \frac{(\sum x_i)^2}{n} = 116,103.03 - \frac{1180.3^2}{12} = 10.689167$$

$$S_{yy} = \sum y_i^2 - \frac{(\sum y_i)^2}{n} = 67,312 - \frac{896^2}{12} = 410.6667$$

Then $r = \dfrac{S_{xy}}{\sqrt{S_{xx} S_{yy}}} = \dfrac{11.5333}{\sqrt{10.689167(410.6667)}} = 0.1741$.

b The test of hypothesis is $H_0: \rho = 0$ versus $H_a: \rho \neq 0$ and the test statistic is

$$t = \frac{r\sqrt{n-2}}{\sqrt{1-r^2}} = \frac{0.1741\sqrt{10}}{\sqrt{1-(0.1741)^2}} = 0.559$$

The rejection region with $\alpha = .05$ is $|t| > t_{.025} = 2.228$ and H$_0$ is not rejected. There is insufficient evidence to indicate a correlation between body temperature and heart rate.

12.6.17 a Use a computer program, your scientific calculator or the computing formulas given in the text to calculate the correlation coefficient r.

$$S_{xy} = \sum x_i y_i - \frac{(\sum x_i)(\sum y_i)}{n} = 1,233,987 - \frac{5028(2856)}{12} = 37,323$$

$$S_{xx} = \sum x_i^2 - \frac{(\sum x_i)^2}{n} = 2,212,178 - \frac{5028^2}{12} = 105,446$$

$$S_{yy} = \sum y_i^2 - \frac{(\sum y_i)^2}{n} = 723,882 - \frac{2856^2}{12} = 44,154$$

Then $r = \dfrac{S_{xy}}{\sqrt{S_{xx} S_{yy}}} = \dfrac{37,323}{\sqrt{105,446(44,154)}} = 0.5470$.

The test of hypothesis is $H_0: \rho = 0$ versus $H_a: \rho > 0$ and the test statistic is

$$t = \frac{r\sqrt{n-2}}{\sqrt{1-r^2}} = \frac{0.5470\sqrt{10}}{\sqrt{1-(0.5470)^2}} = 2.066$$

with p-value $= P(t > 2.066)$ bounded as $0.05 < p$-value < 0.10.

If the experimenter is willing to tolerate a p-value this large, then H$_0$ can be rejected. Otherwise, you would declare the results not significant; there is insufficient evidence to indicate that bending stiffness and twisting stiffness are positively correlated.

b $r^2 = (0.5470)^2 = 0.2992$ so that 29.9% of the total variation in y can be explained by the independent variable x.

Reviewing What You've Learned

12.R.1 a The calculations shown next are done using the computing formulas. An appropriate computer program will provide identical results to within rounding error.

$$\sum x_i = 720 \qquad \sum y_i = 324 \qquad \sum x_i y_i = 17{,}540$$

$$\sum x_i^2 = 49{,}200 \qquad \sum y_i^2 = 9540 \qquad n = 12$$

Then

$$S_{xy} = \sum x_i y_i - \frac{(\sum x_i)(\sum y_i)}{n} = 17{,}540 - \frac{720(324)}{12} = -1900$$

$$S_{xx} = \sum x_i^2 - \frac{(\sum x_i)^2}{n} = 49{,}200 - \frac{720^2}{12} = 6000$$

$$S_{yy} = \sum y_i^2 - \frac{(\sum y_i)^2}{n} = 9540 - \frac{324^2}{12} = 792$$

$b = \dfrac{S_{xy}}{S_{xx}} = \dfrac{-1900}{6000} = -0.317$ and $a = \bar{y} - b\bar{x} = 27 - (-0.317)(60) = 46.000$ and the least squares line is

$$\hat{y} = a + bx = 46 - 0.317x.$$

b The graph of the least squares line and the 12 data points are shown next.

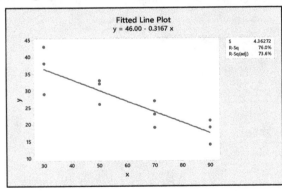

c Since Total SS $= S_{yy} = 792$ and

$$\text{SSR} = \frac{(S_{xy})^2}{S_{xx}} = \frac{(-1900)^2}{6000} = 601.6667$$

Then $\quad \text{SSE} = \text{Total SS} - \text{SSR} = S_{yy} - \dfrac{(S_{xy})^2}{S_{xx}} = 190.3333$

The ANOVA table with 1 df for regression and $n - 2$ df for error follows. Remember that the mean squares are calculated as $\text{MS} = \text{SS}/df$.

Source	df	SS	MS
Regression	1	601.6667	601.6667
Error	10	190.3333	19.0333
Total	11	792.0000	

d The diagnostic plots for the regression analysis follow. Although the second plot indicates that the responses are slightly more variable when $x = 30$, and the normality plot is slightly irregular, these irregularities are probably not significant, given the small sample size.

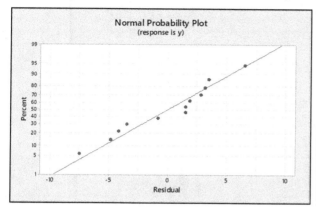
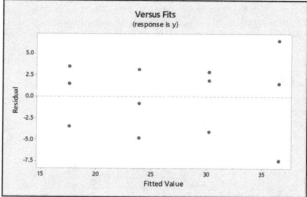

e The 95% confidence interval for β is

$$b \pm t_{.025} \sqrt{\frac{MSE}{S_{xx}}} \Rightarrow -0.317 \pm 2.228 \sqrt{\frac{19.0333}{6000}} \Rightarrow -0.317 \pm 0.125$$

or $-0.442 < \beta < -0.192$.

f When $x = 50$, the estimate of mean potency $E(y)$ is $\hat{y} = 46.00 - 0.317(50) = 30.167$ and the 95% confidence interval is

$$\hat{y} \pm t_{.025} \sqrt{MSE\left(\frac{1}{n} + \frac{(x_0 - \bar{x})^2}{S_{xx}}\right)}$$

$$30.167 \pm 2.228 \sqrt{19.0333\left(\frac{1}{12} + \frac{(50-60)^2}{6000}\right)}$$

$$30.167 \pm 3.074$$

or $27.09 < E(y) < 33.24$.

g The predictor for y when $x = 50$ is $\hat{y} = 30.167$ and the 95% prediction interval is

$$\hat{y} \pm t_{.025}\sqrt{MSE\left(1+\frac{1}{n}+\frac{(x_0-\bar{x})^2}{S_{xx}}\right)}$$

$$30.167 \pm 2.228\sqrt{19.0333\left(1+\frac{1}{12}+\frac{(50-60)^2}{6000}\right)}$$

$$30.167 \pm 10.195$$

or $19.97 < y < 40.36$.

12.R.3 The calculations shown next are done using the computing formulas. An appropriate computer program will provide identical results to within rounding error.

$$S_{xy} = \sum x_i y_i - \frac{(\sum x_i)(\sum y_i)}{n} = 3343 - \frac{42(891)}{12} = 224.5$$

$$S_{xx} = \sum x_i^2 - \frac{(\sum x_i)^2}{n} = 162 - \frac{42^2}{12} = 15$$

$$S_{yy} = \sum y_i^2 - \frac{(\sum y_i)^2}{n} = 69,653 - \frac{891^2}{12} = 3496.25$$

a The correlation coefficient is $r = \frac{S_{xy}}{\sqrt{S_{xx}S_{yy}}} = \frac{224.5}{\sqrt{15(3496.25)}} = .9803$.

b The coefficient of determination is $r^2 = 0.9610$ (or 96.1%).

c Calculate $b = \frac{S_{xy}}{S_{xx}} = \frac{224.5}{15} = 14.9666667$ and $a = \bar{y} - b\bar{x} = 74.25 - (14.9666667)(3.5) = 21.8667$

and the least squares line is

$$\hat{y} = a + bx = 21.8667 + 14.9667x.$$

d We wish to estimate the mean percentage of kill for an application of 4 pounds of nematicide per acre. Since the percent kill y is actually a binomial percentage, the variance of y will change depending on the value of p, the proportion of nematodes killed for a particular application rate. The residual plot versus the fitted values shows this phenomenon as a "football-shaped" pattern. The normal probability plot also shows some deviation from normality in the tails of the plot. A transformation may be needed to assure that the regression assumptions are satisfied.

12.R.5 a Let x = EL and y = API. Then

$$S_{xy} = \sum x_i y_i - \frac{(\sum x_i)(\sum y_i)}{n} = 241,125 - \frac{6236(319)}{8} = -7535.5$$

$$S_{xx} = \sum x_i^2 - \frac{(\sum x_i)^2}{n} = 4,878,636 - \frac{6236^2}{8} = 17,674$$

$$S_{yy} = \sum y_i^2 - \frac{(\sum y_i)^2}{n} = 17,141 - \frac{319^2}{8} = 4420.875$$

Then $r = \frac{S_{xy}}{\sqrt{S_{xx}S_{yy}}} = \frac{-7535.5}{\sqrt{17674(4420.875)}} = -.852$.

The test of hypothesis is $H_0 : \rho = 0$ versus $H_a : \rho \neq 0$ and the test statistic is

$$t = \frac{r\sqrt{n-2}}{\sqrt{1-r^2}} = \frac{0.852\sqrt{6}}{\sqrt{1-(0.852)^2}} = -3.99$$

with p-value $= 2P(t > 3.99)$ bounded as p-value $< 2(.005) = .01$. There is evidence of a significant correlation between the two variables.

12.R.7 The relationship between $y = $ penetrability and $x = $ number of days is apparently nonlinear, as seen by the strong curvilinear pattern in the residual plot. The regression analysis discussed in this chapter is not appropriate; we will discuss the appropriate model in Chapter 13.

12.R.9 **a**   The calculations shown next are done using the computing formulas. An appropriate computer program will provide identical results to within rounding error.

$$\sum x_i = 150 \qquad \sum y_i = 91 \qquad \sum x_i y_i = 986$$

$$\sum x_i^2 = 2750 \qquad \sum y_i^2 = 1120.04 \qquad n = 10$$

$$S_{xy} = \sum x_i y_i - \frac{(\sum x_i)(\sum y_i)}{n} = 986 - \frac{150(91)}{10} = -379$$

$$S_{xx} = \sum x_i^2 - \frac{(\sum x_i)^2}{n} = 2750 - \frac{150^2}{10} = 500$$

$$S_{yy} = \sum y_i^2 - \frac{(\sum y_i)^2}{n} = 1120.04 - \frac{91^2}{10} = 291.94$$

Then $b = \dfrac{S_{xy}}{S_{xx}} = \dfrac{-379}{500} = -.758$ and $a = \bar{y} - b\bar{x} = 9.1 - (-.758)(15) = 20.47$

and the least squares line is $\hat{y} = a + bx = 20.47 - .758x$.

b   Since Total SS $= S_{yy} = 291.94$ and SSR $= \dfrac{(S_{xy})^2}{S_{xx}} = \dfrac{(-379)^2}{500} = 287.282$, then

$$\text{SSE} = \text{Total SS} - \text{SSR} = S_{yy} - \frac{(S_{xy})^2}{S_{xx}} = 4.658$$

The ANOVA table with 1 *df* for regression and $n - 2$ *df* for error follows. Remember that the mean squares are calculated as MS $=$ SS/df.

Source	df	SS	MS
Regression	1	287.282	287.282
Error	8	4.658	.58225
Total	9	291.940	

c   To test $H_0 : \beta = 0, H_a : \beta \neq 0$, the test statistic is

$$t = \frac{b - \beta_0}{s/\sqrt{S_{xx}}} = \frac{-.758}{\sqrt{.58225/500}} = -22.21$$

The rejection region for $\alpha = 0.05$ is $|t| > t_{.025} = 2.306$ and we reject H_0. There is sufficient evidence to indicate that x and y are linearly related.

d   The 95% confidence interval for the slope β is

$$b \pm t_{\alpha/2}\sqrt{MSE/S_{xx}} \Rightarrow -.758 \pm 2.896\sqrt{.58225/500} \Rightarrow -.758 \pm .099$$

or $-.857 < \beta < -.659$.

e When $x = 14$, the estimate of expected freshness $E(y)$ is $\hat{y} = 20.47 - .758(14) = 9.858$ and the 95% confidence interval is

$$\hat{y} \pm t_{.025}\sqrt{MSE\left(\frac{1}{n} + \frac{(x_0 - \bar{x})^2}{S_{xx}}\right)}$$

$$9.858 \pm 2.306\sqrt{.58225\left(\frac{1}{10} + \frac{(14-15)^2}{500}\right)}$$

$$9.858 \pm .562$$

or $9.296 < E(y) < 10.420$.

f Calculate

$$r^2 = \frac{SSR}{\text{Total SS}} = \frac{287.282}{291.94} = 0.984$$

The total variation has been reduced by 98.4%% by using the linear model.

12.R.11 a-b Answers will vary. The *Minitab* printout is shown, along with two diagnostic plots. These plots show one unusual observation, giving it an unusually large influence on the estimate of the regression line. The printout indicates a significant linear regression ($t = 36.21$ with *p*-value = .000) and the regression line is given as $\hat{y} = 0.546 + 0.9741x$.

Regression Analysis: Actual versus Estimated

Analysis of Variance

Source	DF	Adj SS	Adj MS	F-Value	P-Value
Regression	1	1164.82	1164.82	1311.09	0.000
Error	8	7.11	0.89		
Total	9	1171.93			

Model Summary

S	R-sq	R-sq(adj)
0.942569	99.39%	99.32%

Coefficients

Term	Coef	SE Coef	T-Value	P-Value
Constant	0.546	0.410	1.33	0.220
Estimated	0.9741	0.0269	36.21	0.000

Regression Equation

Actual = 0.546 + 0.9741 Estimated

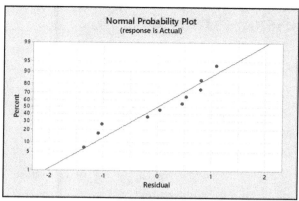 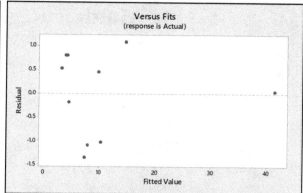

12.R.13 Answers will vary from student to student.

13: Multiple Regression Analysis

Section 13.2

13.2.1 For the equation $E(y) = 3 + x_1 - 2x_2$:

When $x_2 = 0$, $E(y) = 3 + x_1 - 2(0) = x_1 + 3$.
When $x_2 = 1$, $E(y) = 3 + x_1 - 2(1) = x_1 + 1$.
When $x_2 = 2$, $E(y) = 3 + x_1 - 2(2) = x_1 - 1$.

When $x_1 = 0$, $E(y) = 3 + 0 - 2x_2 = 3 - 2x_2$.
When $x_1 = 1$, $E(y) = 3 + 1 - 2x_2 = 4 - 2x_2$.
When $x_1 = 2$, $E(y) = 3 + 2 - 2x_2 = 5 - 2x_2$.

These three straight lines parallel (they have the same slope) and are graphed as follows.

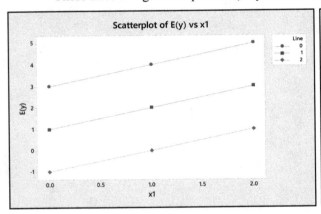
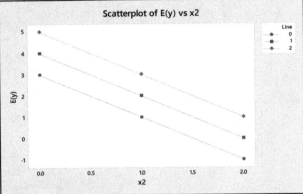

13.2.3 For the equation $E(y) = 2 + 2x_1 - x_2$:

When $x_2 = 0$, $E(y) = 2 + 2x_1 - 2(0) = 2 + 2x_1$.
When $x_2 = 1$, $E(y) = 2 + 2x_1 - 2(1) = 2x_1$.
When $x_2 = 2$, $E(y) = 2 + 2x_1 - 2(2) = -2 + 2x_1$.

When $x_1 = 0$, $E(y) = 2 + 2(0) - x_2 = 2 - x_2$.
When $x_1 = 1$, $E(y) = 2 + 2(1) - x_2 = 4 - x_2$.
When $x_1 = 2$, $E(y) = 2 + 2(2) - x_2 = 6 - x_2$.

These three straight lines parallel (they have the same slope) and are graphed as follows.

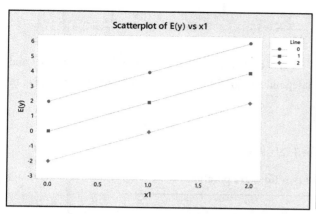
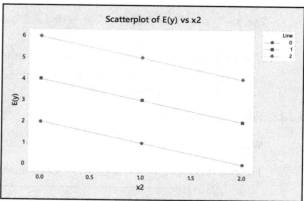

13.2.5 The second portion of the printout shows the terms in the model and the partial regression coefficients. The model is $E(y) = \beta_0 + \beta_1 x_1 + \beta_2 x_2 + \beta_3 x_3$ and the least squares prediction equation is

$$\hat{y} = 9.22 + .225 x_1 - 5.29 x_2 - .059 x_3$$

Using the additivity of the sum of squares and the degrees of freedom and the fact that MS = SS/df, the analysis of variance table can be filled in as follows:

Source	DF	Adj SS	Adj MS	F-Value	P-Value
Regression	3	20.17	6.723	1.33	0.348
Error	6	30.33	5.038		
Total	9	50.40			

To test the hypothesis $H_0 : \beta_1 = \beta_2 = \beta_3 = 0$ H_a: at least one β_i differs from zero, we use the test statistic

$$F = \frac{\text{MSR}}{\text{MSE}} = 1.33 \text{ with P-Value} = 0.348$$

Since the p-value is much larger than .10, we do not reject H_0. The regression is not significant.

13.2.7 The significance of the partial regression coefficients is measured by the p-values associated with the individual t tests. For this exercise, all of the p-values are greater than .10, so that none of the variables—x_1, x_2 and x_3—are significant in the presence of the other predictors already in the model.

The percentage of the total variation in the experiment explained by the model is measured using

$$R^2 = \frac{SSR}{TotalSS} = \frac{20.17}{50.40} = .40$$

Since only 40% of the variation is explained by the model, the model is not very effective.

13.2.9 The hypothesis to be tested is $H_0 : \beta_1 = \beta_2 = \beta_3 = 0$ H_a: at least one β_i differs from zero

and the test statistic is $F = \frac{\text{MSR}}{\text{MSE}} = 57.44$ which has an F distribution with $df_1 = k = 3$ and $df_2 = n - k - 1 = 15 - 3 - 1 = 11$. The rejection region for $\alpha = .05$, which is found in the upper tail of the F distribution, is $F > 3.59$ and H_0 is rejected. Alternatively, the p-value with $df_1 = k = 3$ and $df_2 = n - k - 1 = 15 - 3 - 1 = 11$ can be bounded (using Table 6 in Appendix I) as p-value < .005. There is evidence that the model contributes information for the prediction of y.

13.2.11 The hypotheses of interest are $H_0 : \beta_i = 0$ $H_a : \beta_i \neq 0$ for $i = 1, 2, 3$ and the test statistic is $t = \frac{b_i - \beta_i}{SE(b_i)}$.

For $i = 1$, $t = \dfrac{1.29}{.42} = 3.071$ For $i = 2$, $t = \dfrac{2.72}{.65} = 4.185$ For $i = 3$, $t = \dfrac{.41}{.17} = 2.412$.

If you choose $\alpha = .05$, the rejection region, with $n - (k+1) = 11$ degrees of freedom is $|t| > t_{.025} = 2.201$ and all three hypotheses are rejected. All of the three independent variables contribute to the prediction of y, in the presence of the other two variables.

13.2.13 When $x_2 = 1$ and $x_3 = 0$, $\hat{y} = 3.76 + 1.29x_1$.

When $x_2 = 1$ and $x_3 = .5$, $\hat{y} = (1.04 + 2.72 + .205) + 1.29x_1$ or $\hat{y} = 3.965 + 1.29x_1$.

The two lines are graphed together. Notice that the two lines are parallel.

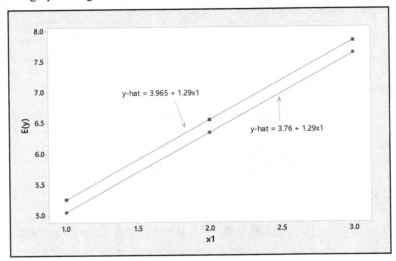

13.2.15 a The *Minitab* printout fitting the model to the data is shown next. The least squares line is

$$\hat{y} = -8.177 + 0.292x_1 + 4.434x_2$$

```
Regression Analysis: y versus x1, x2

Analysis of Variance
Source      DF   Adj SS   Adj MS   F-Value   P-Value
Regression   2   355.22   177.61    16.28     0.002
Error        7    76.38    10.91
Total        9   431.60

Model Summary
   S      R-sq   R-sq(adj)
3.30335   82.30%   77.25%

Coefficients
Term       Coef    SE Coef   T-Value   P-Value
Constant  -8.177    4.21      -1.94     0.093
x1         0.292    0.136      2.15     0.068
x2         4.434    0.800      5.54     0.001

Regression Equation
y  =  -8.18 + 0.292 x1 + 4.434 x2
```

b The F test for the overall utility of the model is $F = 16.28$ with $P = .002$. The results are highly significant; the model contributes significant information for the prediction of y.

c To test the effect of advertising expenditure, the hypothesis of interest is

$$H_0 : \beta_2 = 0, \quad H_a : \beta_2 \neq 0$$

and the test statistic is $t = 5.54$ with p-value $= .001$. Since $\alpha = .01$, H_0 is rejected. We conclude that advertising expenditure contributes significant information for the prediction of y, given that capital investment is already in the model.

d From the ***Minitab*** printout, R-Sq = 82.30%, which means that 82.3% of the total variation can be explained by the quadratic model. The model is very effective.

13.2.17 a The values of R^2(adj) should be used to compare several different regression models. For the nine possible models given in the *Minitab* output, the largest value of R^2(adj) is 40.3% which occurs when the following model is fit to the data:

$$y = \beta_0 + \beta_1 x_2 + \beta_2 x_3 + \beta_3 x_5 + \varepsilon$$

b Since the values of R^2 and R^2(adj) are both very small, even the best of the models is not very useful for predicting overall score based on these three independent variables.

Section 13.3

13.3.1 The model is quadratic.

13.3.3 The hypothesis to be tested is $H_0 : \beta_1 = \beta_2 = 0 \quad H_a$: at least one β_i differs from zero and the test statistic is

$$F = \frac{\text{MSR}}{\text{MSE}} = 989.89$$

which, when H_0 is true, has an F distribution has an F distribution with $df_1 = k = 2$ and $df_2 = n - k - 1 = 20 - 2 - 1 = 17$. The p-value given in the printout is P = .000 and H_0 is rejected. There is evidence that the model contributes information for the prediction of y.

13.3.5 Refer to the printout. The prediction equation is $\hat{y} = 12.48 + 9.89x - 2.329x^2$ and is graphed as follows.

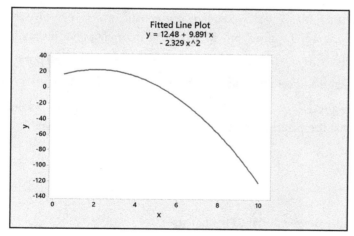

13.3.7 Since $E(y) = \beta_0 + \beta_1 x + \beta_2 x^2$, when $x = 0$, $E(y) = \beta_0$. A test of $E(y$ given $x = 0) = 0$ is equivalent to a test of $H_0 : \beta_0 = 0 \quad H_a : \beta_0 \neq 0$. The individual t-test is

$$t = \frac{b_0}{SE(b_0)} = 3.68 \text{ with } p\text{-value} = .002$$

and H₀ is rejected. The mean value of y differs from zero when $x = 0$.

13.3.9 The hypothesis of interest is $H_0 : \beta_2 = 0$, $H_a : \beta_2 \neq 0$ and the individual t-test is

$$t = \frac{b_2}{SE(b_2)} = -16.91$$

with p-value $= .000$ and H₀ is rejected. There is evidence to indicate curvature.

13.3.11 The individual t-test is $t = -16.91$ as in Exercise 13.3.9. However, the test is one-tailed, which means that the p-value is half of the amount given in the printout. That is, $p\text{-value} = \frac{1}{2}(.000) = .000$. Hence, H₀ is again rejected. There is evidence to indicate a decreasing rate of increase.

13.3.13 Refer to Exercise 13.3.12.

    a From the printout, SSR $= 234.955$ and Total SS $= S_{yy} = 236.015$. Then

$$R^2 = \frac{SSR}{\text{Total SS}} = \frac{234.955}{236.015} = .9955$$

which agrees with the printout.

    b Calculate $R^2(\text{adj}) = \left(1 - \frac{MSE}{\text{Total SS}/(n-1)}\right)100\% = \left(1 - \frac{.353}{236.015/5}\right)100\% = 99.25\%$

    c The value of $R^2(\text{adj})$ can be used to compare two or more regression models using different numbers of independent predictor variables. Since the value of $R^2(\text{adj}) = 99.25\%$ is just slightly larger than the value of $R^2(\text{adj}) = 95.66\%$ for the linear model, the quadratic model fits just slightly better.

13.3.15 a $R^2 = .9985$. Hence, 99.85% of the total variation is accounted for by using x and x^2 in the model.

    b The hypothesis of interest is $H_0 : \beta_1 = \beta_2 = 0$ and the test statistic is $F = 1676.610$ with p-value $= .000$. H₀ is rejected and we conclude that the model provides valuable information for the prediction of y.

    c The hypothesis of interest is $H_0 : \beta_1 = 0$ and the test statistic is $t = -2.652$ with p-value $= .045$. H₀ is rejected and we conclude that the linear regression coefficient is significant when x^2 is in the model.

    d The hypothesis of interest is $H_0 : \beta_2 = 0$ and the test statistic is $t = 15.138$ with p-value $= .000$. H₀ is rejected and we conclude that the quadratic regression coefficient is significant when x is in the model.

Section 13.4

13.4.1 Quantitative

13.4.3 Qualitative ($x_1 = 1$ if B; 0 otherwise $x_2 = 1$ if C; 0 otherwise)

13.4.5 Qualitative ($x_1 = 1$ if day shift; 0 otherwise)

13.4.7 Refer to Exercise 13.4.6,

    When $x_2 = 0$, $E(y) = 3 + x_1 - 2(0) = x_1 + 3$.

When $x_2 = 2$, $E(y) = 3 + x_1 - 2(2) = x_1 - 1$.

When $x_2 = -2$, $E(y) = 3 + x_1 - 2(-2) = x_1 + 7$.

These three straight lines are graphed next.

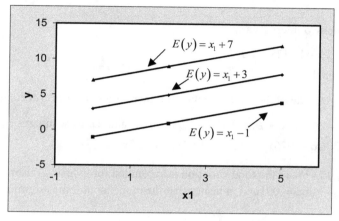

13.4.9 Notice that the lines are no longer parallel. The inclusion of the term $x_1 x_2$ allows the slope of the line to changes as x_2 changes. This allows for "interaction", which is a difference in the relationship between x_1 and y for different values of x_2.

13.4.11 When $x_1 = 0$, $\hat{y} = 12.6 + 3.9x_2^2$ while when $x_1 = 1$,

$$\hat{y} = 12.6 + .54(1) - 1.2x_2 + 3.9x_2^2$$
$$= 13.14 - 1.2x_2 + 3.9x_2^2$$

13.4.13 The basic response equation for a specific type of bonding compound would be

$$E(y) = \beta_0 + \beta_1 x_1 + \beta_2 x_1^2$$

Since the qualitative variable "bonding compound" is at two levels, one dummy variable is needed to incorporate this variable into the model. Define the dummy variable x_2 as follows:

$x_2 = 1$ if bonding compound 2
$\quad = 0$ otherwise

The expanded model is now written as

$$E(y) = \beta_0 + \beta_1 x_1 + \beta_2 x_1^2 + \beta_3 x_2 + \beta_4 x_1 x_2 + \beta_5 x_1^2 x_2 \text{ or}$$
$$y = \beta_0 + \beta_1 x_1 + \beta_2 x_1^2 + \beta_3 x_2 + \beta_4 x_1 x_2 + \beta_5 x_1^2 x_2 + \varepsilon$$

13.4.15 **a** From the *Minitab* printout, $F = 69.83$ (with P-Value $= .000$) and $R^2 = .9544$ can be used to test the overall utility of the model, and we conclude that model contributes significant information for the prediction of y. There are no obvious violations of the regression assumptions, since the patterns in the diagnostic plots are as expected when the regression assumptions are satisfied.

b The equation for chicken uses $x_1 = 0$ and the prediction equation is

$$\hat{y} = 23.57 + 7.75x_2$$

while the equation for beef uses $x_1 = 1$ and the prediction equation is

$$\hat{y} = (23.57 + 69.00) + (7.750 - 12.29)x_2 = 92.57 - 4.54x_2$$

c Using the second prediction equation from part **b**, the point estimate is

$$\hat{y} = 92.57 - 4.54(8) = 56.25$$

Because of rounding error, this value is slightly different from the value marked "Fit" in the computer printout.

d From the printout, the two intervals are

$$48.4387 < E(y) < 64.1328 \text{ and}$$
$$44.1291 < y < 68.4423$$

Since you are predicting outside of the experimental unit, there is a danger of inaccurate predictions!

13.4.17 a The model is $E(y) = \beta_0 + \beta_1 x_1 + \beta_2 x_2 + \beta_3 x_3 + \beta_4 x_4 + \beta_5 x_5$ and the least squares equation is found from the printout: $\hat{y} = 15 - .306 x_1 + .076 x_2 - 48.1 x_3 + 1.93 x_4 + 1.127 x_5$

b $R^2 = .9541$. Hence, 95.41% of the total variation is accounted for by using this model. The F-test yields $F = 24.94$ with p-value $= .001$ and indicates that the model contributes significant information for the prediction of y.

c Scan down the list of p-values for the individual t-tests. The only variable with p-values less than .05 is $x_5 =$ API score from the previous year. This is a useful independent variable, in the presence of the others.

d The best model has the largest value of $R^2 (\text{adj}) = 93.3\%$ and includes variables x_3, x_4, and x_5. It would not be wise to "extrapolate" to next year, since the appropriate model may be quite different.

Reviewing What You've Learned

13.R.1 a From the printout, the prediction equation is $\hat{y} = 8.59 + 3.821x - 0.2166x^2$.

b R^2 is labeled "R-sq" or $R^2 = .9436$. Hence 94.36% of the total variation in y is accounted for by using x and x^2 in the model.

c The hypothesis of interest is $H_0 : \beta_1 = \beta_2 = 0$ H_a: at least one β_i differs from zero and the test statistic is $F = 33.44$ with p-value $= .003$. Hence, H_0 is rejected, and we conclude that the model contributes significant information for the prediction of y.

d The hypothesis of interest is $H_0 : \beta_2 = 0$ $H_a: \beta_2 \neq 0$ and the test statistic is $t = -4.93$ with p-value $= .008$. Hence, H_0 is rejected, and we conclude that the quadratic model provides a better fit to the data than a simple linear model.

e The pattern of the diagnostic plots does not indicate any obvious violation of the regression assumptions.

13.R.3 The *Minitab* printout for the data is shown next.

Regression Analysis: y versus x1, x2, x3, x1x2, x1x3

Analysis of Variance

Source	DF	Adj SS	Adj MS	F-Value	P-Value
Regression	5	74.830	14.9661	110.50	0.000
Error	9	1.219	0.1354		
Total	14	76.049			

Model Summary

S	R-sq	R-sq(adj)
0.368028	98.40%	97.51%

Coefficients

Term	Coef	SE Coef	T-Value	P-Value
Constant	14.100	0.386	36.53	0.000
x1	1.040	0.116	8.94	0.000
x2	3.530	0.546	6.47	0.000
x3	4.760	0.546	8.72	0.000
x1x2	-0.430	0.165	-2.61	0.028
x1x3	-0.080	0.165	-0.49	0.639

Regression Equation

y = 14.100 + 1.040 x1 + 3.530 x2 + 4.760 x3 - 0.430 x1x2 - 0.080 x1x3

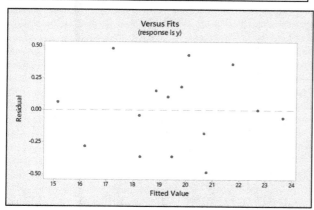
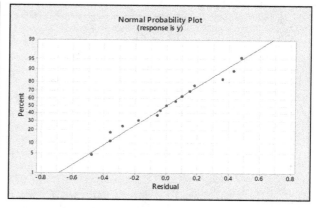

a The model fits very well, with an overall $F = 110.50$ (P-Value = .000) and $R^2 = .9840$. The diagnostic plots indicate no violations of the regression assumptions.

b The parameter estimates are found in the column marked "Coef" and the prediction equation is

$$\hat{y} = 14.100 + 1.040x_1 + 3.530x_2 + 4.760x_3 - 0.430x_1x_2 - 0.080x_1x_3$$

Using the dummy variables defined in Exercise 13.R.2, the coefficients can be combined to give the three lines that are graphed in the figure that follows.

Men: $\hat{y} = 14.100 + 1.040x_1$

Children: $\hat{y} = 17.630 + 0.610x_1$

Women: $\hat{y} = 18.860 + 0.960x_1$

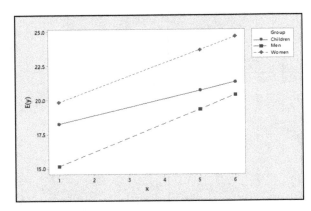

c The hypothesis of interest is $H_0: \beta_4 = 0$ $H_a: \beta_4 \neq 0$ and the test statistic is $t = -2.61$ with P-Value $= .028$. Since this value is less than .05, the results are significant at the 5% level of significance and H_0 is rejected. There is a difference in the slopes.

d The hypothesis of interest is $H_0: \beta_4 = \beta_5 = 0$ $H_a:$ at least one β_i differs from zero for $i = 4, 5$

Using the methods of Section 13.5 and the *Minitab* printout above, $SSE_2 = 1.219$ with 9 degrees of freedom, while the printout that follows, fit using the reduced model gives $SSE_1 = 2.265$ with 11 degrees of freedom.

Regression Analysis: y versus x1, x2, x3

Analysis of Variance

Source	DF	Adj SS	Adj MS	F-Value	P-Value
Regression	3	73.784	24.5948	119.44	0.000
Error	11	2.265	0.2059		
Total	14	76.049			

Model Summary

S	R-sq	R-sq(adj)
0.453772	97.02%	96.21%

Coefficients

Term	Coef	SE Coef	T-Value	P-Value
Constant	14.610	0.321	45.53	0.000
x1	0.8700	0.0828	10.50	0.000
x2	2.240	0.287	7.81	0.000
x3	4.520	0.287	15.75	0.000

Regression Equation

y = 14.610 + 0.8700 x1 + 2.240 x2 + 4.520 x3

Hence, the degrees of freedom associated with $SSE_1 - SSE_2 = 1.046$ is $11 - 9 = 2$. The test statistic is

$$F = \frac{(SSE_1 - SSE_2)/2}{SSE_2/9} = \frac{1.046/2}{.1354} = 3.86$$

The rejection region with $\alpha = .05$ is $F > F_{.05} = 4.26$ (with 2 and 9 df) and H_0 is not rejected. The interaction terms in the model are not significant. The experimenter should consider eliminating these terms from the model.

e Answers will vary.

13.R.5 a The model is $y = \beta_0 + \beta_1 x_1 + \beta_2 x_2 + \beta_3 x_1^2 + \beta_4 x_1 x_2 + \beta_5 x_1^2 x_2 + \varepsilon$ and the *Minitab* printout follows.

> **Regression Analysis: y versus x1, x2, x1-sq, x1x2, x1-sqx2**
>
> Analysis of Variance
>
> | Source | DF | Adj SS | Adj MS | F-Value | P-Value |
> |---|---|---|---|---|---|
> | Regression | 5 | 664164 | 132833 | 25.85 | 0.000 |
> | Error | 39 | 200434 | 5139 | | |
> | Total | 44 | 864598 | | | |
>
> Model Summary
>
> | S | R-sq | R-sq(adj) |
> |---|---|---|
> | 71.6891 | 76.82% | 73.85% |
>
> Coefficients
>
> | Term | Coef | SE Coef | T-Value | P-Value |
> |---|---|---|---|---|
> | Constant | 4.509 | 42.2 | 0.11 | 0.916 |
> | x1 | 6.394 | 5.78 | 1.11 | 0.275 |
> | x2 | -50.854 | 56.2 | -0.90 | 0.371 |
> | x1-sq | 0.132 | 0.169 | 0.78 | 0.439 |
> | x1x2 | 17.064 | 7.10 | 2.40 | 0.021 |
> | x1-sqx2 | -0.502 | 0.199 | -2.52 | 0.016 |
>
> Regression Equation
>
> y = 4.5 + 6.39 x1 - 50.9 x2 + 0.132 x1-sq + 17.06 x1x2 - 0.502 x1-sqx2

b The fitted prediction model uses the coefficients given in the column marked "Coef" in the printout:

$$\hat{y} = 4.509 + 6.394x_1 - 50.854x_2 + 17.064x_1x_2 + .132x_1^2 - .502x_1^2x_2$$

The F test for the model's utility is $F = 25.85$ with P = .000 and $R^2 = .7682$. The model fits quite well.

c If the dolphin is female, $x_2 = 0$ and the prediction equation becomes

$$\hat{y} = 4.509 + 6.394x_1 + .132x_1^2$$

d If the dolphin is male, $x_2 = 1$ and the prediction equation becomes

$$\hat{y} = -46.345 + 23.458x_1 - .370x_1^2$$

e The hypothesis of interest is

$$H_0: \beta_4 = 0 \quad H_a: \beta_4 \neq 0$$

and the test statistic is $t = .78$ with p-value $= .439$. H_0 is not rejected and we conclude that the quadratic term is not important in predicting mercury concentration for female dolphins.

13.R.7 a The correlation matrix for Cost, y, x_1, x_2, x_3, x_4, and x_5 was obtained using *Minitab* and is shown as follows:

Correlation: Price($), y, x1, x2, x3, x4, x5

Correlations

	Price($)	y	x1	x2	x3	x4
y	0.682 0.004					
x1	0.562 0.024	0.446 0.083				
x2	0.275 0.302	0.312 0.239	-0.200 0.458			
x3	0.582 0.018	0.553 0.026	0.124 0.647	0.372 0.156		
x4	-0.020 0.941	0.399 0.126	-0.149 0.582	-0.149 0.582	0.277 0.298	
x5	0.472 0.065	0.601 0.014	0.432 0.095	0.011 0.968	0.241 0.370	0.322 0.224

Cell Contents
  Pearson correlation
  P-Value

Scan down the first column labeled "Price($)" and look for correlations with p-values less than .05. You will see that Price is significantly correlated with y = overall score (p-value = .004), x_1 (p-value = .024) and x_3 (p-value = .018) and *almost* significantly correlated with x_5 (p-value = .065). Now scan down the second column labeled "y", and you will find that x_3 and x_5 both have p-values less than 0.05, indicating a significant correlation with y.

b-c Using all five independent variables, the model is $E(y) = \beta_0 + \beta_1 x_1 + \beta_2 x_2 + \beta_3 x_3 + \beta_4 x_4 + \beta_5 x_5$

and the *Minitab* printout follows.

Regression Analysis: y versus x1, x2, x3, x4, x5

Analysis of Variance

Source	DF	Adj SS	Adj MS	F-Value	P-Value
Regression	5	390.0	77.99	5.34	0.012
Error	10	146.0	14.60		
Total	15	536.0			

Model Summary

S	R-sq	R-sq(adj)
3.82133	72.76%	59.13%

Coefficients

Term	Coef	SE Coef	T-Value	P-Value
Constant	-31.7	24.2	-1.31	0.220
x1	11.22	5.07	2.21	0.051
x2	4.74	2.41	1.97	0.077
x3	2.70	2.98	0.91	0.386
x4	5.44	2.78	1.96	0.079
x5	0.0868	0.0827	1.05	0.318

Regression Equation

y = -31.7 + 11.22 x1 + 4.74 x2 + 2.70 x3 + 5.44 x4 + 0.0868 x5

Notice that $R^2 = 72.76\%$ and $R^2(adj) = 59.13\%$. Hence, 72.76% of the total variation is accounted for by using this regression model, so that the model does a reasonable job of explaining the inherent variability in y. The overall model is significant, with $F = 5.34$ and p-value = .012.

d The *Minitab* Best Subsets program was run to isolate the best model and the printout follows:

Response is y

Vars	R-Sq	R-Sq (adj)	R-Sq (pred)	Mallows Cp	S	x1	x2	x3	x4	x5
1	36.1	31.5	18.0	11.5	4.9474					X
1	30.6	25.7	11.6	13.5	5.1540			X		
2	53.8	46.7	28.0	7.0	4.3639			X		X
2	45.4	37.0	16.6	10.0	4.7443		X			X
3	67.6	59.5	*	3.9	3.8040	X	X		X	
3	57.7	47.1	*	7.5	4.3465	X		X		X
4	70.5	59.8	*	4.8	3.7903	X	X		X	X
4	69.7	58.7	*	5.1	3.8393	X	X	X	X	
5	72.8	59.1	*	6.0	3.8213	X	X	X	X	X

The model with the largest value for R^2(adj) is a model including all but x_3 (R^2(adj) = 59.8%). The reduced model is

$$E(y) = \beta_0 + \beta_1 x_1 + \beta_2 x_2 + \beta_3 x_4 + \beta_4 x_5$$

and the *Minitab* printout follows:

Regression Analysis: y versus x1, x2, x4, x5

Analysis of Variance

Source	DF	Adj SS	Adj MS	F-Value	P-Value
Regression	4	378.0	94.49	6.58	0.006
Error	11	158.0	14.37		
Total	15	536.0			

Model Summary

S	R-sq	R-sq(adj)
3.79025	70.52%	59.80%

Coefficients

Term	Coef	SE Coef	T-Value	P-Value
Constant	-34.4	23.8	-1.44	0.177
x1	12.48	4.84	2.58	0.026
x2	5.81	2.08	2.79	0.018
x4	6.42	2.54	2.52	0.028
x5	0.0855	0.0820	1.04	0.319

Regression Equation

y = -34.4 + 12.48 x1 + 5.81 x2 + 6.42 x4 + 0.0855 x5

Notice that $R^2(adj) = 59.80\%$, which means that the reduced model is *slightly* better than the full model $R^2(adj) = 59.13\%$ for explaining the inherent variability in *y*. The overall model is now *highly* significant, with $F = 6.58$ and *p*-value = .006.

13.R.9 The hypothesis of interest is $H_0: \beta_1 = \beta_2 = \beta_3 = 0$ and the test statistic is $F = 1.49$ with *p*-value = .235. H_0 is not rejected and we conclude that the model does not provide valuable information for the prediction of *y*. This matches the results of the analysis of variance F-test.

14: Analysis of Categorical Data

Section 14.2

14.2.1 See Section 14.1 of the text.

14.2.3-5 Index Table 5, Appendix I, with χ^2_α and the appropriate degrees of freedom.

    3. $\chi^2_{.01} = 20.0902$ **5.** $\chi^2_{.05} = 11.0705$

14.2.7-9 For a test of specified cell probabilities, the degrees of freedom are $k-1$. Use Table 5, Appendix I:

    7. $df = 9$; $\chi^2_{.01} = 21.666$; reject H₀ if $X^2 > 21.666$

    9. $df = 4$; $\chi^2_{.05} = 9.48773$; reject H₀ if $X^2 > 9.48773$

14.2.11-13 Refer to Table 5 in Appendix I. Since the chi-square test is one-tailed, the *p*-value is the probability in the tail of the chi-square distribution ($df = k-1$) to the right of the observed value of the test statistic X^2.

    11. Since $X^2 = 20.62$ is greater than $\chi^2_{.005} = 18.5476$, the *p*-value $= P(X^2 > 20.62)$ is less than .005.

    13. Since $X^2 = 9.56$ is between $\chi^2_{.01} = 9.21034$ and $\chi^2_{.005} = 10.5966$, the *p*-value $= P(X^2 > 9.56)$ is between .005 and .01.

14.2.15 We are given three of the four probabilities with $p_1 = .25, p_2 = .15, p_3 = .10$. Since we must have $\sum p_i = 1$, the fourth probability must be $p_4 = 1 - .25 - .15 - .10 = .50$ and the expected cell counts are calculated as

$$E_1 = np_1 = 250(.25) = 62.5 \qquad E_2 = np_2 = 250(.15) = 37.5$$

$$E_3 = np_3 = 250(.10) = 25 \qquad E_4 = np_4 = 250(.50) = 125$$

14.2.17 Three hundred responses were each classified into one of three categories. The hypothesis of interest is $H_0 : p_1 = .4; p_2 = .3; p_3 = .3$ versus the alternative that at least one of the p_i is different from the value specified in H₀.

The number of degrees of freedom is equal to the number of cells, k, less one degree of freedom for each linearly independent restriction placed on $p_1, p_2, \ldots, p_k$. For this exercise, $k = 3$ and one degree of freedom is lost because of the restriction that $\sum p_i = 1$. Hence, X^2 has $k - 1 = 2$ degrees of freedom.

The rejection region for this test is located in the upper tail of the chi-square distribution with $df = 2$. From Table 5, the appropriate upper-tailed rejection region is $X^2 > \chi^2_{.05} = 5.99147$.

The test statistic is $X^2 = \sum \dfrac{(O_i - E_i)^2}{E_i}$ which, when n is large, possesses an approximate chi-square distribution in repeated sampling. The values of O_i are the actual counts *observed* in the experiment, and

$$E_i = np_i = 300 p_i \quad \text{for } i = 1, 2, 3.$$

A table of observed and expected cell counts follows:

Category	1	2	3
O_i	130	98	72
E_i	120	90	90

The test statistic is

$$X^2 = \frac{(130-120)^2}{120} + \frac{(98-90)^2}{90} + \frac{(72-90)^2}{90} = 5.14$$

which has an approximate chi-square (if H_0 is true) with $df = k-1 = 2$. From Table 5, with 2 degrees of freedom, the observed value of $X^2 = 5.14$ falls between $\chi^2_{.05}$ and $\chi^2_{.10}$. Hence, $.05 < p\text{-value} < .10$. Since the p-value is not less than .05 and the test statistic is not in the rejection region, the results are not statistically significant (although some researchers might say they are *tending toward significance*). That is, we do not have sufficient evidence to conclude that the cell probabilities are different from the hypothesized values.

14.2.19 The null hypothesis to be tested is $H_0: p_1 = \frac{1}{8}; p_2 = \frac{1}{8}; p_3 = \frac{1}{8}; p_4 = \frac{1}{8}; p_5 = \frac{2}{8}; p_6 = \frac{2}{8}$ against the alternative that at least one of these probabilities is incorrect. A table of observed and expected cell counts follows:

Day	Monday	Tuesday	Wednesday	Thursday	Friday	Saturday
O_i	95	110	125	75	181	214
E_i	100	100	100	100	200	200

The test statistic is

$$X^2 = \frac{(95-100)^2}{100} + \frac{(110-100)^2}{100} + \cdots + \frac{(214-200)^2}{200} = 16.535$$

The number of degrees of freedom is $k-1=5$ and the rejection region with $\alpha = .05$ is $X^2 > \chi^2_{.05} = 11.07$ and H_0 is rejected. The manager's claim is refuted.

14.2.21 The flower fall into one of four classifications, with theoretical ratio 9:3:3:1.

Converting these ratios to probabilities,

$p_1 = 9/16 = .5625$ $\qquad p_2 = 3/16 = .1875$

$p_3 = 3/16 = .1875$ $\qquad p_4 = 1/16 = .0625$

We will test the null hypothesis that the probabilities are as above against the alternative that they differ. The table of observed and expected cell counts follows:

	AB	Ab	aB	aa
O_i	95	30	28	7
E_i	90	30	30	10

The test statistic is

$$X^2 = \frac{(95-90)^2}{90} + \frac{(30-30)^2}{30} + \frac{(28-30)^2}{30} + \frac{(7-10)^2}{10} = 1.311$$

The number of degrees of freedom is $k-1=3$ and the rejection region with $\alpha = .01$ is $X^2 > \chi^2_{.01} = 11.3449$. Since the observed value of X^2 does not fall in the rejection region, we do not reject H_0. We do not have enough information to contradict the theoretical model for the classification of flower color and shape.

14.2.23 It is necessary to determine whether proportion of non-accidental deaths at a given hospital differ from the known population proportions. A table of observed and expected cell counts follows:

Disease	HD	C	RD	S	Other	Totals
O_i	78	81	28	16	105	308
E_i	72.072	69.300	17.248	15.708	133.672	308

The null hypothesis to be tested is

$$H_0 : p_1 = .234;\ p_2 = .225;\ p_3 = .056;\ p_4 = .051$$

against the alternative that at least one of these probabilities is incorrect. The test statistic is

$$X^2 = \frac{(78-72.072)^2}{72.072} + \frac{(81-69.300)^2}{69.300} + \cdots + \frac{(105-133.672)^2}{133.672} = 15.321$$

The number of degrees of freedom is $k-1=4$ and, since the observed value of $X^2 = 15.321$ is greater than $\chi^2_{.005}$, the p-value is less than .005 and the results are declared highly significant. We reject H₀ and conclude that the proportions of non-accidental deaths from certain diseases at this hospital differ from the proportions for the larger population.

14.2.25 Similar to previous exercises. The peas fall into one of four classifications, with theoretical ratio 9:3:3:1. Converting these ratios to probabilities,

$$p_1 = 9/16 = .5625 \qquad p_2 = 3/16 = .1875$$
$$p_3 = 3/16 = .1875 \qquad p_4 = 1/16 = .0625$$

We will test the null hypothesis that the probabilities are as above against the alternative that they differ. The table of observed and expected cell counts follows:

Peas	Round Yellow	Wrinkled Yellow	Round Green	Wrinkled Green
O_i	56	19	17	8
E_i	56.25	18.75	18.75	6.25

The test statistic is

$$X^2 = \frac{(56-56.25)^2}{56.25} + \frac{(19-18.75)^2}{18.75} + \frac{(17-18.75)^2}{18.75} + \frac{(8-6.25)^2}{6.25} = .658$$

The number of degrees of freedom is $k-1=3$ and the rejection region with $\alpha = .01$ is $X^2 > \chi^2_{.01} = 11.3449$. Since the observed value of X^2 does not fall in the rejection region, we do not reject H₀. We do not have enough information to contradict the theoretical model for the classification of peas.

14.2.27 It is necessary to determine whether proportions given by the Mars Company are correct. The null hypothesis to be tested is

$$H_0 : p_1 = .12;\ p_2 = .15;\ p_3 = .12;\ p_4 = .23;\ p_5 = .23;\ p_6 = .15$$

against the alternative that at least one of these probabilities is incorrect. A table of observed and expected cell counts follows:

Color	Brown	Yellow	Red	Blue	Orange	Green
O_i	70	87	64	115	106	85
E_i	63.24	79.05	63.24	121.21	121.21	79.05

The test statistic is

$$X^2 = \frac{(70-63.24)^2}{63.24} + \frac{(87-79.05)^2}{79.05} + \cdots + \frac{(85-79.05)^2}{79.05} = 4.206$$

The number of degrees of freedom is $k-1=5$ and, since the observed value of $X^2 = 4.206$ is less than $\chi^2_{.10} = 9.24$, the p-value is greater than .10 and the results are not significant. We conclude that the proportions reported by the Mars Company are substantiated by our sample.

Section 14.3

14.3.1-3 Refer to Section 14.3 of the text. For an $r \times c$ contingency table, there are $(r-1)(c-1)$ degrees of freedom.

1. For a 3×5 contingency table, there are $(r-1)(c-1) = (2)(4) = 8$ degrees of freedom.

3. For a 3×3 contingency table, there are $(r-1)(c-1) = (2)(2) = 4$ degrees of freedom.

14.3.5-7 Refer to Table 5 in Appendix I. Since the chi-square test is one-tailed, the rejection region is in the right tail of the chi-square distribution with $df = (r-1)(c-1)$) and area α to its right.

5. $df = 3$; $\chi^2_{.05} = 7.81473$; reject H₀ if $X^2 > 7.81473$.

7. $df = 4$; $\chi^2_{.10} = 7.77944$; reject H₀ if $X^2 > 7.77944$.

14.3.9 The experiment is analyzed as a 3×4 contingency table. Hence, the expected cell counts must be obtained for each of the cells. Since values for the cell probabilities are not specified by the null hypothesis, they must be estimated, and the appropriate estimator is

$$\hat{E}_{ij} = \frac{r_i c_j}{n},$$

where r_i is the total for row i and c_j is the total for column j (see Section 14.3). The contingency table, including column and row totals and the estimated expected cell counts (in parentheses) follows.

	Column				
Row	1	2	3	4	Total
1	120 (67.68)	70 (66.79)	55 (67.97)	16 (58.56)	261
2	79 (84.27)	108 (83.17)	95 (84.64)	43 (72.91)	325
3	31 (78.05)	49 (77.03)	81 (78.39)	140 (67.53)	301
Total	230	227	231	199	887

The test statistic is
$$X^2 = \sum \frac{(O_{ij} - \hat{E}_{ij})^2}{\hat{E}_{ij}} = \frac{(120-67.68)^2}{67.68} + \cdots + \frac{(140-67.53)^2}{67.53} = 211.71$$

using the two-decimal accuracy given above. The degrees of freedom are $(r-1)(c-1) = (3-1)(4-1) = 6$.

14.3.11 Since $r = 2$ and $c = 3$, the total degrees of freedom are $(r-1)(c-1) = (1)(2) = 2$.

14.3.13 With $\alpha = .01$, a one-tailed rejection region is found using Table 5 to be $X^2 > \chi^2_{.01} = 9.21$.

14.3.15 The contingency table (from Exercise 12), including column and row totals and the estimated expected cell counts, follows.

	Column			
Row	1	2	3	Total
1	37 (42.23)	34 (37.31)	93 (84.46)	164
2	66 (60.77)	57 (53.69)	113 (121.54)	236
Total	103	91	206	400

and $X^2 = \dfrac{(37-42.23)^2}{42.23} + \dfrac{(34-37.31)^2}{37.31} + \cdots + \dfrac{(113-121.54)^2}{121.54} = 3.059$

From Table 5, the value $X^2 = 3.059$ is smaller than $\chi^2_{.10} = 4.605$ so that p-value $> .10$. Since the p-value is larger than .05, H_0 is not rejected. There is no reason to expect a dependence between rows and columns.

14.3.17 The hypothesis to be tested is

H_0: participation is independent of state

H_a: participation is dependent on state

	Rhode Island	Colorado	California	Florida	Total
Participate	46 (63.620)	63 (78.627)	108 (97.876)	121 (97.876)	338
Do not participate	149 (131.380)	178 (162.373)	192 (202.124)	179 (202.124)	698
Total	195	241	300	300	1036

The test statistic is $X^2 = \dfrac{(46-63.620)^2}{63.62} + \dfrac{(63-78.627)^2}{78.627} + \cdots + \dfrac{(179-202.124)^2}{202.124} = 21.52$

With $df = 3$, the p-value is less than .005 and H_0 is rejected. There is a difference in the proportions for the four states. The difference can be seen by considering the proportion of people participating in each of the four states:

	Rhode Island	Colorado	California	Florida
Participate	$\dfrac{46}{195} = .24$	$\dfrac{63}{241} = .26$	$\dfrac{108}{300} = .36$	$\dfrac{121}{300} = .40$

14.3.19 a The hypothesis of independence between attachment pattern and child care time is tested using the chi-square statistic. The contingency table, including column and row totals and the estimated expected cell counts, follows.

	Child Care			
Attachment	Low	Moderate	High	Total
Secure	24 (24.086)	35 (30.968)	5 (8.946)	64
Anxious	11 (10.914)	10 (14.032)	8 (4.054)	29
Total	111	51	297	459

The test statistic is

$$X^2 = \dfrac{(24-24.086)^2}{24.086} + \dfrac{(35-30.968)^2}{30.968} + \cdots + \dfrac{(8-4.054)^2}{4.054} = 7.27$$

and the rejection region is $X^2 > \chi^2_{.05} = 5.99$ with 2 df. H_0 is rejected. There is evidence of a dependence between attachment pattern and child care time.

b The value $X^2 = 7.27$ is between $\chi^2_{.05}$ and $\chi^2_{.025}$ so that $.025 < p$-value $< .05$. The results are significant.

14.3.21 a The hypothesis of no difference in the proportions of men and women in each hair color category is equivalent to testing the hypothesis of independence between hair color and gender, and is tested using the chi-square statistic. The contingency table, including column and row totals and the estimated expected cell counts, follows.

	Hair Color						
Gender	Light Blonde	Blonde	Light Brown	Brown	Black	Red	Total
Male	4 (4.385)	46 (50.667)	45 (55.538)	176 (165.641)	23 (17.051)	10 (10.718)	304
Female	5 (4.615)	58 (53.333)	69 (58.462)	164 (174.359)	12 (17.949)	12 (11.282)	320
Total	9	104	114	340	35	22	624

The test statistic is

$$X^2 = \frac{(4-4.385)^2}{4.385} + \frac{(46-50.667)^2}{50.667} + \cdots + \frac{(12-11.282)^2}{11.282} = 10.208 \text{ (10.207 using full accuracy)}$$

and the rejection region is $X^2 > \chi^2_{.05} = 11.0705$ with 5 df. Alternatively, we can bound the p-value as

$$.05 < p\text{-value} = P(X^2 > 10.208) < .10$$

and H_0 is not rejected at the 5% level. There is insufficient evidence to indicate a difference in the proportion of individuals with these hair colors between men and women.

b The contingency table is collapsed to combine "light blonde" and "blonde" into one category. The table follows, along with the estimated expected cell counts.

	Hair Color					
Gender	Blonde	Light Brown	Brown	Black	Red	Total
Male	50 (55.051)	45 (55.538)	176 (14.14)	23 (17.051)	10 (10.718)	304
Female	63 (57.949)	69 (58.462)	164 (174.359)	12 (17.949)	12 (11.282)	320
Total	113	114	340	35	22	624

The test statistic is $X^2 = \frac{(50-55.051)^2}{55.051} + \cdots + \frac{(12-11.282)^2}{11.282} = 10.207$ and the rejection region is

$X^2 > \chi^2_{.05} = 9.48773$ with 4 df. Alternatively, we can bound the p-value as

$$.025 < p\text{-value} = P(X^2 > 10.207) < .05$$

and H_0 *is* rejected at the 5% level. There *is* sufficient evidence to indicate a difference in the proportion of individuals with these hair colors between men and women.

14.3.23 a The hypothesis of independence between salary and number of workdays at home is tested using the chi-square statistic. The contingency table, including column and row totals and the estimated expected cell counts, generated by *Minitab* follows.

```
Tabulated Statistics: Worksheet rows, Worksheet columns

Rows: Worksheet rows   Columns: Worksheet columns

              At least
       Less   one, not   All at
       than one   all    home    All

  1      38      16      14      68
       36.27   21.08   10.65

  2      54      26      12      92
       49.07   28.52   14.41

  3      35      22       9      66
       35.20   20.46   10.34

  4      33      29      12      74
       39.47   22.94   11.59

  All   160      93      47     300

  Cell Contents
       Count
       Expected count

Chi-Square Test
                    Chi-Square   DF   P-Value
         Pearson      6.447       6    0.375
  Likelihood Ratio    6.421       6    0.378
```

The test statistic is

$$X^2 = \frac{(38-36.27)^2}{36.27} + \frac{(16-21.08)^2}{21.08} + \cdots + \frac{(12-11.59)^2}{11.59} = 6.447$$

and the rejection region with $\alpha = .05$ and $df = 3(2) = 6$ is $X^2 > \chi^2_{.05} = 12.59$ and the null hypothesis is not rejected. There is insufficient evidence to indicate that salary is dependent on the number of workdays spent at home.

b The observed value of the test statistic, $X^2 = 6.447$, is less than $\chi^2_{.10} = 10.6446$ so that the p-value is more than .10. This would confirm the non-rejection of the null hypothesis from part **a**.

14.3.25 The null hypothesis is that the two methods of classification are independent. The 2×2 contingency table with estimated expected cell counts in parentheses follows.

	Infection	No Infection	Total
Antibody	4	78	82
	(8.913)	(73.087)	
No antibody	11	45	56
	(6.087)	(49.913)	
Total	15	123	138

The test statistic is

$$X^2 = \frac{(4-8.913)^2}{8.913} + \frac{(78-73.087)^2}{73.087} + \cdots + \frac{(45-49.913)^2}{49.913} = 7.487$$

($X^2 = 7.488$ using computer accuracy). The rejection region with $\alpha = .05$ and 1 df is $X^2 > 3.84$ (alternatively, the p-value is bounded as $.005 < p\text{-value} < .01$) and H_0 is rejected. There is evidence that the injection of antibodies affects the likelihood of infections.

Section 14.4

14.4.1-3 Similar to previous exercises, except that the number of observations per row were selected prior to the experiment. The test procedure is identical to that used for an $r \times c$ contingency table. The contingency table, including column and row totals and the estimated expected cell counts, follows.

	Category			
Population	1	2	3	Total
1	108 (102.33)	52 (47.33)	40 (50.33)	200
2	87 (102.33)	51 (47.33)	62 (50.33)	200
3	112 (102.33)	39 (47.33)	49 (50.33)	200
Total	307	142	151	600

1. The test statistic is

$$X^2 = \frac{(108-102.33)^2}{102.33} + \frac{(52-47.33)^2}{47.33} + \cdots + \frac{(49-50.33)^2}{50.33} = 10.597$$

using calculator accuracy.

3. The null hypothesis is not rejected. There is insufficient evidence to indicate that the proportions in each of the three categories depend upon the population from which they are drawn.

14.4.5-7 The test procedure is identical to that used for an $r \times c$ contingency table. The contingency table, including column and row totals and the estimated expected cell counts, follows.

	Population			
	A	B	C	Total
Number of successes	24 (25.33)	19 (25.33)	33 (25.33)	76
Number of failures	76 (74.67)	81 (74.67)	67 (74.67)	224
Total	100	100	100	300

5. If we define p_A, p_B, and p_C as the probability of success for each of the three binomial populations, then the null hypothesis of independence of rows and columns is the same as a test of the equality of the three binomial proportions: $H_0 : p_A = p_B = p_C$

7. Since the p-value is greater than .05, the null hypothesis is not rejected. There is insufficient evidence to indicate that the proportions depend upon the population from which they were drawn.

14.4.9 Because a set number of chickens were selected for each group of varying contact, we have a contingency table with fixed columns. The table, with estimated expected cell counts appearing in parentheses, follows.

	No contact	Moderate contact	Heavy contact	Total
Diseased	87 (100)	89 (100)	124 (100)	300
Not diseased	9913 (9900)	9911 (9900)	9876 (9900)	29,700
Total	10,000	10,000	10,000	30,000

The test statistic is

$$X^2 = \frac{(87-100)^2}{100} + \frac{(89-100)^2}{100} + \cdots + \frac{(9876-9900)^2}{9900} = 8.75$$

and the rejection region with 2 df is $X^2 > 5.99$. H_0 is rejected and we conclude that the incidence of the disease is dependent on the amount of contact.

14.4.11 Similar to previous exercises, except that the observed cell counts must be obtained from the given information as

$$O_i = \frac{(\text{number of samples}) \times (\text{percentage})_i}{100}.$$

The approximate observed and estimated expected cell counts are shown in the following table.

	1	2	3	4	5	6	7	Total
Have nodules	23 (54.796)	25 (19.721)	35 (30.145)	18 (11.833)	52 (34.793)	159 (157.767)	11 (13.945)	323
No nodules	366 (334.204)	115 (120.279)	179 (183.855)	66 (72.167)	195 (212.207)	961 (962.233)	88 (85.055)	1970
Total	389	140	214	84	247	1120	99	2293

The test statistic is

$$X^2 = \frac{(23-54.796)^2}{54.796} + \frac{(25-19.721)^2}{19.721} + \cdots + \frac{(88-85.055)^2}{85.055} = 38.41$$

and the *p*-value with $df = 6$ is less than .005. The null hypothesis is rejected, and we conclude that age and probability of finding nodules are dependent.

14.4.13 a The proportions (in italics) in each of the six categories are shown in the table, along with the observed and expected cell counts.

	Access	Cost	Substance Abuse	Cancer	Obesity	Other	Total
2016	.20 40 (44)	.27 54 (43)	.03 6 (17)	.12 24 (23)	.08 16 (15)	.30 60 (58)	1.00 200
2017	.24 48 (44)	.16 32 (43)	.14 28 (17)	.11 22 (23)	.07 14 (15)	.28 56 (58)	1.00 200
Total	88	86	34	46	30	116	400

The test statistic is

$$X^2 = \frac{(40-44)^2}{44} + \frac{(54-43)^2}{43} + \cdots + \frac{(56-58)^2}{58} = 20.949$$

and the *p*-value with $df = 5$ is less than .005. The null hypothesis is rejected, and we conclude that there is a significant change in proportions from 2016 to 2017.

b In order to test specifically for a change in the proportion of adults whose concern was substance abuse, use the two-sample procedure in Section 9.5. The hypothesis of interest is

$$H_0: p_1 - p_2 = 0 \quad \text{versus} \quad H_a: p_1 - p_2 \neq 0$$

Calculate $\hat{p}_1 = \frac{6}{200} = .03$, $\hat{p}_2 = \frac{28}{200} = .14$, and $\hat{p} = \frac{x_1 + x_2}{n_1 + n_2} = \frac{6+28}{200+200} = .085$.

The test statistic is then

$$z = \frac{\hat{p}_1 - \hat{p}_2}{\sqrt{\hat{p}\hat{q}\left(\frac{1}{n_1} + \frac{1}{n_2}\right)}} = \frac{.03 - .14}{\sqrt{.085(.915)(1/200 + 1/200)}} = -3.94$$

The rejection region, with $\alpha = .05$, is $|z| > 1.96$ and H_0 is rejected. There is sufficient evidence to indicate a difference in the proportion of adults whose concern was substance abuse from 2016 to 2017.

14.4.15 a Similar to previous exercises. The contingency table, including column and row totals and the estimated expected cell counts, follows.

Condition	Treated	Untreated	Total
Improved	117 (95.5)	74 (95.5)	191
Not improved	83 (104.5)	126 (104.5)	209
Total	200	200	400

The test statistic is

$$X^2 = \frac{(117-95.5)^2}{95.5} + \frac{(74-95.5)^2}{95.5} + \cdots + \frac{(126-104.5)^2}{104.5} = 18.527$$

To test a one-tailed alternative of "effectiveness", first check to see that $\hat{p}_1 > \hat{p}_2$. Then the rejection region with 1 *df* has a right-tail area of $2(.05) = .10$ or $X^2 > \chi^2_{2(.05)} = 2.706$. H_0 is rejected and we conclude that the serum is effective.

b Consider the treated and untreated patients as comprising random samples of two hundred each, drawn from two populations (i.e., a sample of 200 treated patients and a sample of 200 untreated patients). Let p_1 be the probability that a treated patient improves and let p_2 be the probability that an untreated patient improves. Then the hypothesis to be tested is

$$H_0 : p_1 - p_2 = 0 \qquad H_a : p_1 - p_2 > 0$$

Using the procedure described in Chapter 9 for testing the hypothesis about the difference between two binomial parameters, the following estimators are calculated:

$$\hat{p}_1 = \frac{x_1}{n_1} = \frac{117}{200} \qquad \hat{p}_2 = \frac{x_2}{n_2} = \frac{74}{200} \qquad \hat{p} = \frac{x_1 + x_2}{n_1 + n_2} = \frac{117+74}{400} = .4775$$

The test statistic is

$$z = \frac{\hat{p}_1 - \hat{p}_2 - 0}{\sqrt{\hat{p}\hat{q}(1/n_1 + 1/n_2)}} = \frac{.215}{\sqrt{.4775(.5225)(.01)}} = 4.304$$

And the rejection region for $\alpha = .05$ is $z > 1.645$. Again, the test statistic falls in the rejection region. We reject the null hypothesis of no difference and conclude that the serum is effective. Notice that

$$z^2 = (4.304)^2 = 18.52 = X^2 \text{ (to within rounding error)}$$

14.4.17 The *Minitab* printout for this 2×4 contingency table follows.

```
Rows: Worksheet rows   Columns: Worksheet columns
                          Wild
           Banana  Cherry  Fruit  Strawberry-Banana  All

   1          14      20      7                  9    50
           12.857  24.286  5.000              7.857

   2           4      14      0                  2    20
            5.143   9.714  2.000              3.143

  All         18      34      7                 11    70

Cell Contents
       Count
       Expected count

Chi-Square Test
                  Chi-Square   DF   P-Value
       Pearson         6.384    3     0.094
       Likelihood Ratio 8.188   3     0.042

2 cell(s) with expected counts less than 5.
```

The test statistic for the test of independence of the two classifications of $X^2 = 6.384$ with p-value $= .094$ and H_0 is not rejected. There is insufficient evidence to indicate a difference in the perception of the best taste between adults and children. If the company intends to use a flavor as a marketing tool, the cherry flavor does not seem to provide an incentive to buy this product.

Reviewing What You've Learned

14.R.1 If the housekeeper actually has no preference, he or she has an equal chance of picking any of the five floor polishes. Hence, the null hypothesis to be tested is

$$H_0 : p_1 = p_2 = p_3 = p_4 = p_5 = \frac{1}{5}$$

The values of O_i are the actual counts observed in the experiment, and $E_i = np_i = 100(1/5) = 20$.

Polish	A	B	C	D	E
O_i	27	17	15	22	19
E_i	20	20	20	20	20

Then $\quad X^2 = \dfrac{(27-20)^2}{20} + \dfrac{(17-20)^2}{20} + \cdots + \dfrac{(19-20)^2}{20} = 4.40$

The p-value with $df = k - 1 = 4$ is greater than .10 and H_0 is not rejected. We cannot conclude that there is a difference in the preference for the five floor polishes. Even if this hypothesis **had** been rejected, the conclusion would be that at least one of the values of the p_i was significantly different from 1/6. However, this does not imply that p_i is necessarily greater than 1/6. Hence, we could not conclude that polish A is superior.

If the objective of the experiment is to show that polish A is superior, a better procedure would be to test the hypothesis as follows:

$$H_0 : p_1 = 1/6 \qquad H_a : p_1 > 1/6$$

From a sample of $n = 100$ housewives, $x = 27$ are found to prefer polish A. A z-test can be performed on the single binomial parameter p_1.

14.R.3 a To test for equality of the two binomial proportions, we use chi-square statistic and a 2×2 contingency table. The null hypothesis is evaluation (positive or negative) is independent of teaching approach; that is, there is no difference in p the proportion of positive evaluations for the discovery based versus the standard teaching approach. The contingency table generated by *Minitab* follows.

```
Rows: Worksheet rows   Columns: Worksheet columns
            Positive   Negative   All

    1          37        11       48
               34        14

    2          31        17       48
               34        14

    All        68        28       96

    Cell Contents
        Count
        Expected count

Chi-Square Test
                    Chi-Square   DF   P-Value
           Pearson     1.815      1    0.178
   Likelihood Ratio    1.826      1    0.177
```

The observed value of the test statistic is $X^2 = 1.815$ with p-value = .178 and the null hypothesis is not rejected at the 5% level of significance. There is insufficient evidence to indicate that there is a difference in the proportion of positive evaluations for the discovery based versus the standard teaching approach.

b Since the observed value of the test statistic is less than $\chi^2_{.10} = 2.71$ with $df = 1$, the approximate p-value is greater than .10. This agrees with the actual p-value = .178 given in the printout.

14.R.5 a Each of the three milestones is tested as a 2×2 contingency table using the chi-square test or the two-sample z test for binomial proportions.

b The researchers are interested in detecting a difference in the proportions of infants achieving a particular milestone at the 4-month checkup.

c It is not clear from the table whether the tests are one- or two-tailed; however, the p-values for each of the three tests are large enough to indicate that none of the proportions are significantly different for the two groups of infants. The *Minitab* chi-square printouts that follow answer these questions.

```
Rows: Worksheet rows   Columns: Worksheet columns
                       Side or
            Stomach     back      All

    1          79        144      223
              76.12     146.88

    2           6         20       26
               8.88      17.12

    All        85        164      249

    Cell Contents
        Count
        Expected count

Chi-Square Test
                    Chi-Square   DF   P-Value
           Pearson     1.579      1    0.209
   Likelihood Ratio    1.674      1    0.196
```

```
Rows: Worksheet rows   Columns: Worksheet columns
                       Side or
            Stomach     back      All

    1         102        167      269
             103.46     165.54

    2           3          1        4
              1.54       2.46

    All       105        168      273

    Cell Contents
        Count
        Expected count

Chi-Square Test
                    Chi-Square   DF   P-Value
           Pearson     2.290      1    0.130
   Likelihood Ratio    2.239      1    0.135

   2 cell(s) with expected counts less than 5.
```

```
Rows: Worksheet rows   Columns: Worksheet columns

                 Side or
      Stomach    back    All

1      107       183     290
       107.05    182.95

2      3         5       8
       2.95      5.05

All    110       188     298

Cell Contents
   Count
   Expected count


Chi-Square Test
                     Chi-Square   DF   P-Value
         Pearson     0.001        1    0.972
   Likelihood Ratio  0.001        1    0.972

1 cell(s) with expected counts less than 5.
```

Notice that the researcher's *p*-values match the values given in the printout, indicating that the researchers are doing two-tailed tests.

d *Minitab* warns you that the assumption that E_i is greater than or equal to five for each cell has been violated. Since the effect of a small expected value is to inflate the value of X^2, you need not be too concerned (X^2 was not big enough to reject H_0 in any of the tests).

14.R.7 a The 2×4 contingency table with observed and estimated expected cell counts follows. Remember that $E_{ij} = \dfrac{(r_i)(c_j)}{n}$.

	Very Confident	Somewhat Confident	Not too Confident	Not at all Confident	Total
Men	210 (164.5)	241 (273.5)	68 (70.5)	5 (10.5)	524
Women	129 (164.5)	306 (273.5)	73 (70.5)	16 (10.5)	524
Total	329	547	141	21	1048

Then the test statistic is

$$X^2 = \frac{(210-164.5)^2}{164.5} + \frac{(241-273.5)^2}{273.5} + \cdots + \frac{(16-10.5)^2}{10.5} = 33.017$$

with $df = (r-1)(c-1) = 3$. With $\alpha = .05$, the rejection region is $X^2 \geq \chi^2_{3,.05} = 7.81$ and H_0 is rejected.

b Since the observed value of $X^2 = 33.017$ is larger than the largest value in Table 5 ($\chi^2_{.005} = 12.8381$), the *p*-value is less than .005.

14.R.9 The null hypothesis to be tested is

H_0 : color distribution is independent of vehicle type

H_a : color distribution is dependent on vehicle type

A table of observed and expected cell counts follows, with $E_{ij} = \dfrac{(r_i)(c_j)}{n}$.

	Silver	Black	Gray	Blue	Red	White	Other	All
1	40	55	39	18	23	51	24	250
	34.50	47.50	33.50	18.50	27.00	65.00	24.00	
2	29	40	28	19	31	79	24	250
	34.50	47.50	33.50	18.50	27.00	65.00	24.00	
All	69	95	67	37	54	130	48	500

Cell Contents
  Count
  Expected count

Then the test statistic is

$$X^2 = \frac{(40-34.5)^2}{34.5} + \frac{(55-47.5)^2}{47.5} + \cdots + \frac{(24-24)^2}{24} = 13.171$$

Since the observed value of X^2 with $df = k-1 = 6$ is between $\chi^2_{.025}$ and $\chi^2_{.05}$, the p-value is between .025 and .05 (a computer printout will show p-value = .040). We can reject H_0. There is sufficient evidence to suggest a difference in color distributions for the two types of vehicles.

14.R.11 The 4×4 contingency table is analyzed using *Minitab* to test

H_0: fast food choice is independent of age group

H_a: fast food choice is dependent on age group

Rows: Worksheet rows   Columns: Fast Food Choice

	McDonalds	Burger King	Wendys	Other	All
1	75	34	10	6	125
	59.75	38.25	16.00	11.00	
2	89	42	19	10	160
	76.48	48.96	20.48	14.08	
3	54	52	28	18	152
	72.66	46.51	19.46	13.38	
4	21	25	7	10	63
	30.11	19.28	8.06	5.54	
All	239	153	64	44	500

Cell Contents
  Count
  Expected count

Chi-Square Test

	Chi-Square	DF	P-Value
Pearson	32.182	9	0.000
Likelihood Ratio	32.125	9	0.000

The *Minitab* printout shows $X^2 = 32.182$ with *p*-value $= .000$, so that H_0 is not rejected. There is a dependence between age group and favorite fast food restaurant. The practical implications of this conclusion can be explored by looking at the conditional distribution of fast food choices for each of the four age groups. Each student will present slightly different conclusions.

14.R.13 The 3×4 contingency table is analyzed as in previous exercises. The *Minitab* printout shows the observed and estimated expected cell counts, the test statistic and its associated *p*-value.

```
Rows: Worksheet rows   Columns: Education
         A      B      C      D     All

   1    55     40     43     30    168
       45.39  43.03  43.03  36.55

   2    15     25     18     22     80
       21.61  20.49  20.49  17.40

   3     7      8     12     10     37
       10.00   9.48   9.48   8.05

  All   77     73     73     62    285

Cell Contents
   Count
   Expected count

Chi-Square Test
                    Chi-Square   DF   P-Value
       Pearson         10.227     6    0.115
    Likelihood Ratio   10.310     6    0.112
```

The results are not significant (p-value $= .115$) and we cannot conclude that number of arrests is dependent on the educational achievement of a criminal offender.

14.R.15 In order to perform a chi-square "goodness of fit" test on the given data, it is necessary that the values O_i and E_i are known for each of the five cells. The O_i (the number of measurements falling in the *i*-th cell) are given. However, $E_i = np_i$ must be calculated. Remember that p_i is the probability that a measurement falls in the *i*-th cell. The hypothesis to be tested is

H_0: the experiment is binomial   versus H_a: the experiment is not binomial

Let x be the number of successes and p be the probability of success on a single trial. Then, assuming the null hypothesis to be true,

$$p_0 = P(x = 0) = C_0^4 p^0 (1-p)^4 \qquad p_1 = P(x = 1) = C_1^4 p^1 (1-p)^3$$
$$p_2 = P(x = 2) = C_2^4 p^2 (1-p)^2 \qquad p_3 = P(x = 3) = C_3^4 p^3 (1-p)^1$$
$$p_4 = P(x = 4) = C_4^4 p^4 (1-p)^0$$

Hence, once an estimate for p is obtained, the expected cell frequencies can be calculated using the above probabilities. Note that each of the 100 experiments consists of four trials and hence the complete experiment involves a total of 400 trials.

The best estimator of p is $\hat{p} = x/n$ (as in Chapter 9). Then,

$$\hat{p} = \frac{x}{n} = \frac{\text{number of successes}}{\text{number of trials}} = \frac{0(11)+1(17)+2(42)+3(12)+4(9)}{400} = \frac{1}{2}$$

The experiment (consisting of four trials) was repeated 100 times. There are a total of 400 trials in which the result "no successes in four trials" was observed 11 times, the result "one success in four trials" was observed 17 times, and so on. Then

$$p_0 = C_0^4 (1/2)^0 (1/2)^4 = 1/16 \quad p_1 = C_1^4 (1/2)^1 (1/2)^3 = 4/16$$
$$p_2 = C_2^4 (1/2)^2 (1/2)^2 = 6/16 \quad p_3 = C_3^4 (1/2)^3 (1/2)^1 = 4/16$$
$$p_4 = C_4^4 (1/2)^4 (1/2)^0 = 1/16$$

The observed and expected cell frequencies are shown in the following table.

x	0	1	2	3	4
O_i	11	17	42	21	9
E_i	6.25	25.00	37.50	25.00	6.25

and the test statistic is

$$X^2 = \frac{(11-6.25)^2}{6.25} + \frac{(17-25.00)^2}{25.00} + \cdots + \frac{(9-6.25)^2}{6.25} = 8.56$$

In order to bound the *p*-value or set up a rejection region, it is necessary to determine the appropriate degrees of freedom associated with the test statistic. Two degrees of freedom are lost because:

1 The cell probabilities are restricted by the fact that $\sum p_i = 1$.

2 The binomial parameter *p* is unknown and must be estimated before calculating the expected cell counts. The number of degrees of freedom is equal to $k-1-1 = k-2 = 3$. With $df = 3$, the *p*-value for $X^2 = 8.56$ is between .025 and .05 and the null hypothesis can be rejected at the 5% level of significance. We conclude that the experiment in question does not fulfill the requirements for a binomial experiment.

15: Nonparametric Statistics

Section 15.1

15.1.1 If distribution 1 is shifted to the right of distribution 2, the rank sum for sample 1 (T_1) will tend to be large. The test statistic will be T_1^*, the rank sum for sample 1 if the observations had been ranked from large to small. The null hypothesis will be rejected if T_1^* is unusually small. From Table 7a with $n_1 = 6$, $n_2 = 8$ and $\alpha = .05$, H_0 will be rejected if $T_1^* \leq 31$. From Table 7c with $n_1 = 6$, $n_2 = 8$ and $\alpha = .01$, H_0 will be rejected if $T_1^* \leq 27$.

15.1.3 The hypothesis to be tested is

H_0 : populations 1 and 2 are identical

H_a : population 1 is shifted to the left of population 2

and the data, with ranks in parentheses, are given next.

Sample 1	Sample 2
1(1)	4(5)
3(3.5)	7(9)
2(2)	6(7.5)
3(3.5)	8(10)
5(6)	6(7.5)

Note that tied observations are given an average rank, the average of the ranks they would have received if they had not been tied. Then

$$T_1 = 1 + 3.5 + 2 + 3.5 + 6 = 16$$
$$T_1^* = n_1(n_1 + n_2 + 1) - T_1 = 5(10+1) - 16 = 39$$

With $n_1 = n_2 = 5$, the one-tailed rejection region with $\alpha = .05$ is found in Table 7a to be $T_1 \leq 19$. The observed value, $T_1 = 16$, falls in the rejection region and H_0 is rejected. We conclude that population 1 is shifted to the left of population 2.

15.1.5 Since n_1 and n_2 are large, the large sample approximation is used to test

H_0 : the populations are identical   versus H_a : the populations differ in location

Calculate

$$\mu_T = \frac{n_1(n_1 + n_2 + 1)}{2} = \frac{20(45+1)}{2} = 460$$

$$\sigma_T^2 = \frac{n_1 n_2 (n_1 + n_2 + 1)}{12} = \frac{20(25)(46)}{12} = 1916.667$$

The test statistic is

$$z = \frac{T_1 - \mu_T}{\sigma_T} = \frac{252 - 460}{\sqrt{1916.667}} = -4.75$$

and the p-value is

$$p\text{-value} = 2P(z > 4.75) < 2(1 - .9998) = .0004.$$

The *p*-value is less than $\alpha = .05$ and the results are highly significant. There is a difference in the two population distributions.

15.1.7 Since the two samples are independent, use the Wilcoxon rank sum test. The data, with corresponding ranks, are shown in the next table.

Stimulus 1	Stimulus 2
1 (2.5)	4 (16)
3 (12.5)	2 (7)
2 (7)	3 (12.5)
1 (2.5)	3 (12.5)
2 (7)	1 (2.5)
1 (2.5)	2 (7)
3 (12.5)	3 (12.5)
2 (7)	3 (12.5)
$T_1 = 53.5$	

Calculate

$$T_1 = 53.5$$
$$T_1^* = n_1(n_1 + n_2 + 1) - T_1 = 8(16+1) - 53.5 = 82.5$$

The test statistic is $T = \min(T_1, T_1^*) = 53.5$. With $n_1 = 8$ and $n_2 = 8$, the two-tailed rejection region with $\alpha = .05$ is found in Table 7b to be $T \le 49$. The observed value, $T = 53.5$, does not fall in the rejection region and H₀ is not rejected. We cannot conclude that the distributions of reaction times for the two stimuli are different.

15.1.9 The data are already in rank form. The "substantial experience" sample is designated as sample 1, and $n_1 = 5, n_2 = 7$. Calculate

$$T_1 = 19$$
$$T_1^* = n_1(n_1 + n_2 + 1) - T_1 = 5(13) - 19 = 46$$

The test statistic is $T = \min(T_1, T_1^*) = 19$. With $n_1 = n_2 = 12$, the one-tailed rejection region with $\alpha = .05$ is found in Table 7a to be $T_1 \le 21$. The observed value, $T = 19$, falls in the rejection region and H₀ is rejected. There is sufficient evidence to indicate that the review board considers experience a prime factor in the selection of the best candidates.

15.1.11 The hypothesis of interest is

H₀: populations 1 and 2 are identical   versus   Hₐ: population 2 is shifted to the right of population 1

The data, with ranks in parentheses, are given next.

20s	11(20)	7(11)	6(7.5)	8(14)	6(7.5)	9(16.5)	2(2)	10(18.5)	3(3.5)	6(7.5)
65-70s	1(1)	9(16.5)	6(7.5)	8(14)	7(11)	8(14)	5(5)	7(11)	10(18.5)	3(3.5)

Then

$$T_1 = 20 + 11 + \cdots + 7.5 = 108$$
$$T_1^* = n_1(n_1 + n_2 + 1) - T_1 = 10(20+1) - 108 = 102$$

The test statistic is $T = \min(T_1, T_1^*) = 102$. With $n_1 = n_2 = 10$, the one-tailed rejection region with $\alpha = .05$ is found in Table 7a to be $T \le 82$ and the observed value, $T = 102$, does not fall in the rejection region; H₀ is not rejected. We cannot conclude that this drug improves memory in mean aged 65 to 70 to that of 20 year olds.

15.1.13 Similar to previous exercises. The data, with corresponding ranks, are shown in the following table.

Deaf (1)	Hearing (2)
2.75 (15)	0.89 (1)
2.14 (11)	1.43 (7)
3.23 (18)	1.06 (4)
2.07 (10)	1.01 (3)
2.49 (14)	0.94 (2)
2.18 (12)	1.79 (8)
3.16 (17)	1.12 (5.5)
2.93 (16)	2.01 (9)
2.20 (13)	1.12 (5.5)
$T_1 = 126$	

Calculate

$$T_1 = 126$$
$$T_1^* = n_1(n_1 + n_2 + 1) - T_1 = 9(19) - 126 = 45$$

The test statistic is $T = \min(T_1, T_1^*) = 45$. With $n_1 = n_2 = 9$, the two-tailed rejection region with $\alpha = .05$ is found in Table 7b to be $T_1^* \leq 62$. The observed value, $T = 45$, falls in the rejection region and H₀ is rejected. We conclude that the deaf children do differ from the hearing children in eye-movement rate.

15.1.15 The data, with corresponding ranks, are shown in the following table.

Lake 2	Lake 1
14.1 (12.5)	12.2 (2)
15.2 (16)	13.0 (6)
13.9 (10)	14.1 (12.5)
14.5 (14)	13.6 (7)
14.7 (15)	12.4 (3)
13.8 (8.5)	11.9 (1)
14.0 (11)	12.5 (4)
16.1 (18)	13.8 (8.5)
12.7 (5)	
15.3 (17)	
	$T_1 = 44$

Calculate

$$T_1 = 44$$
$$T_1^* = n_1(n_1 + n_2 + 1) - T_1 = 8(18+1) - 44 = 108$$

The test statistic is $T = \min(T_1, T_1^*) = 44$. With $n_1 = 8$ and $n_2 = 10$, the two-tailed rejection region with $\alpha = .05$ is found in Table 7b to be $T \leq 53$. The observed value, $T = 44$, falls in the rejection region and H₀ is rejected. We conclude that the distribution of weights for the tagged turtles exposed to the two lake environments were different.

Section 15.2

15.2.1 If the sign test is two-tailed $(H_a : p \neq .5)$, then the experimenter would like to show that one population of measurements lies above or below the other population. An exact practical statement of the alternative hypothesis would depend on the experimental situation. It is necessary that α (the probability of rejecting the null hypothesis when it is true) take values less than or equal to $\alpha = .10$. Assuming the null hypothesis to be true, the two populations are identical and consequently, $p = P(A$ exceeds B for a given pair of observations$)$ is 1/2. The binomial probability was discussed in Chapter 5. In particular, it was noted that the distribution of the random variable x is symmetrical about the mean np when $p = 1/2$. For example, with $n = 15$, $P(x = 0) = P(x = 15)$. Similarly, $P(x = 1) = P(x = 14)$ and so on. Hence, the lower portion of the two-tailed rejection region must have probability less than or equal to $\alpha/2 = .05$. From Table 1, Appendix I we look probabilities less than .05 to determine the rejection regions.

Rejection Region $n = 15$	α	Rejection Region $n = 25$	α
$x \leq 2; x \geq 13$	.008	$x \leq 5; x \geq 20$	.004
$x \leq 3; x \geq 12$	.036	$x \leq 6; x \geq 19$	.014
		$x \leq 7; x \geq 18$	.044

15.2.3 Similar to Exercise 15.2.1. If the sign test is lower-tailed $(H_a : p < .5)$, then the experimenter would like to show that one population of measurements lies below the other population. The values of α available for this lower tailed test ($\alpha \leq .10$) and the corresponding rejection regions are shown next.

Rejection Region $n = 10$	α	Rejection Region $n = 15$	α
$x \leq 0$	.001	$x \leq 2$	.004
$x \leq 1$	.011	$x \leq 3$	.018
$x \leq 2$	.055	$x \leq 4$	.059

15.2.5 Define $x =$ number of positive differences, $x_1 - x_2$, and $p = P($positive difference$)$. The hypothesis of interest is

$$H_0 : p = 1/2 \quad \text{versus} \quad H_a : p \neq 1/2$$

With $n = 10$, the rejection region must be calculated using the binomial formula or the binomial tables with $n = 10$ and $p = .5$ (Table 1, Appendix I). If we choose to use $\{x \leq 1, x \geq 9\}$ as the rejection region, then

$$\alpha_1 = .011 + .011 = .022$$

If we choose $\{x \leq 2, x \geq 8\}$ as the rejection region, then

$$\alpha_2 = .055 + .055 = .11$$

which is too large. Hence, the rejection region will be $\{x \leq 1, x \geq 9\}$ with $\alpha = .022$.

There is $x = 1$ positive differences. Since this observed value of x falls in the rejection region, H_0 is rejected. We conclude that there is a difference between the populations.

15.2.7 Refer to Exercise 15.2.6.

a The hypothesis of interest is now $H_0 : p = .5$ versus $H_a : p > .5$ where p is the probability that the store brand is judged to be better than the national brand.

b The test statistic is $x = 8$, the number of times that the store brand is judged better. . Since there are 10 ties, the sample size is reduced to $n = 15$ and the one-tailed rejection region with $\alpha \approx .05$ is $x \geq 11$ with $\alpha = .059$ and H_0 is not rejected. There is insufficient evidence to indicate that the store brand is preferred to the national brand. Apparently, it is very hard to tell the difference between the store and the national brand.

15.2.9 Define p to be the probability that a vaccinated monkey showed no evidence of the SIV infection after a year and a half and x to be number of monkeys showing no evidence of infection. Then $n = 6$ and $x = 2$. The hypothesis to be tested is

$$H_0 : p = 1/2 \quad \text{versus} \quad H_a : p > 1/2$$

Since, if the vaccine is effective, the number of monkeys showing no evidence of infection should be large. The one-tailed rejection region with $\alpha \approx .10$ is $x \geq 5$ with $\alpha = .109$. Hence, H_0 is not rejected. There is no evidence to indicate that the vaccine is effective.

15.2.11 a Define $p = P(\text{chef A's rating exceeds chef B's rating for a given meal})$ and $x =$ number of meals for which chef A exceeds B. The hypothesis to be tested is

$$H_0 : p = 1/2 \quad \text{versus} \quad H_a : p \neq 1/2$$

using the sign test with x as the test statistic. Notice that for this exercise $n = 20$, since a tie rating was given to meals 7 and 14. The observed value of the test statistic is $x = 11$.

Critical value approach: Various two tailed rejection regions are tried in order to find a region with $\alpha \approx .05$. These are shown in the following table.

Rejection Region	α
$x \leq 3; x \geq 17$	.002
$x \leq 4; x \geq 16$	.012
$x \leq 5; x \geq 15$	.042
$x \leq 6; x \geq 14$	.116

We choose to reject H_0 if $x \leq 5$ or $x \geq 15$ with $\alpha = .042$. Since $x = 11$, H_0 is not rejected. There is insufficient evidence to indicate a difference between the two gourmets.

p-value approach: For the observed value $x = 11$, calculate the two-tailed p-value:

$$p\text{-value} = 2P(x \geq 11) = 2(1 - .588) = .824$$

b The large sample z statistic, with $n = 20$ and $p = .5$ is $z = \dfrac{x - .5n}{.5\sqrt{n}} = \dfrac{11 - 10}{.5\sqrt{20}} = .45$ and the two-tailed rejection region with $\alpha = .05$ is $|z| > 1.96$. The null hypothesis is not rejected.

c The results are the same.

15.2.13 a Since we are interested in a difference in recovery rates, let p be the probability that the recovery rate for A exceeds B at a given hospital. Let x be the number of times that A exceeds B. The hypothesis to be tested is

$$H_0 : p = 1/2 \quad \text{versus} \quad H_a : p \neq 1/2$$

and the data are shown in the following table.

Hospital	A	B	Sign of (A − B)
1	75.0	85.4	−
2	69.8	83.1	−
3	85.7	80.2	+
4	74.0	74.5	−
5	69.0	70.0	−
6	83.3	81.5	+
7	68.9	75.4	−
8	77.8	79.2	−
9	72.2	85.4	−
10	77.4	80.4	−

Various two tailed rejection regions are tried in order to find a region with $\alpha \approx .10$. These are shown in the table.

Rejection Region	α
$x=0; x=10$	.002
$x \leq 1; x \geq 9$	.022
$x \leq 2; x \geq 8$	.110

We choose to reject H_0 if $x \leq 2$ or $x \geq 8$ with $\alpha = .110$. Since $x = 2$, H_0 is rejected. There is sufficient evidence to indicate a difference between the two drugs.

b In the above analysis, we made no assumptions concerning the underlying distributions of the data. To use the t test we must be able to assume normality of the distributions and equal variances for the two populations. Since the observations given above are percentages, their distributions may be almost mound-shaped, but the variances will not be equal unless H_0 is true.

Section 15.4

15.4.1 The three possible tests are the paired-difference Student's t test (which requires that the population of differences is normal with mean 0 when H_0 is true), the sign test (no requirements about the population of differences) and the Wilcoxon signed-rank test (requires that the population of differences is symmetric with mean 0 when H_0 is true).

15.4.3 The hypothesis of interest is:

H_0: populations 1 and 2 are identical versus H_a: population 1 is shifted to the right of population 2

15.4.5 The hypothesis of interest is

H_0: population distributions 1 and 2 are identical versus H_a: the distributions differ in location

Since Table 8, Appendix I gives critical values for rejection in the lower tail of the distribution, we use the smaller of T^+ and T^- as the test statistic. From Table 8 with $n = 30$, $\alpha = .05$ and a two-tailed test, the rejection region is $T \leq 137$. Since $T^+ = 249$, we can calculate

$$T^- = \frac{n(n+1)}{2} - T^+ = \frac{30(31)}{2} - 249 = 216.$$

The test statistic is the smaller of T^+ and T^- or $T = 216$ and H_0 is not rejected. There is no evidence of a difference between the two distributions.

15.4.7 Since $n > 25$, the large sample approximation to the signed rank test can be used to test the hypothesis given in Exercise 15.4.5. Calculate

$$E(T) = \frac{n(n+1)}{4} = \frac{30(31)}{4} = 232.5$$

$$\sigma_T^2 = \frac{n(n+1)(2n+1)}{24} = \frac{30(31)(61)}{24} = 2363.75$$

The test statistic is

$$z = \frac{T - E(T)}{\sigma_T} = \frac{216 - 232.5}{\sqrt{2363.75}} = -.34$$

with p-value = $P(|z| < .34) = 2(.3669) = .7338$ and H_0 is not rejected. The results agree with Exercise 15.4.5.

15.4.9 a The hypothesis to be tested is

H_0: population distributions 1 and 2 are identical

H_a: the distributions differ in location

and the test statistic is T, the rank sum of the positive (or negative) differences. The ranks are obtained by ordering the differences according to their absolute value. Define d_i to be the difference between a pair in populations 1 and 2 (i.e., $x_{1i} - x_{2i}$). The differences, along with their ranks (according to absolute magnitude), are shown in the following table.

d_i	.1	.7	.3	−.1	.5	.2	.5
Rank $\lvert d_i \rvert$	1.5	7	4	1.5	5.5	3	5.5

The rank sum for positive differences is $T^+ = 26.5$ and the rank sum for negative differences is $T^- = 1.5$ with $n = 7$. Consider the smaller rank sum and determine the appropriate lower portion of the two-tailed rejection region. Indexing $n = 7$ and $\alpha = .05$ in Table 8, the rejection region is $T \leq 2$ and H_0 is rejected. There is a difference in the two population locations.

b The results do not agree with those obtained in Exercise 15.2.4. We are able to reject H_0 with the more powerful Wilcoxon test.

15.4.11 a The Wilcoxon signed rank test is used, and the differences, along with their ranks (according to absolute magnitude), are shown in the following table.

d_i	−4	2	−2	−5	−3	0	1	1	−6
Rank $\lvert d_i \rvert$	6	3.5	3.5	7	5	--	1.5	1.5	8

The sixth pair is tied and is hence eliminated from consideration. Pairs 7 and 8, 2 and 3 are tied and receive an average rank. Then $T^+ = 6.5$ and $T^- = 29.5$ with $n = 8$. Indexing $n = 8$ and $\alpha = .05$ in Table 8, the lower portion of the two-tailed rejection region is $T \leq 4$ and H_0 is not rejected. There is insufficient evidence to detect a difference in the two machines.

b If a machine continually breaks down, it will eventually be fixed, and the breakdown rate for the following month will decrease.

15.4.13 a The paired data are given in the exercise. The differences, along with their ranks (according to absolute magnitude), are shown in the following table.

d_i	1	2	−1	1	3	1	−1	3	−2	3	1	0
Rank $\lvert d_i \rvert$	3.5	7.5	3.5	3.5	10	3.5	3.5	10	7.5	10	2.5	--

Let $p = P(A \text{ exceeds B for a given intersection})$ and $x =$ number of intersections at which A exceeds B. The hypothesis to be tested is

$$H_0: p = 1/2 \quad \text{versus} \quad H_a: p \neq 1/2$$

using the sign test with x as the test statistic.

Critical value approach: Various two tailed rejection regions are tried in order to find a region with $\alpha \approx .05$. These are shown in the following table.

Rejection Region	α
$x \leq 1; x \geq 10$	.012
$x \leq 2; x \geq 9$	.066
$x \leq 3; x \geq 8$	.226

We choose to reject H_0 if $x \leq 2$ or $x \geq 9$ with $\alpha = .066$. Since $x = 8$, H_0 is not rejected. There is insufficient evidence to indicate a difference between the two methods.

p-value approach: For the observed value $x = 8$, calculate the two-tailed p-value:

$$p\text{-value} = 2P(x \geq 8) = 2(1-.887) = .226$$

Since the *p*-value is greater than .10, H₀ is not rejected.

b To use the Wilcoxon signed rank test, we use the ranks of the absolute differences shown in the table above. Then $T^+ = 51.5$ and $T^- = 14.5$ with $n = 11$. Indexing $n = 11$ and $\alpha = .05$ in Table 8, the lower portion of the two-tailed rejection region is $T \leq 11$ and H₀ is not rejected, as in part **a**.

15.4.15 a Since the experiment has been designed as a paired experiment, there are three tests available for testing the differences in the distributions with and without imagery – (1) the paired difference *t* test; (2) the sign test or (3) the Wilcoxon signed rank test. In order to use the paired difference *t* test, the scores must be approximately normal; since the number of words recalled has a binomial distribution with $n = 25$ and unknown recall probability, this distribution may not be approximately normal.

b Using the **sign test**, the hypothesis to be tested is

$$H_0: p = 1/2 \quad \text{versus} \quad H_a: p > 1/2$$

For the observed value $x = 0$ we calculate the two-tailed *p*-value:

$$p\text{-value} = 2P(x \leq 0) = 2(.000) = .000$$

The results are highly significant; H₀ is rejected and we conclude there is a difference in the recall scores with and without imagery.

Using the **Wilcoxon signed-rank test**, the differences will all be positive ($x = 0$ for the sign test), so that and

$$T^+ = \frac{n(n+1)}{2} = \frac{20(21)}{2} = 210 \quad \text{and} \quad T^- = 210 - 210 = 0$$

Indexing $n = 20$ and $\alpha = .01$ in Table 8, the lower portion of the two-tailed rejection region is $T \leq 37$ and H₀ is rejected.

Section 15.5

15.5.1 The Kruskal-Wallis *H* test provides a nonparametric analog to the analysis of variance *F* test for a completely randomized design presented in Chapter 11. The hypothesis of interest is

H₀: the three population distributions are identical

Hₐ: at least two of the three population distributions differ in location

The test statistic, based on the rank sums, is

$$H = \frac{12}{n(n+1)} \sum \frac{T_i^2}{n_i} - 3(n+1)$$

$$= \frac{12}{15(26)} \left[\frac{(35)^2}{5} + \frac{(63)^2}{5} + \frac{(22)^2}{5} \right] - 3(16) = 8.78$$

The rejection region with $\alpha = .05$ and $k - 1 = 2\ df$ is based on the chi-square distribution, or $H > \chi^2_{.05} = 5.99$. The null hypothesis is rejected and we conclude that there is a difference among the three treatments.

15.5.3 Similar to Exercise 15.5.1. The hypothesis of interest is

H₀: the three population distributions are identical

Hₐ: at least two of the three population distributions differ in location

The test statistic, based on the rank sums, is

$$H = \frac{12}{n(n+1)}\sum \frac{T_i^2}{n_i} - 3(n+1)$$

$$= \frac{12}{17(18)}\left[\frac{21^2}{6} + \frac{60^2}{5} + \frac{72^2}{6}\right] - 3(18) = 11.000$$

The rejection region with $\alpha = .05$ and $k-1 = 2\ df$ is based on the chi-square distribution, or $H > \chi^2_{.05} = 5.99$. The null hypothesis is rejected and we can conclude that there is a difference among the three treatments.

15.5.5 The data are jointly ranked from smallest to largest, with ties treated as in the Wilcoxon rank sum test. The data with corresponding ranks in parentheses are shown next.

Treatment			
1	2	3	4
124 (9)	147 (20)	141 (17)	117 (4.5)
167 (26)	121 (7)	144 (18.5)	128 (10.5)
135 (14)	136 (15)	139 (16)	102 (1)
160 (24)	114 (3)	162 (25)	119 (6)
159 (23)	129 (12)	155 (22)	128 (10.5)
144 (18.5)	117 (4.5)	150 (21)	123 (8)
133 (13)	109 (2)		
$T_1 = 127.5$	$T_2 = 63.5$	$T_3 = 119.5$	$T_4 = 40.5$
$n_1 = 7$	$n_2 = 7$	$n_3 = 6$	$n_4 = 6$

The test statistic, based on the rank sums, is

$$H = \frac{12}{n(n+1)}\sum \frac{T_i^2}{n_i} - 3(n+1)$$

$$= \frac{12}{26(27)}\left[\frac{(127.5)^2}{7} + \frac{(63.5)^2}{7} + \frac{(119.5)^2}{6} + \frac{(40.5)^2}{6}\right] - 3(27) = 13.90$$

The rejection region with $\alpha = .05$ and $k-1 = 3\ df$ is based on the chi-square distribution, or $H > \chi^2_{.05} = 7.81$. The null hypothesis is rejected and we conclude that there is a difference among the four treatments.

15.5.7 Similar to Exercise 15.5.5. The data with corresponding ranks in parentheses are shown next.

Location			
1	2	3	4
5.7 (15.5)	6.2 (22.5)	5.4 (12)	3.7 (4)
6.3 (24)	5.3 (11)	5.0 (8)	3.2 (1)
6.1 (21)	5.7 (15.5)	6.0 (19)	3.9 (5)
6.0 (19)	6.0 (19)	5.6 (14)	4.0 (6)
5.8 (17)	5.2 (9.5)	4.9 (7)	3.5 (2)
6.2 (22.5)	5.5 (13)	5.2 (9.5)	3.6 (3)
$T_1 = 119.0$	$T_2 = 90.5$	$T_3 = 69.5$	$T_4 = 21.0$
$n_1 = 6$	$n_2 = 6$	$n_3 = 6$	$n_4 = 6$

a The test statistic, based on the rank sums, is

$$H = \frac{12}{n(n+1)}\sum\frac{T_i^2}{n_i} - 3(n+1)$$

$$= \frac{12}{24(25)}\left[\frac{(119.0)^2}{6} + \frac{(90.5)^2}{6} + \frac{(69.5)^2}{6} + \frac{(21.0)^2}{6}\right] - 3(25) = 17.075$$

The rejection region with $\alpha = .05$ and $k-1 = 3$ df is based on the chi-square distribution, or $H > \chi^2_{.05} = 7.81$. The null hypothesis is rejected and we conclude that there is a difference in location among the four locations.

b Since the observed value $H = 17.075$ exceeds $\chi^2_{.005} = 12.8381$, the p-value is less than .005.

c-d From Exercise 11.2.11, $F = 57.38$ with 3 and 20 df. Since $F_{.005} = 5.82$, the p-value is less than .005 as in part **b**. The nonparametric test allows us to draw the same conclusions with less restrictive assumptions.

15.5.9 Similar to previous exercises. The ranks of the data are shown next.

Region		
Northeast	Middle Atlantic	Southeast
19	24	23
5	11	14.5
7	14.5	29
1	2	26
18	21	12
3	10	30
8	22	27
4	28	25
6	16.5	16.5
13	9	20
$T_1 = 84$	$T_2 = 158$	$T_3 = 223$
$n_1 = 10$	$n_2 = 10$	$n_3 = 10$

a The test statistic is

$$H = \frac{12}{n(n+1)}\sum\frac{T_i^2}{n_i} - 3(n+1)$$

$$= \frac{12}{30(31)}\left[\frac{(84)^2}{10} + \frac{(158)^2}{10} + \frac{(223)^2}{10}\right] - 3(31) = 12.48$$

The rejection region with $\alpha = .05$ and $k-1 = 2$ df is based on the chi-square distribution, or $H > \chi^2_{.05} = 5.99$. The null hypothesis is rejected and we conclude that there is a difference in the levels of acidity for the three locations.

b Since the observed value $H = 12.48$ exceeds $\chi^2_{.005} = 10.5966$, the p-value is less than .005.

Section 15.6

15.6.1 In using the Friedman F_r test, data are ranked **within a block** from 1 to k. The treatment rank sums are then calculated as usual. The data and their corresponding ranks are shown next.

	Treatment		
Block	1	2	3
1	3.2 (3)	3.1 (2)	2.4 (1)
2	2.8 (2)	3.0 (3)	1.7 (1)
3	4.5 (2)	5.0 (3)	3.9 (1)
4	2.5 (1)	2.7 (3)	2.6 (2)
5	3.7 (2)	4.1 (3)	3.5 (1)
6	2.4 (2.5)	2.4 (2.5)	2.0 (1)
	$T_1 = 12.5$	$T_2 = 16.5$	$T_3 = 7$

The test statistic is

$$F_r = \frac{12}{bk(k+1)} \sum T_i^2 - 3b(k+1)$$

$$= \frac{12}{6(3)(4)}\left[(12.5)^2 + (16.5)^2 + 7^2\right] - 3(6)(4) = 7.58$$

and the rejection region is $F_r > \chi^2_{.05} = 5.99$. The observed value, $F_r = 7.58$, falls between $\chi^2_{.025}$ and $\chi^2_{.01}$. Hence, $.01 < p\text{-value} < .025$. H_0 is rejected and we conclude that there is a difference among the three treatments.

15.6.3 The analysis of variance is performed as in Chapter 11. The ANOVA table is shown next.

Source	df	SS	MS	F
Treatments	2	1.56	0.78	10.833
Blocks	5	10.965	2.193	30.458
Error	10	0.72	0.072	
Total	17	13.245		

The analysis of variance F test for treatments is $F = 10.833$ and the approximate p-value with 2 and 10 df is $p\text{-value} < .005$. The analysis of variance F test for blocks is $F = 30.458$ and the approximate p-value with 5 and 10 df is $p\text{-value} < .005$. Both tests are highly significant. There are significant differences among the treatment means and among the block means (blocking was effective).

15.6.5 Refer to Exercises 15.6.1 and 15.6.3. In both instances, H_0 is rejected. However, the p-value for the parametric test is smaller. Hence, if all of the parametric assumptions are met, the parametric test will be more powerful than its nonparametric analog.

15.6.7 a The design is a randomized block design, with conditions as treatments and employees as blocks.

b Use the Friedman's F_r Test. The ranks within each block are shown next.

	Conditions		
Employees	1	2	3
1	3	1	2
2	2	1	3
3	3	2	1
4	2	1	3
5	1	2	3
6	1	2	3
7	3	1	2
8	3	1	2
9	2	1	3
10	1.5	1.5	3
11	1	2	3
12	2	1	3
Totals	24.5	16.5	31

a Use the Friedman F_r statistic calculated as

$$F_r = \frac{12}{12(3)(4)}\left[24.5^2 + 16.5^2 + 31^2\right] - 3(12)(4) = 8.79$$

and the rejection region is $F_r > \chi^2_{.05} = 5.99$. Hence, H$_0$ is rejected and we conclude that there is a difference in employee fatigue as measured by stoppages for these three conditions.

15.6.9 Similar to previous exercises, with rats as blocks. The data are shown next along with corresponding ranks within blocks. Note that we have rearranged the data to eliminate the random order of presentation in the display.

	Treatment		
Rat	A	B	C
1	6 (3)	5 (2)	3 (1)
2	9 (2.5)	9 (2.5)	4 (1)
3	6 (2)	9 (3)	3 (1)
4	5 (1)	8 (3)	6 (2)
5	7 (1)	8 (2.5)	8 (2.5)
6	5 (1.5)	7 (3)	5 (1)
7	6 (2)	7 (3)	5 (1)
8	6 (1)	7 (2.5)	7 (2.5)
	$T_1 = 14$	$T_2 = 21.5$	$T_3 = 12.5$

a The test statistic is

$$F_r = \frac{12}{bk(k+1)}\sum T_i^2 - 3b(k+1)$$

$$= \frac{12}{8(3)(4)}\left[(14)^2 + (21.5)^2 + (12.5)^2\right] - 3(8)(4) = 5.81$$

and the rejection region is $F_r > \chi^2_{.05} = 5.99$. Hence, H$_0$ is not rejected and we cannot conclude that there is a difference among the three treatments.

b The observed value, $F_r = 5.81$, falls between $\chi^2_{.05}$ and $\chi^2_{.10}$. Hence, $.05 < p$-value $< .10$.

15.6.11 Similar to previous exercises. The ranks within each block are shown next.

Varieties	1	2	3	4	5	6	T_i
A	4	2	3	3	3	3	18
B	1	3	2	2	2	2	12
C	5	4	4	4	4	5	26
D	3	5	5	5	5	4	27
E	2	1	1	1	1	1	7

a Use the Friedman F_r statistic calculated as

$$F_r = \frac{12}{30(6)}\left[18^2 + 12^2 + \cdots + 7^2\right] - 3(36) = 20.13$$

and the rejection region is $F_r > \chi^2_{.05} = 9.49$. Hence, H$_0$ is rejected and we conclude that there is a difference in the levels of yield for the five varieties of wheat.

b From Chapter 11, $F = 18.61$ with p-value $= .000$ and H$_0$ is rejected. The results are the same.

Section 15.7

15.7.1 Table 9, Appendix I gives critical values r_0 such that $P(r_s \geq r_0) = \alpha$. Hence, for an upper-tailed test, the critical value for rejection can be read directly from the table.

For $\alpha = .05$, $r_s \geq .425$ while for $\alpha = .01$, $r_s \geq .601$.

15.7.3 For a two-tailed test of correlation, the value of α given along the top of the table is doubled to obtain the **actual** value of α for the test.

To obtain $\alpha = .05$, index .025 and the rejection region is $|r_s| \geq .400$. To obtain $\alpha = .01$, index .005 and the rejection region is $|r_s| \geq .526$.

15.7.5 Since there are no tied observations, the simpler formula for r_s is used, and

$$r_s = 1 - \frac{6\sum d_i^2}{n(n^2 - 1)} = 1 - \frac{6\left[(-6)^2 + (-3)^2 + \cdots + (4)^2\right]}{8(63)}$$

$$= 1 - \frac{840}{504} = -0.6667$$

15.7.7 Use the definition formula for Spearman's rank correlation coefficient and calculate

$$r_s = \frac{S_{xy}}{\sqrt{S_{xx} S_{yy}}} = \frac{74.5}{\sqrt{(82.5)(82.5)}} = 0.903$$

15.7.9 To calculate the Spearman's rank correlation coefficient, the data are ranked separately according to the variables x and y.

Rank x	4.5	1	3	6	4.5	2
Rank y	3.5	6	3.5	1	5	2

Calculate $\sum x_i y_i = 64.75$, $\sum x_i^2 = 90.5$, $\sum y_i^2 = 90.5$

$n = 6$, $\sum x_i = 21$, $\sum y_i = 21$

Then

$$S_{xy} = 64.75 - \frac{21^2}{6} = -8.75 \qquad S_{xx} = S_{yy} = 90.5 - \frac{21^2}{6} = 17$$

and

$$r_s = \frac{S_{xy}}{\sqrt{S_{xx} S_{yy}}} = \frac{-8.75}{\sqrt{17(17)}} = -.515.$$

To test for correlation with $\alpha = .05$, index .025 in Table 9 and the rejection region is $|r_s| \geq .886$. The null hypothesis is not rejected and we cannot conclude that there is a correlation between the two variables.

15.7.11 a The two variables (rating and distance) are ranked from low to high, and the results are shown in the following table.

Voter	x	y	Voter	x	y
1	7.5	3	7	6	4
2	4	7	8	11	2
3	3	12	9	1	10
4	12	1	10	5	9
5	10	8	11	9	5.5
6	7.5	11	12	2	5.5

Calculate $\quad \sum x_i y_i = 442.5 \quad \sum x_i^2 = 649.5 \quad \sum y_i^2 = 649.5$
$\quad\quad\quad\quad n = 12 \quad\quad\quad \sum x_i = 78 \quad\quad \sum y_i = 78$

Then

$$S_{xy} = 422.5 - \frac{78^2}{12} = -84.5 \quad S_{xx} = 649.5 - \frac{78^2}{12} = 142.5 \quad S_{yy} = 649.5 - \frac{78^2}{12} = 142.5$$

and $\quad r_s = \frac{S_{xy}}{\sqrt{S_{xx}S_{yy}}} = \frac{-84.5}{142.5} = -.593.$

b The hypothesis of interest is H_0: no correlation versus H_a: negative correlation. Consulting Table 9 for $\alpha = .05$, the critical value of r_s, denoted by r_0 is $-.497$. Since the value of the test statistic is less than the critical value, the null hypothesis is rejected. There is evidence of a significant negative correlation between rating and distance.

15.7.13 a The data are ranked separately according to the variables x and y.

Rank x	7	6	5	4	1	12	8	3	2	11	10	9
Rank y	7	8	4	5	2	10	12	3	1	6	11	9

Since there were no tied observations, the simpler formula for r_s is used, and

$$r_s = 1 - \frac{6\sum d_i^2}{n(n^2-1)} = 1 - \frac{6\left[(0)^2 + (-2)^2 + \cdots + (0)^2\right]}{12(143)}$$

$$= 1 - \frac{6(54)}{1716} = .811$$

b To test for positive correlation with $\alpha = .05$, index .05 in Table 9 and the rejection region is $r_s \geq .497$. Hence, H_0 is rejected, there is a positive correlation between x and y.

15.7.15 The data given in this exercise have already been ranked and may be substituted into the simpler formula since no ties exist in either the x or y rankings.

$$r_s = 1 - \frac{6\sum d_i^2}{n(n^2-1)} = 1 - \frac{6\left[(1)^2 + (-2)^2 + \cdots + (1)^2\right]}{10(99)} = .903$$

To test for positive correlation with $\alpha = .05$, index .05 in Table 9 and the rejection region is $r_s \geq .564$. We reject the null hypothesis of no association and conclude that a positive association exists.

15.7.17 a The ranks of the two variables are shown next.

Subject	1	2	3	4	5	6	7
Rank x	5	1.5	7	6	3	4	1.5
Rank y	6	1	7	3.5	5	2	3.5

Calculate $\quad \sum x_i y_i = 129.75 \quad \sum x_i^2 = 139.5 \quad \sum y_i^2 = 139.5$
$\quad\quad\quad\quad n = 7 \quad\quad\quad\quad \sum x_i = 28 \quad\quad \sum y_i = 28$

Then

$$S_{xy} = 129.75 - \frac{28^2}{7} = 17.75 \quad S_{xx} = 139.5 - \frac{28^2}{7} = 27.5 \quad S_{yy} = 139.5 - \frac{28^2}{7} = 27.5$$

and

$$r_s = \frac{S_{xy}}{\sqrt{S_{xx}S_{yy}}} = \frac{17.75}{\sqrt{27.5(27.5)}} = .645.$$

To test for positive correlation with $\alpha = .05$, index .05 in Table 9, and the rejection region is $r_s \geq .714$. The null hypothesis is not rejected and we cannot conclude that there is a positive correlation between the two variables.

b In Chapter 12, $r = .760$ and there was evidence of significant positive correlation. Notice that the parametric test allows rejection of H₀ while the nonparametric test does not. Assuming that the assumptions for the parametric test have been met, the parametric test is more powerful.

Reviewing What You've Learned

15.R.1 a Define $p = P(\text{response for stimulus 1 exceeds that for stimulus 2})$ and x = number of times the response for stimulus 1 exceeds that for stimulus 2. The hypothesis to be tested is

$$H_0 : p = 1/2 \quad \text{versus} \quad H_a : p \neq 1/2$$

using the sign test with x as the test statistic. Notice that for this exercise, $n = 9$, and the observed value of the test statistic is $x = 2$. Various two tailed rejection regions are tried in order to find a region with $\alpha \approx .05$. These are shown in the following table.

Rejection Region	α
$x = 0; x = 9$	.004
$x \leq 1; x \geq 8$	.040
$x \leq 2; x \geq 7$	.180

We choose to reject H₀ if $x \leq 1$ or $x \geq 8$ with $\alpha = .040$. Since $x = 2$, H₀ is not rejected. There is insufficient evidence to indicate a difference between the two stimuli.

b The experiment has been designed in a paired manner, and the paired difference test is used. The differences are: d_i $-.9$ -1.1 1.5 -2.6 -1.8 -2.9 -2.5 2.5 -1.4

The hypothesis to be tested is

$$H_0 : \mu_1 - \mu_2 = 0 \qquad H_a : \mu_1 - \mu_2 \neq 0$$

Calculate

$$\bar{d} = \frac{\sum d_i}{n} = \frac{-9.2}{9} = -1.022 \qquad s_d^2 = \frac{\sum d_i^2 - \frac{(\sum d_i)^2}{n}}{n-1} = \frac{37.14 - 9.404}{8} = 3.467$$

and the test statistic is $t = \dfrac{\bar{d}}{\sqrt{\dfrac{s_d^2}{n}}} = \dfrac{-1.022}{\sqrt{\dfrac{3.467}{9}}} = -1.646$.

The rejection region with $\alpha = .05$ and 8 df is $|t| > 2.306$ and H₀ is not rejected.

15.R.3 a Define $p = P(\text{school A exceeds school B in test score for a pair of twins})$ and x = number of times the score for school A exceeds the score for school B. The hypothesis to be tested is

$$H_0 : p = 1/2 \quad \text{versus} \quad H_a : p \neq 1/2$$

using the sign test with x as the test statistic. Notice that for this exercise, $n = 10$, and the observed value of the test statistic is $x = 7$.

Critical value approach: Various two tailed rejection regions are tried in order to find a region with $\alpha \approx .05$. These are shown in the following table.

Rejection Region	α
$x=0; x=10$	.002
$x \leq 1; x \geq 9$	.022
$x \leq 2; x \geq 8$	.110

We choose to reject H_0 if $x \leq 1$ or $x \geq 9$ with $\alpha = .022$. Since $x = 7$, H_0 is not rejected. There is insufficient evidence to indicate a difference between the two schools.

p-value approach: For the observed value $x = 7$, calculate the two-tailed p-value:

$$p\text{-value} = 2P(x \geq 7) = 2(1-.828) = .344$$

and H_0 is not rejected. There is insufficient evidence to indicate a difference between the two schools.

b Consider the one-tailed test of hypothesis as follows:

$$H_0 : p = 1/2 \quad \text{versus} \quad H_a : p > 1/2$$

This alternative will imply that school A is superior to school B. From Table 1, the one-tailed rejection region with $\alpha \approx .05$ is $x \geq 8$ with $\alpha = .055$. The null hypothesis is still not rejected, since $x = 7$. (The one-tailed p-value $= .172$.)

15.R.5 a-b The measurements are ordered according to magnitude, and ranked "from the outside in" as described in part (ii). The resulting ranks are

Instrument	Response	Rank
A	1060.21	1
B	1060.24	3
A	1060.27	5
B	1060.28	7
B	1060.30	9
B	1060.32	8
A	1060.34	6
A	1060.36	4
A	1060.40	2

c The hypothesis to be tested is $H_0 : \sigma_A^2 = \sigma_B^2$ versus $H_a : \sigma_A^2 > \sigma_B^2$ and the test statistic is the Wilcoxon rank sum. If the alternative hypothesis is true (that is, the variance for instrument A is greater than the variance for instrument B), then the measurements for instrument A should be very low and very high in the sequence of measurements. Hence, they will be assigned the lower ranks, and the "sum of ranks" for the A observations will be small; the rank sum of the B observations will be large. A one-tailed test of hypothesis is required, with α near .05. Since there are fewer measurements in sample B, you let $n_B = 4$ and $n_A = 5$. Calculate

$$T_B = 3+7+9+8 = 27$$
$$T_B^* = n_B(n_B + n_A + 1) - T_B = 4(10) - 27 = 13$$

The test statistic is $T = \min(T_B, T_B^*) = 13$. With $n_B = 4$ and $n_A = 5$, the one-tailed rejection region with $\alpha = .05$ is found in Table 7b to be $T_B^* \leq 12$. The observed value, $T = 13$, does not fall in the rejection region and H_0 is not rejected. We cannot conclude that the more expensive instrument B is more precise than A.

d The samples variances, s_1^2 and s_2^2, must be calculated in order to use the F test of Section 10.6.

$$s_1^2 = \frac{\sum x_{1j}^2 - \frac{\left(\sum x_{1j}\right)^2}{n_1}}{n_1 - 1} = \frac{.0230}{4} = .00575 \qquad s_2^2 = \frac{\sum x_{2j}^2 - \frac{\left(\sum x_{2j}\right)^2}{n_2}}{n_2 - 1} = \frac{.0035}{3} = .00117$$

Then the test statistic is $F = \frac{s_1^2}{s_2^2} = \frac{.00575}{.00117} = 4.914$. The rejection region with 4 and 3 df and $\alpha = .05$ is $F > 9.12$ and the null hypothesis is not rejected.

15.R.7 The hypothesis to be tested is

H₀: population distributions 1 and 2 are identical

Hₐ: the distributions differ in location

and the test statistic is T, the rank sum of the positive (or negative) differences. The ranks are obtained by ordering the differences according to their absolute value. Define d_i to be the difference between a pair in populations 1 and 2 (i.e., $x_{1i} - x_{2i}$). The differences, along with their ranks (according to absolute magnitude), are shown in the following table.

d_i	−31	−31	−6	−11	−9	−7	7
Rank $|d_i|$	14.5	14.5	4.5	12.5	10.5	7	7

d_i	−11	7	−9	−2	−8	−1	−6	−3
Rank $|d_i|$	12.5	7	10.5	2	9	1	4.5	3

The rank sum for positive differences is $T^+ = 14$ and the rank sum for negative differences is $T^- = 106$ with $n = 15$. Consider the smaller rank sum and determine the appropriate lower portion of the two-tailed rejection region. Indexing $n = 15$ and $\alpha = .05$ in Table 8, the rejection region is $T \leq 25$ and H₀ is rejected. We conclude that there is a difference between math and art scores.

15.R.9 The experiment is completely randomized, and the Kruskal Wallis H test is used. The data is jointly ranked as shown next.

Training Program			
1	2	3	4
9.5	27	8	13
15	18	20	1
18	21	12	5
23	24	14	2
3	25	9.5	18
6.5	16		11
	26		6.5
	22		4
$T_1 = 75$	$T_2 = 179$	$T_3 = 63.5$	$T_4 = 60.5$
$n_1 = 6$	$n_2 = 8$	$n_3 = 5$	$n_4 = 8$

a The test statistic, based on the rank sums, is

$$H = \frac{12}{n(n+1)} \sum \frac{T_i^2}{n_i} - 3(n+1)$$

$$= \frac{12}{27(28)} \left[\frac{(75)^2}{6} + \frac{(179)^2}{8} + \frac{(63.5)^2}{5} + \frac{(60.5)^2}{8} \right] - 3(28) = 14.52$$

With $df = k-1 = 3$, the observed value $H = 14.52$ exceeds $\chi^2_{.005}$ so that p-value $< .005$. The null hypothesis is rejected and we conclude that there is a difference from one training program to another.

    b From Chapter 11, $F = 9.84$, and H_0 was rejected. The results are the same.

15.R.11 a Neither of the two plots follow the general patterns for normal populations with equal variances.

    b Use the Friedman F_r test for a randomized block design. The *Minitab* printout follows.

Friedman Test: Cadmium vs Rate, Harvest

Method

Treatment = Rate
Block = Harvest

Descriptive Statistics

Rate	N	Median	Sum of Ranks
1	3	155.633	3.0
2	3	199.333	8.0
3	3	214.000	14.0
4	3	194.867	9.0
5	3	207.517	12.0
6	3	230.950	17.0
Overall	18	200.383	

Test

Null hypothesis H_0: All treatment effects are zero
Alternative hypothesis H_1: Not all treatment effects are zero

DF	Chi-Square	P-Value
5	11.57	0.041

Since the *p*-value is .041, the results are significant. There is evidence of a difference among the responses to the three rates of application.

15.R.13 Similar to previous exercises. The data, with corresponding ranks, are shown in the following table.

Supplier 1	Supplier 2
.86 (20)	.55 (14)
.69 (18)	.40 (10)
.72 (19)	.22 (7)
1.18 (23)	.09 (2)
.45 (12)	.16 (4.5)
1.41 (24)	.26 (8)
.65 (16.5)	.58 (15)
1.13 (22)	.16 (4.5)
.65 (16.5)	.07 (1)
.50 (13)	.36 (9)
1.04 (21)	.20 (6)
.41 (11)	.15 (3)
$T_1 = 216$	

a Calculate

$$T_1 = 216$$
$$T_1^* = n_1(n_1 + n_2 + 1) - T_1 = 12(25) - 216 = 84$$

The test statistic is $T = \min(T_1, T_1^*) = 84$. With $n_1 = n_2 = 12$, the two-tailed rejection region with $\alpha = .05$ is found in Table 7b to be $T_1^* \leq 115$. The observed value, $T = 84$, falls in the rejection region and H_0 is rejected. There is sufficient evidence to indicate a difference in the contaminant percentages for the two suppliers.

b Calculate the mean and variance as

$$\mu_T = \frac{n_1(n_1 + n_2 + 1)}{2} = \frac{12(25)}{2} = 150 \text{ and } \sigma_T^2 = \frac{n_1 n_2 (n_1 + n_1 + 1)}{12} = \frac{12(12)(25)}{12} = 300$$

$$z = \frac{T - \frac{n_1(n_1 + n_2 + 1)}{2}}{\sqrt{\frac{n_1 n_2 (n_1 + n_2 + 1)}{12}}} = \frac{84 - 150}{\sqrt{300}} = -3.81$$

The two-tailed rejection region with $\alpha = .05$ is $|z| > 1.96$ and H_0 is rejected. The conclusion is identical to the conclusion in part a.